HYBRID ENERGY SYSTEM MODELS

HYBRID ENERGY SYSTEM MODELS

Edited by

ASMAE BERRADA
Professor of Energy, College of Engineering, LERMA Lab,
International University of Rabat, Sala El Jadida, Morocco

RACHID EL MRABET
Senior Expert and General Director of the Public Utility
Development Company, Morocco

ELSEVIER

ACADEMIC PRESS
An imprint of Elsevier

Academic Press is an imprint of Elsevier
125 London Wall, London EC2Y 5AS, United Kingdom
525 B Street, Suite 1650, San Diego, CA 92101, United States
50 Hampshire Street, 5th Floor, Cambridge, MA 02139, United States
The Boulevard, Langford Lane, Kidlington, Oxford OX5 1GB, United Kingdom

Library of Congress Cataloging-in-Publication Data
A catalog record for this book is available from the Library of Congress

British Library Cataloguing-in-Publication Data
A catalogue record for this book is available from the British Library

ISBN 978-0-12-821403-9

For information on all Academic Press publications
visit our website at https://www.elsevier.com/books-and-journals

Publisher: Brian Romer
Acquisitions Editor: Lisa Reading
Editorial Project Manager: Naomi Robertson
Production Project Manager: Prem Kumar Kaliamoorthi
Cover Designer: Victoria Pearson

Typeset by SPi Global, India

Contents

Contributors

Raymond Adomatis
Department of Chemical and Biomolecular Engineering, Institute for Systems Research, University of Maryland, College Park, MD, United States

Sung-Hoon Ahn
Department of Mechanical and Aerospace Engineering, Seoul National University, Seoul, South Korea

Loiy Al-Ghussain
Mechanical Engineering Department, University of Kentucky, Lexington, KY, United States

Arechkik Ameur
Faculty of Science, Ibn Tofail University, Kenitra; School of Science and Engineering, Al Akhawayn University, Ifrane, Morocco

Somayeh Asadi
Department of Architectural Engineering, Pennsylvania State University, University Park, PA, United States

Mohamed Bakhouya
College of Engineering, LERMA Lab, International University of Rabat, Sala El Jadida, Morocco

Soukayna Berrabah
College of Engineering, LERMA Lab, International University of Rabat, Sala El Jadida, Morocco

Asmae Berrada
College of Engineering, LERMA Lab, International University of Rabat, Sala El Jadida, Morocco

Binayak Bhandari
Department of Railroad Engineering and Transport Management, Woosong University, Daejeon, South Korea

Zineb Bouhssine
College of Engineering, LERMA Lab, International University of Rabat, Sala El Jadida, Morocco

Sofia Boulmrharj
College of Engineering, LERMA Lab, International University of Rabat, Sala El Jadida; CUR"EnR&SIE, Faculty of Sciences, University Research Center on Renewable Energies & Intelligent Systems for Energy - Chouaïb Doukkali University, El Jadida, Morocco

Mohammed El Ibrahimi
College of Engineering, LERMA Lab, International University of Rabat, Sala El Jadida, Morocco

Anas El Maakoul
College of Engineering, LERMA Lab, International University of Rabat, Sala El Jadida, Morocco

Rachid El Mrabet
Senior Expert and General Director of the Public Utility Development Company, Morocco

Amir Fakhim-Babaei
Faculty of Electrical and Computer Engineering, University of Tabriz, Tabriz, Iran

Mohammed Khaidar
CUR"EnR&SIE, Faculty of Sciences, University Research Center on Renewable Energies & Intelligent Systems for Energy - Chouaïb Doukkali University, El Jadida, Morocco

Ismail Khay
College of Engineering, LERMA Lab, International University of Rabat, Sala El Jadida, Morocco

Khalid Loudiyi
School of Science and Engineering, Al Akhawayn University, Ifrane, Morocco

Behnam Mohammadi-Ivatloo
Faculty of Electrical and Computer Engineering, University of Tabriz, Tabriz, Iran

Morteza Nazari-Heris
Department of Architectural Engineering, Pennsylvania State University, University Park, PA, United States

Meng Ni
Building Energy Research Group, Department of Building and Real Estate; Environmental Energy Research Group, Research Institute for Sustainable Urban Development (RISUD), The Hong Kong Polytechnic University, Kowloon, Hong Kong, China

Radouane Ouladsine
College of Engineering, LERMA Lab, International University of Rabat, Sala El Jadida, Morocco

Onur Taylan
Mechanical Engineering Program, Middle East Technical University, Northern Cyprus Campus, Mersin; Center for Solar Energy Research and Applications (GÜNAM), Middle East Technical University, Ankara, Turkey

Haoran Xu
Department of Chemical Engineering, Loughborough University, Loughborough, United Kingdom

CHAPTER 1

Introduction to hybrid energy systems

Asmae Berrada[a], Khalid Loudiyi[b], and Rachid El Mrabet[c]
[a]College of Engineering, LERMA Lab, International University of Rabat, Sala El Jadida, Morocco
[b]School of Science and Engineering, Al Akhawayn University, Ifrane, Morocco
[c]Senior Expert and General Director of the Public Utility Development Company, Morocco

1.1 Introduction

In recent years, access to energy has been improved significantly. The number of people without access to electricity has increased from one billion in 2016 (and 1.2 billion in 2010) to about 840 million today. About 600 million people in Africa live without electricity, while for hundreds of millions more, the electricity supply is insufficient or random. In addition, nearly 3 billion people use polluting fuels such as wood and other forms of biomass for cooking and heating purposes. This lack of continuous energy provision costs the country (Africa) about 2%–4% GDP growth per year [1]. During blackouts, large industries often depend on expensive backup generators.

The global energy system is undergoing a major transformation, where renewable energy systems play a critical role in the development of modern and robust energy systems. The transition is being driven by a sharp decline in the cost of clean energy, whereas the deployment of disruptive technologies—smart grids and meters or geolocated data systems—contributes to a radical improvement in planning.

In many countries, new large-scale solutions combining stand-alone installations and grid-based systems improve access to energy. Other countries depend on minigrid. At the same time, domestic solar installations are witnessing an increase in performance and a decrease in cost, making them affordable in sub-Saharan Africa and South Asia, the two regions where access to energy is significantly deficient.

With a significant increase in energy requirement and its continuous fluctuation along with the environmental concerns associated with the use of fossil fuels, various clean energy generation technologies have emerged, which are based on renewable energy sources such as wind,

Hybrid Energy System Models
https://doi.org/10.1016/B978-0-12-821403-9.00001-9

solar, and biomass and therefore are considered harmless to the environment compared to the conventional ways of producing energy.

The trend toward increasing the deployment of renewable energy systems is expected to grow in the coming years. A high contribution of variable renewable energy resources disturbs the reliability and the correct operation of the utility network. Electric grids are now at the starting point of a new revolution where expensive requirements have been enforced to secure a proper supply of electricity. This shift to more installed renewables combined with other energy technologies requires new ways of thinking about the electricity system. System planners and utility grid operators face various challenges such as integration of intermittent generation, weak grid infrastructure, multiple gas turbine starts per day, and increased need for spinning reserve. Many recent studies have shown that there is no single form of power generation that is optimal in all applications. Solar and wind energy production are variable, but they neither consume fuel nor emit greenhouse gases. On the other hand, natural gas-fueled generation systems have low CAPEX ($/kW) and are dispatchable, but they do emit greenhouse gases. To solve these issues, several solutions are being investigated. The deployment of hybrid energy system technologies is one among them.

1.2 Hybridization: A key solution to energy transition

The need to mitigate global warming and the inevitable decline in conventional energy sources have encouraged several countries to introduce new energy policies that support the use of renewable energy. An energy transition is occurring in developing countries with an increase in the utilization of solar and wind energies globally. To connect these fluctuating renewable energy sources into the electric grid at the scale necessary to reduce climate change, hybrid systems including energy storage are the key solution.

Renewable energy systems offer many advantages, particularly in terms of the environment and growth potential. On the other hand, they pose serious challenges, particularly in terms of costs and service continuity (as they are intermittent sources). The combination of two or more energy sources to form a multisource system (hybrids) transcends these problems by providing economical and reliable electricity while meeting ecological requirements. The main concern in the design of such systems is the precise selection (optimal sizing) of their components. The aim is to minimize the cost of generating electricity as well as the use of the national grid and/or conventional energy sources while ensuring optimal service continuity (reliability).

A hybrid energy production system, in its most general view, is one that combines and exploits several easily accessible energy sources. Hybridization is defined as the increasingly frequent coupling of different energy sources at different levels of an energy system. Hybrid energy system solutions are very well positioned to address the challenges of managing a transformable power system as more renewable energy technologies are integrated into a grid that does not have adequate flexible resources to guarantee reliability. Hybrid power systems typically combine multiple sources of energy generation with a control system to overcome the deficiencies of a specific generation type. These systems may include energy storage technologies. This combination will provide the power that is reliable, sustainable, and cost-effective. In fact, various gas/renewable/energy storage hybrid systems have been deployed worldwide. Research is needed to investigate such hybrid energy systems.

Hybrid systems can be divided into two groups. In the first group, we find hybrid systems, working in parallel with the electric grid. These systems help satisfy the electrical system's load. The hybrid systems, in the second group, operate in an isolated or autonomous mode. They must meet the needs of consumers located in sites that are far from the electric grid: mountain huts, islands, or isolated villages.

Three criteria can be considered in this classification according to the structure of a hybrid system. The first criterion is whether or not there is a conventional energy source. This conventional source can be a diesel generator, a microgas turbine, and, in the case of a study of the entire power grid, an entire power plant. A second possible criterion is whether or not a storage device is present. The presence of storage ensures better satisfaction of electrical loads during periods in the absence of a primary resource to be converted into electricity. Storage devices can be rechargeable batteries, electrolysis with hydrogen tanks, etc.

The last possible classification is the type of renewable energy source used. Indeed, the structure of the hybrid system may contain one or more combination of renewable energy sources. An important criterion for the selection of the source used is the availability of the energy potential, which depends on where the hybrid system is installed. Another determining factor is the powered electric consumer. Its importance determines the need for an additional source, a storage device, and/or a conventional source.

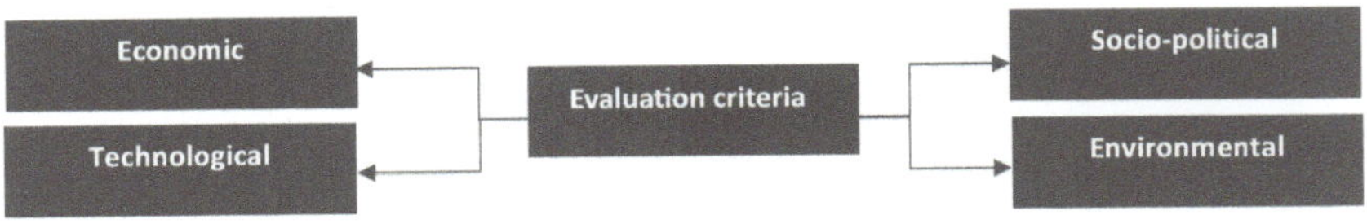

Fig. 1.1 Criteria for designing and evaluating a hybrid system.

The combination of several renewable energy sources makes it possible to optimize electricity generation systems as much as possible, from both a technical and economic point of view.

The selection of evaluation criteria is an important element in the design of a multisource system for a given locality. Fig. 1.1 presents various criteria for the design and evaluation of these multisource hybrid systems (technological, economic, socio-political, and environmental factors).

A major concern related to the planning of renewable energy projects is based on economic efficiency. Low economic efficiency is considered as one of the main disadvantages compared to multisource energy systems. An energy solution with the highest economic benefits is typically the one that is achieved. Technical or ecological aspects are often considered to be of inferior importance.

Components of a multisource system are subject to:

- Minimizing the cost of generating electricity,
- The Insurance of the service of the load according to certain criteria of reliability (loss of load probability),
- Minimizing the energy purchased from the grid (probability of loss of supply for systems connected to the grid).

To achieve these goals, data about the user's load demand, energy resources, as well as technical and economic information should be available.

1.2.1 The role of digitalization in hybrid energy systems

Hybrid energy systems have become more connected, reliable, and intelligent thanks to digital technologies. Striking progress in connectivity and analytics is supporting the development of new digital systems such as smart machines. Digitalization is enhancing the accessibility, safety, and productivity of energy technologies.

Besides the fact that all energy demand sectors are witnessing the impact of digitalization such as planes, cars, and their supporting infrastructure, energy use in buildings could be reduced by about 10% using digitalization to enhance operational efficiency. Digital transformation is projected to have a wide impact on the future design of energy technologies.

Digitally interconnected hybrid energy systems could fundamentally transform electricity markets. The integration of intermittent renewable energy system to the electric grid can be done with the help of digitalization. The latter provides a better matching of energy demand and supply. This would enable more flexibility to the electric grid while saving money in avoiding further investment of new energy infrastructure. The development of distributed energy resources can also be facilitated using digitalization. New incentives and devices would be created to support energy trading within local hybrid energy districts.

The development and the implementation of integrated and smart hybrid energy technologies require reconfigured and new value chains, regulatory and organizational innovation, business models and landscape of energy services, and new research actors, in addition to an optimized integration of end users into the electricity system.

1.3 Hybrid energy systems with conventional technologies

1.3.1 PV with conventional sources

This type of hybrid system is most often implemented in sites that are characterized by a hot climate and have significant solar potentials such as Saudi Arabia [2], Morocco [3], the Maldives [4], and Corsica [5]. The aim of these systems, when operated in autonomous mode, is to supply energy without interruption to a house [5], residential/administrative buildings [2], or villages [6]. Other hybrid systems supply energy to research centers [7] or are connected to the electric grid [4].

Autonomous systems often incorporate batteries [8, 9] as well as other energy storage systems. The batteries and the PV field produce direct current. On the other hand, diesel engines can drive continuous or alternating generators. Most often, consumers demand alternating current; a distinction is then made between different system structures depending on the type of electric machine coupled with the diesel engine. These structures have been described and classified according to the type of energy flow by the authors in Ref. [10].

A number of studies have investigated existing hybrid systems composed of photovoltaic systems coupled with a conventional source [3,6]. Others studied the possibility of integrating PV panels as an additional energy source in existing installations with a conventional source [6,9,11]. The authors have carried out studies on the analysis of the processes that occur within the hybrid system [12]. Other authors have investigated the optimization of the dimensioning of the hybrid system [13], or the energy management strategy used by this system [14].

1.3.2 Hybrid solar-Stirling engine

Recent studies have shown that there is big potential in thermosolar generation systems. In fact, these systems might be efficiently useful in thermal storage and hybridization to reduce solar dependability. The use of these hybrid technologies is interesting for isolated units targeted in distributed generation systems. Dish-Stirling technologies are a type of thermosolar system; They have great potential to be used as power feeding systems thanks to the units' modularity that varies between 3 and 25 kWe [15].

The analysis presented in this section focuses on the integration of hybridization and thermal storage within an isolated system of a dish-Stirling system. Hybridization could particularly enable continuous system operation. Hybridizations are analyzed for both renewable energy sources and conventional fuel. The dish-Stirling analysis will determine the interest of introducing complementary technologies.

Solar-Stirling dish engines demonstrate interesting potential in regions with a large amount of solar radiation. Additionally, these systems outperform parabolic troughs in producing power at high efficiency and are more cost-effective. Nevertheless, it is important to mention that Stirling dish systems, in comparison with solar technologies, have not received as much attention as other technologies. Stirling systems are most appropriate for hybrid combinations due to their efficient ability to integrate various heat sources into one application. Indeed, parabolic dishes concentrate direct radiations only; hence, trackers are mandatory for a continuous orientation toward the sun. To allow high-efficient solar-electric conversion, high temperatures at the receiver are highly recommended. For practical systems, the solar to electric efficiency varies between 16% and 30%.

Concentrated solar power is used by high concentration solar thermal to convert the captured solar radiations into mechanical energy by means of thermodynamic cycling; This is then converted to electrical power. Most known solar thermal technologies are central receivers and parabolic troughs. Dish-Stirling systems are in the first phase of industrialization. To focus solar radiations into the receiver, these systems make use of mirrors positioned on the parabolic surface. The receiver is responsible for transferring energy to the Stirling engine. Two main advantages of these systems include modularity and high efficiency. Its modularity enables the system to be used individually in a remote location or to operate in small connected groups linked to the utility grid. The basic components of a hybrid system include a tracking system, concentrator shell, cooling system, receiver, structural framework, Stirling engine, battery, and controller [16].

One negative aspect associated with the use of solar-Stirling technology is its dependence on solar availability; however, the use of both hybridization and thermal storage may solve this issue, as it provides flexibility and more operating hours. There are a lot of research studies investigating components, which could be utilized for hybridization such as the receiver. A number of advantages are provided by hybridization such as the enhancement of power management and the adaptation to weather transitions. The hybrid system also improves the operation time and the investment recovery of the plant. Different energy sources can be provided to the Stirling engine as it is considered an external combustion engine. The application of this hybrid system means the utilization of substitute energy sources combined with solar power. Because of the availability of this fossil fuel, the aforementioned hybridization is feasible and requires the addition of a combustion system to the power plant. The implementation of this hybrid system could reach high potential in regions where fossil fuels are produced such as the Middle East and North Africa. However, the hybridization of natural gas and solar power can face some limitations such as national regulations, which depend on the type of fuel used. For example, Spain has limited the annual electricity produced from fossil fuels and natural gas to 12% and 15%, respectively. During high energy demand, especially in the morning, hybridization of the plant with natural gas will be turned on. As an alternative to using natural gas, it is also interesting to make use of biogas as a source of fuel as it has a lower environmental impact. Furthermore, it is particularly attractive for isolated systems in which a natural gas supply is not feasible. Compared to natural gas, the imposed Spanish regulations, for example, are higher for biogas as they are limited to 50%. That is, biogas combustion can provide 50 % of the output of the hybridized plant. To ensure continuous operation of the power plant, thermal energy storage can also be utilized as an auxiliary energy system. Energy is stored in the latter during periods of low energy demand. This energy is discharged when needed during peak periods. However, because of the limited capacity of the intrinsic storage system, continuous operation of the facility will not be guaranteed. The power generated from the thermal energy storage will be typically offered during the evening or in the cloudy weather periods when solar energy production falls. Therefore, the implementation of a solar dish-Stirling hybrid system with thermal storage will increase the electricity output of the plant.

Solar Stirling dishes can produce electricity ranging from kW to MW. There have been many recent studies dealing with new technologies that could produce optimal power from the use of this hybrid system. In 2012, Wu et al. [16] assessed the thermoelectrical conversion performance of a solar parabolic dish. The obtained possible output power was about 19 kW with an efficiency of 20%. In 2013, Kleih [17] discussed a new HIMAP measuring system using a camera-video to test a parabolic dish system ranging from 5 to 25 kW. In 2015, Nepveu et al. [18] proposed a thermal power model of Eurodish Stirling unit of 10 kWel.

Several solar dish models were discussed by researchers for the possibility of using hybrid systems in cooking. A solar cooking unit using vacuum tube collectors with combined water pipes under hot and cold conditions has been investigated by Balzar (2011) et al. [19]. In 2015, Grupp et al. [20] proposed an automatic synopsis user meter for solar cooking models. In 2007, the authors in Ref. [21] described a new system of solar cooker using PCM storage. Ultimately, Badran et al. [22] proposed a design of a portable solar cooker integrated with a water heater.

Solar-Stirling dish systems can be used for various applications, among them is water heating. In 2009, Mohammed [23] designed a hybrid solar parabolic dish for water heating where the water can heat up to 100%. Similarly, in 2010, Akinbode and Manukaji built a hybrid solar dish concentrator used for water heating and cooking in addition to many other applications [24]. In 2014, Dafle and Shinde [25], using 16 m^2 Scheffler reflector, constructed a water heater system working at 2 bar and 110°C; it was reported that the system can also be used for cooking.

Parabolic dish system is composed of a parabolic reflector in the shape of dish supported by a structure, a Stirling engine, parabolic solar receiver, a solar tracking system, and a generator to produce electricity. A tracking control system is used to direct the solar parabolic dish, throughout the day, toward the sun. There exist two types of solar parabolic dish systems; the Eurodish, a 10 kW electrical solar-Stirling dish system, and the SunCatcher, a 25 kWe solar-Stirling dish system [26–27].

The aim of combining solar dishes with a Stirling engine is to create large electricity-generating plants that are connected to the transmission network. Since similar technologies used by the hybrid solar-Stirling engine are utilized by solar plants, the investment costs in Table 1.1 could be used to

Table 1.1 Performance/cost indicators.

Component cost	Units	Higher production 2030	Higher production 2020	Commercial engine 2005	Hybrid system 2000
Plant size	MW	30	30	30	1
Concentrator	$/kW	300	400	1550	2800
Hybrid		250	270	400	500
Generator		40	40	45	50
Electrical		25	25	35	45
Receiver		70	80	80	120
Engine		90	90	260	800
Cooling system		30	30	40	65
Balance of plant		240	240	300	425
General plant facilities		110	110	150	190
Start-up cost		18	18	35	70
Engineering fee		115	128	286	500
Init Cat & Chem. Inventory		6	6	12	60
Inventory capital		4	4	12	40
Land (16,250/ha)		26	26	26	26
Total capital		1324	1476	3231	5691
Total capital w/o hybrid		1074	1197	2831	5191
Labor O&M costs	¢/kW	0.55	0.55	1.20	2.10
Material O&M costs	¢/kW	0.50	0.50	1.10	1.60
Total cost	¢/kW	1.05	1.05	2.30	3.70

estimate the cost of the Stirling engine in combination with a solar dish. The initial investment is the major cost of this system. Table 1.1 presents the performance and cost indicators of the different components [28–29].

Dish-Stirling engine technologies have few environmental impacts. As compared to internal combustion diesel and gasoline engines, Stirling engines are quiet. The radiator cooling fan is the main source of noise from a dish-Stirling system. This technology has not been developed enough to determine its visual impact objectively. The structure can be elevated in profile and reach up to 15 m. However, from an esthetic point of view, the dish-engine systems look like satellite dishes, which are publically accepted. The system emissions are also very low. Other than the likelihood of spilling small amounts of coolant and motor oil, such devices do not generate any waste as they are operated using a solar source. The combustion mechanisms used in both Brayton and Stirling processes result in particularly low emissions even when fossil fuel is used.

Dish-Stirling engine systems have high versatility and efficiency, as well as interesting hybrid functionality attributes. Good efficiency results in high power density. Among all other solar systems, the dish-Stirling engine technology has demonstrated the highest energy conversion coefficient, which is about 29%. Hence, it has great potential compared to other renewable electricity sources. The principal advantages of dish-engine systems are modularity and high performance with devices.

Due to the use of heat engines by dish-engine systems, it is inherently capable of working with fossil fuels. A hybrid capability only requires the addition of a fossil-fuel combustor, as the system makes use of the same conversion components such as the engine, generator, cable, and switching gear. Therefore, the addition of a hybrid capacity is simple for dish-Brayton systems. A fossil fuel combustion system able to operate continually at full power can be supplied with minimal barriers and expenses. This hybrid burning device is located downstream of the receiver and has no negative effect on performance. The total efficiency of the system is increased because gas turbine can operate at its optimum efficiency. The efficiency of the hybrid-mode dish-Brayton system is expected to be around 30%, based on the higher heating value (HHV).

The introduction of a hybrid capacity is difficult for dish-Stirling devices. In Stirling engines, the isothermal heat addition is easily integrated with solar thermal power compared to combustion heat. Geometric limitations make it even more challenging to integrate simultaneously. As a consequence, costs are projected to increase by $250/kWe for large-scale production dish-Stirling hybrid technology. The addition of a separate diesel generator set or large-scale gas turbine requires more than the aforementioned cost. Even though the cost of these technologies is projected to be significantly lower than a continuously variable hybrid receiver, their operating flexibility will decrease significantly. The efficiency of hybrid dish-Stirling systems based on HHV is projected to be around 33%.

Hybridization of solar thermal energy systems has received considerable interest because of the intermittent and variable characteristics of solar sources. It is a practical way of increasing the system operating period and investment turnover. The integration of subsystems enables hybrid technology to benefit from the power provided by other sources. However, the introduction of these devices must be balanced by an increased performance advantage. Both renewable and nonrenewable fuels can be used for hybridization. Most of the initial plant components remain the same; only a few additions or modifications are made such as the incorporation of a fuel supply system or a burner. For this technology, volume and weight restrictions are therefore important. Furthermore, a burner's form and weight are important to achieve maximum compactness and minimal effect on system parameters.

1.3.3 Wind systems with conventional technologies

These systems are more prevalent on islands, where sea breezes and wind favor the use of wind power for power generation. Studies have been carried out on systems installed on islands of various sizes—small islands such as the Canary Islands [30], medium islands such as Corsica [31], and large islands such as England [32].

Some studies examined existing hybrid systems and presented measurement results [32]. The authors in Ref. [33] used a wind generator as an additional source of energy to diversify production sources. Other works have investigated the political aspects of hybrid systems [34], about the analysis of energy flows [35] or about optimizing its structure [36,31].

The loads, supplied by this hybrid system, are of different types: isolated houses [31], residential buildings [37], public buildings [35], villages [38], or even islands [30]. In these cases, the hybrid system operates in autonomous mode. The authors in Ref. [39] studied a system connected to the eclectic network. When the hybrid system operates in an autonomous mode, it often includes an energy storage system such as batteries [32] or an electrolyzer and a fuel cell with hydrogen storage [36].

1.3.3.1 Wind energy system with diesel generator and battery storage

Diesel or fuel generators are widely adopted in the agricultural sector and remote locations in most of the developing countries. The main drawback of such energy resources is mediocre energy productivity even with important fuel consumptions. In addition to that, the long under loading of such generators leads to several costly repairs. To solve these problems, a set of batteries joined with a power converter can be used due to the many advantages they provide. The particularity of the system is that the power converter can be either used to charge the set of batteries or used as a power inverter. In fact, the last function enables the structure to avoid damaging the generators at insignificant loads and to withstand loads larger than the generator capacity. Consequently, this combination increases significantly the system efficiency. However, since sustainable energy sources are an interesting solution for the reduction of the use of fuels such as diesel, another combination of renewable and nonrenewable resources is reasonable to ensure both reliability and sustainability.

The development of renewable energy resources such as solar and wind energies had led to an important drift toward hybrid systems combining renewable and nonrenewable resources. The optimal sustainable energy source to be used in a hybrid system is the one that provides the least intermittency as well as the most improved efficiency and energy production in the studied area or region. However, wind energy joined with diesel generators represents a favorable solution for diesel consumption with additional features such as carbon footprint reduction compared to the use of fossil fuel by itself. On the other hand, even wind energy presents numerous drawbacks due to the intermittent nature of the wind. These disadvantages reside in the fact of generating too much agitation regarding the frequency and the voltage.

The issue with such systems is the ability to size a suitable production to match the energy demand. In contrast to fossil fuel power plants that have a fixed output, the nature of renewable energy resources is highly intermittent, and thus, their output is not constant. As a solution, an additional energy system whose main function is the storing of energy is very necessary to deal with the inequality between energy production and consumption by either storing energy or supplying it. The main challenge of sizing a wind-diesel hybrid system with storage is the selection of the most suitable storage technology considering all the aspects related to the storage system such as cost and autonomy period. The batteries as a storage technology remain one of the best choices for operating those hybrid systems owing to their large energy content compared to their size.

The combination of wind energy or any other renewable energy with a diesel generator and a battery storage system increases the effectiveness of the overall system. This combination joined to a power inverter has been named by too many researches as the Static Power Pack.

The power converter is considered an important part of the system. Its main purpose is the conversion of the power signal into a waveform, which acts as the one supplied from the network. The most operated converters make use of semiconductor technologies such as MOSFETs and commutation techniques. A suitable inverter should present at least an exit voltage regulated by more or less than 3% and should have an efficiency higher than 85%. It should also withstand high operating temperatures in addition to operating at a critical condition such as operating at almost three times its rated power. The power converter should possess good conversion qualities, a significant voltage regulation, and a limited reactive power, which has various negative effects on the system. A control method is also crucial for the hybrid system. The role of the control system is to perform automatically the switching operations and to select a suitable energy mode while keeping the energy supply in continuity.

Different battery technologies can be used in this hybrid system. However, lead acid technology remains one of the most adopted technologies since it requires less O&M costs, which is indispensable in such projects that usually take place in remote areas and can be suitable for rural regions. By performing a proper sizing of the system's components such as the generator and the batteries, the maximum loads can be satisfied by conventional and nonconventional resources. The generator can supply up to 1.1 times its rated output, and the wind-battery-inverter set can supply twice its rated output. Thus, for an equal share of the diesel and wind-battery-inverter set, the system can withstand up to 3.1 times the rated overall capacity. As a result, the remote areas could purchase reduced sized generator units and consume less fuel while keeping the generator in a good state. The operation of the hybrid system can have three different states. As a function of the load, the system will either be charging or discharging energy from the batteries. In addition, the control system is responsible for regulating the voltage and for the synchronization and the management of the different aspects of the hybrid system so that the change between the different charging and discharging states will not be noticed.

The main purpose of the addition of a wind turbine or any other renewable to the diesel and battery hybrid system is the reduction of the amount of diesel burned inside the generator as well as the operation period of the latter.

The benefits that can be obtained from the diesel/wind/storage hybrid system are the optimization of the system efficiency, better use of the diesel generator, and the ability to reduce the size of the diesel set, which reduces the bulkiness and the noise from the latter in addition to the possibility to purchase fewer batteries compared to the diesel-battery only. These benefits are supported by the simplicity of renewable energy penetration in the isolated systems, which highly encourages their use. The incorporation of small wind generators has shown interesting results and reduced payback periods.

1.3.3.2 Wind/PV/diesel hybrid systems

The objective of having such a combination of systems is to diversify renewable energy sources. This system results in a more significant reduction in the amount of fuel consumed since renewable sources can complement each other and provide a greater amount of energy.

Some research works have studied existing installed systems [40]. There are also studies on system interconnection in the power grid, in New York State [41] or in Saudi Arabia [42]. Other authors optimized the configuration of this hybrid system according to various criteria such as the probability of energy losses [43], minimization of system cost and emissions [44], and minimization of unsatisfied load [45]. The optimal system has been obtained using Pareto algorithms in Ref. [44], genetic and evolutionary algorithms in Ref. [45], or fuzzy logic in Ref. [43].

These systems supply homes and shelters [46] or villages [47]. They can be autonomous [48, 44] or interconnected with the electrical grid [43].

Energy storage can be in the form of batteries [44], hydrogen tank with electrolyzer and fuel cell [46], or a combination of the latter [45]. The authors developed an optimal management strategy for an energy storage system in Ref. [49].

1.4 Hybrid systems without conventional technologies

These hybrid systems operate mainly in autonomous mode in regions where the supply of diesel fuel or where the interconnection with the electric network is difficult, if not impossible.

1.4.1 Hybrid PV system with energy storage

The photovoltaic system must be connected with another energy source to meet the energy demand of the load at night or in cloudy weather. These systems provide energy to either residential houses [50] or villages [51]. Some studies dealt with modeling and analysis of this hybrid system [52, 53, 54]. Others examined the optimal dimensioning of the system components [50, 55]. The most commonly used optimization criterion is the probability of power supply losses [56]. Vosseler et al. presented the results obtained from analyzing multiple systems with a total power of 86.3 kWp [51]. The energy management strategy has been studied in Ref. [57]. Algorithms for tracking the maximum power point have been developed from fuzzy logic in Ref. [58]. Photovoltaic hydrogen production is also studied using electrolyzers in Ref. [59] and using different technologies in Ref. [60].

Energy storage is sometimes the ultimate [50]. Examples of energy storage that can be used are a battery bank [50], an electrolyzer with hydrogen tank [61], or a combination of the two aforementioned energy storage devices or electrolyzer and supercapacitor [62]. Supercapacitors are used for very short storage period with the reactivity of a few seconds, while electrolyzer and the hydrogen allow more important storage in terms of quantity stored but are slower [63].

1.4.2 Hybrid wind-PV system

This hybrid power generation consists of the combination and the exploitation of two renewable energy technologies, which include photovoltaic and wind turbines. The system is meant to capture energy in a complementary and combined way through the use of solar photovoltaic panels during daytime and the wind turbines during periods of potential wind speed. The generated energy can be stored in a battery bank.

In this hybrid system configuration, the power sources and the storage means have to meet two objectives, which include the provision of appropriate production to cover all users' energy consumption and cost-effectiveness. Energy storage is often used in small hybrid systems to power the load for a relatively long time (hours or even days). It is used to eliminate power fluctuations in the short term.

Energy storage is usually achieved through batteries. An example of batteries includes lead-acid and nickel-cadmium batteries. There exist other forms of storage, such as pumped hydro, flywheels, and hydrogen storage.

Although the first two storage methods are currently the most used in a hybrid PV/wind system, they are far from satisfactory [64].

The advantage of coupling PV-wind hybrid systems with an energy storage device includes:

- System operation without interruption;
- Ability to preserve the excess energy generated by the hybrid system;
- Security of energy supply under diverse weather conditions.

This hybrid system can be used in large applications as well as for residential ones. It has other different usages such as cell phone recipient stations, hospitals, hotels, laboratories, and rural village electrification. In addition, this system can be helpful for street lightning. There is a need for much research and development effort in this field including solar, wind, and storage systems to improve the performance of the hybrid system, determine methodologies for accurately forecasting their output, and integrate the aforementioned technologies with other conventional generating sources.

The incorporation of an energy storage system with a wind generator can have two main objectives. It can either be used as a buffer when the system operates in parallel with the electrical network. The storage device allows the smoothing of the rapid variations in the electric power coming from the wind turbine [65, 66]. It can also be used as a longer-term storage system in an autonomous regime to allow consumers to be supplied power during periods of low wind speed [67]. The electrical network can be large, as in the United States [68], or small (island network) [69]. As with other hybrid systems previously discussed, this system can provide energy to either a household or a building and even a village [65].

The storage methods studied can be a battery [65], an electrolyzer with a hydrogen reservoir [70], a combination of the two aforementioned systems [67], compressed air energy storage [68], superconductive magnets [71], or a combination of an electrolyzer with a hydrogen reservoir and a supercapacitor [66].

1.4.3 Hybrid wind/PV system without energy storage

This type of hybrid system is used very rarely because it does not provide security of energy supply—it lacks either a conventional source or an energy storage device. For this reason, research studies on these systems are limited. Work has been found on an existing system [72] used for educational purposes in Northern Iowa, USA. The authors in Ref. [73] studied a hybrid system project for the electrical network of a Croatian island; others have estimated the performance of this hybrid system [74].

1.4.4 Wind energy and hydropower

The wind is a clean and inexpensive source of renewable energy, but it is unstable because it depends on the weather that is variable. However, engineers have found a way to overcome this problem by combining it with hydropower. Many countries in the world, which have significant wind potential, are interested in wind-hydro systems.

Many studies have provided simulations of how the wind-hydro system does work and its benefits since the twenties. However, a real installation was not implemented until 2016. In March 2016, Max Boegl Wind AG, a German company, established an agreement with GE Renewable Energy company to create onshore wind farm combined with hydropower plant as the first renewable hybrid in the world and their expectations were to start using the wind-hydro system in 2018 [75–76].

Wind-hydropower system consists of connecting a wind plant to a pumped-hydro storage system that stores the excess wind generation by pumping water into an elevated reservoir and resupplying this energy to the electric grid during high demand of electricity through its hydroturbines. A case study a hybrid wind-hydro power plants in Crete Island, Greece, has been presented by the authors in Ref. [77]. The obtained results have shown that pumped storage systems are less expensive and more efficient than other storage technologies. This study used a computer algorithm to simulate the data of one year of the hybrid plant operation. The performance of the reversible pump-turbine machines and double penstock was included in the study as two different configurations of the pumped storage system. In addition, the possibility of installing a large hybrid power plant has been explained by the authors. The latter also provided details about the design parameters of the plant, such as the reservoir capacity. The results of this study have shown that the hybrid plant can replace fossil fuel plants effectively in Crete Island [77].

The authors in Ref. [78] performed a study in Rio Grande do Sul, Brazil. This analysis compares the operation of the hybrid hydroelectric wind system, which includes conventional hydropower using a reservoir, and a pumped-storage hydroelectric plant. HOMER software was used in this study to evaluate the technical and financial solutions for hydroelectric plants with pumped hydro storage. The results of this study show that pumped storage hydropower system requires high initial installation cost, while the hybrid system using hydropower requires high operating cost per year. However, this comparative study shows that it is difficult to decide whether the conventional reservoir or pumped storage facility is better because they both depend on some factors, such as the average load to be served, the hydrological barriers, and the energy costs [78].

A. Hamann and G. Hug studied the ability of hydropower to offset the variability of wind energy. A case study has been performed by the aforementioned authors in the Pacific Northwest region of the United States. MATLAB was used in this case study for two simulations for a period of five days. The first simulation deals with the hydro-wind combination, and the second simulation investigates the generation of electricity without wind while using only hydropower. The obtained results show that the generated power output is higher with hydro plus wind than hydro without wind. According to this work, there are two criteria to evaluate the performance of a hydro-wind system. The first criterion is related to how much hydropower plant can increase its production when integrating wind turbines, while the second one deals with how much load or wind can be decreased when ramping is not distributed equally in the hydro-wind system. The results of this study give a clear conclusion that the ramping score is higher in the first simulation (hydro-wind system) than in the second one (hydropower only) in all five selected regions including Wells, Rocky Reach, Rock Island, Wanapum, and Priest Rapids [79].

The author in Ref. [80] investigates the profitability and the value of combining wind energy with hydropower. Different case studies have been performed in various countries, such as Germany, Sweden, and Denmark to compare the revenues that can be obtained from the use of wind energy with those that can be gained from operating a hybrid wind plant with hydropower. This article reveals that wind power should not be used in areas where the cost of electricity is low, but its use should be based on the value consideration or on the benefits of wind power in general. Benefits can be obtained from the use of hydro reservoir power with wind power despite the change in weather conditions [80].

The authors in Ref. [81] studied the feasibility of a hybrid wind-hydro system for the production of low-cost energy. A case study has been performed on the island of Ikaria, Greece. The latter is known to have great potential in wind energy. The government and the local authorities in Greece encourage the use of multiple renewable energy sources so that they can reduce the utilization of diesel engines and limit their use only during periods when the energy production from renewable sources is not enough to meet the demand. The operation of wind energy and hydropower system on the island of Ikaria has been investigated in this case study using the Monte Carlo simulation program. This study shows that the cost of electric energy production in the country is 0.19 \$/kWh. The obtained results reveal that even the combination of wind and hydropower is not sufficient during the summer period in the island because of tourism. Therefore, to meet this high demand for energy, the use of diesel engines is also needed during the aforementioned period [81]. The simulation results demonstrate that the hybrid wind-hydro system requires less cost than an autonomous power station. It has been shown that using 14 wind turbines costs 0.12 \$/kWh as a minimum and 0.14 \$/kWh as a maximum. On the other hand, using 23 wind turbines costs 0.13 \$/kWh as a minimum and 0.15 \$/kWh as a maximum. However, in both cases, the cost is lower than that obtained from an autonomous power station, which is about 0.19 \$/kWh. Therefore, the hybridization of hydro and wind power is economically beneficial to the country [81]. Martínez-Lucas et al. analyzed the frequency regulation of a hybrid wind–hydro system. The authors aim to evaluate the ability of the hybrid power plant in providing frequency regulation to an isolated power system. A variable speed wind turbine (VSWT) is a turbine that is mainly designed to operate over a wide range of rotor speeds and can provide primary frequency regulation. This study was conducted using control strategies using the kinetic energy stored in the variable speed wind turbine (proportional, inertial, and their combination as well) in El Hierro. The latter is a Canary Island, which aims to prevent the carbon dioxide emissions in its area. The results of the study confirm that the VSWT contributes to the frequency regulation, and thus, it enhances the lifetime of the hydropower turbine control system by reducing its movement at the expense of an increased and permissible workload imposed on wind turbine blades [82]. The authors in Ref. [83] studied a wind hydro system as a hybrid system that uses the squirrel cage induction generator (SCIG). In this analysis, the hybrid hydroelectric wind system includes a 50 kW wind turbine and a 50 kW hydroturbine. In addition, the nominal power of the squirrel cage induction generator is the same as that of the wind turbine with a power of 50 kW. The results of this study showed that renewable energy sources, such as hydropower, solar, and wind energy, complement each other. However, there are constraints for many isolated sites because they cannot be connected to the grid, while they have interesting renewable energy potentials. For such sites, a three-phase four-wire wind hydro and wind solar hybrid system was modeled using cage generators and a photovoltaic system (PV), as well as a battery energy storage [83]. Castronuovo et al. investigated the optimal methodology to coordinate wind power and hydro pumping storage. To increase the profit that could be generated from the use of wind energy, optimization is necessary.

This could be achieved by the incorporation of hydro storage in the plant. The authors describe an approach to analyze the wind storage combination as a solution to overcome wind variability. This approach is based on the probabilistic forecasting and optimization to make decisions regarding wind-storage system. A study in the Iberian Peninsula was conducted and described by the authors in Ref. [84]. The performed case study includes two wind farms with a total capacity of 352 MW and a pumped-storage hydroelectric power station with a generation capacity of 90 MW and a pumping capacity of 184 MW. The efficiency of the hydro station is 92%, while the pumping efficiency is 84%. The electricity supply can be injected directly from the two wind farms or from the stored energy. To predict wind power production of each wind farm, the ANEMOS platform was used by the probabilistic forecasting model. This study has shown that the combination of wind energy and hydro storage can increase profit by 11% [84].

The combination of wind energy and hydroelectric power is beneficial for obtaining electricity at low prices and for avoiding energy shortage when the wind does not blow. A large area for the reservoirs is needed, which is considered a constraint in most cases. However, if the plant is installed close enough to the sea, this problem is solved. This hybrid system should mainly be used in areas where there is wind potential in exploiting this renewable source in an efficient and effective way. The results of the case studies presented in these sections encourage the development of wind-hydro systems. This combination of renewable energies will lead to the production of electricity from 100% renewable energies.

1.5 Hybrid systems with concentrated solar power (CSP)

One of the promising and fast-growing energy generation technologies, which relies on solar energy, is concentrated solar power or concentrating solar power (CSP). The basic working principle of this technology consists of harnessing thermal energy from the sun and converting the captured heat into mechanical energy to generate electricity. The technology is, therefore, only suitable and best suited for regions with high solar direct normal irradiation. Moreover, electricity production with CSP plants is also sensitive to other weather conditions including the ambient temperature, humidity, and wind speed [85].

The world's total installed capacity of CSP increased from around 354 MW back in 2005 to around 5500 MW in 2018, with Spain being the world leader and accounting for almost half that capacity with its

2.3 GW, then comes the United States with a total installed capacity of CSP of approximately 1.7 GW. Morocco is also contributing to that total installed capacity with its world's biggest CSP Project—Noor Power Station located in Ouarzazate, Morocco. The total installed capacity in Morocco exceeds 500 MW. The country has also set an ambitious target consisting of producing 2 GW from solar power by 2020 [86].

The major challenge associated with the integration of renewable technologies, which prevents their wide deployment, is the intermittent nature of renewable energy sources. Maintaining the balance between energy supply and energy demand at all times becomes a great issue with the fluctuations and the unpredictability in the energy input. Appropriate energy storage increases the operational flexibility of such energy systems and enables producing energy around the clock, i.e., from wind and solar technologies even when the wind is not blowing and the sun is not shining.

1.5.1 CSP working principle

All existing CSP systems use lenses or mirror configurations called reflectors to concentrate the sun's rays onto a small area, referred to as a solar receiver, where the sun's energy is converted into high-temperature heat. The absorbed heat is then exploited to produce mechanical work that will drive a power cycle. The latter is similar to the one used in conventional power plants and, ultimately, produces electric power (solar thermal electricity—STE) [85]. CSP systems consist of two main parts. The first one is used to collect solar thermal energy and convert it into heat. The second one converts that heat into electricity. There are four types of CSP systems that include the parabolic trough, the power tower, the Fresnel mirror system, and the dish parabolic system. In this section, further details about each CSP technology will be provided.

1.5.1.1 Parabolic trough

Parabolic trough systems, also known as linear concentrator systems, use U-shaped mirrors to focus most of the thermal energy they receive from the sun onto a receiver; called a heat absorber or a heat collector. The latter is in the form of a long pipe located at the center, specifically at the focal line of the parabolically curved reflectors. This is filled with a heat transfer fluid, which holds the heat well, such as synthetic oil or molten salt. The fluid absorbs heat from the reflected sun rays and becomes superhot, reaching temperatures of around 390°C. It will then go through a heat exchanger to heat water that circulates in the latter and turns it into steam. The steam will expand into a standard steam turbine that will spin a generator and generate electricity. After driving the turbine, the steam is recovered to be condensed and recycled over and over in the power generation systems. The same applies to the working fluid, which also gets recycled and is used again after transferring its heat to water. A typical solar collector field comprises multiple parallel rows of parabolic troughs aligned and oriented along a north-south axis in a way that allows for tracking the sun's east-to-west movement during the day. This positioning increases the exposure time of the solar reflectors to solar radiation and ensures that the sun is continuously focused on the absorber tube. Coatings on the latter allow for maximizing the solar radiation absorption and the use of an evacuated glass envelope around the pipe. Thus, it reduces heat losses to the surrounding.

1.5.1.2 Power tower systems

Power tower systems rely on large flat mirrors, called heliostats, which track the sun throughout the day and focus its rays onto a receiver located at the top of a centrally located and elevated tower (central receiver system). Inside the receiver, a heat transfer fluid, which retains heat efficiently, gets heated to around 600°C and produces steam. The latter expands into a turbine to, eventually, produce electric power. Earliest power tower systems used water as the working fluid. Water gets pumped to the receiver at the top of the centered tower, it gets heated into steam by the concentrated solar thermal energy, and then, if no storage is desired, the steam is directly used to drive the traditional power cycle that generates electricity as done in conventional power plants (i.e., Rankine power cycle). The superhot steam can also be stored in a tank for later use when the clouds block the sun or for a few hours after dark, for instance. Due to the low heat transfer and weak energy storage capabilities of water, other fluids, such as molten salt, were then preferred over water in this type of CSP system.

The working process is quite similar to the direct steam design of power tower technology except that water is replaced by cold molten salt. The latter gets pumped to the receiver to be heated; it then flows back and through a heat exchanger, and its heat gets transferred to water to produce steam. Electricity is then produced in the power generation block of the system using steam. The system also allows for storing the heated salt in a hot storage tank instead of using it immediately for making steam that will be used in the conventional Rankine power cycle.

1.5.1.3 Compact linear Fresnel reflector

Linear Fresnel technology uses flat mirrors that receive solar rays and concentrate them into a tube running through the focal point of the reflectors. The pipe has water flowing inside it, and hence, steam is directly generated and used for power generation. Linear Fresnel collectors are arranged the same way as parabolic trough CSP systems, i.e., in the form of a large number of long arrays, which are aligned in a north-south orientation to maximize the collection of solar energy. The Linear Fresnel technology offers the possibility to store energy for later use instead of directly using the steam for generating power.

1.5.1.4 Parabolic dish systems

Parabolic dish systems use mirrors that are mounted over a parabolic-shaped dish to focus the sun's rays onto a receiver. The latter is mounted at the focal point of the dish along with a heat engine (Stirling or Brayton cycle engine), which has thin tubes inside it. The tubes contain a gas, such as helium or hydrogen, or even air. As the collected thermal energy from the sun falls on the receiver, it gets heated to very high temperatures reaching approximately 750–1000°C. The hot gas inside the heat engine tubes expands inside the cylinders and drives their pistons, which then turn a crankshaft that drives an electric generator producing electricity. The working principle of this system is similar to the one used by cars' internal heat engine except that in the case of dish engines, the heat that is applied to the gases inside the engine comes from an external source, that is the sun. This is why in the case of parabolic dish systems, the heat engine is referred to as an external combustion heat engine.

As is the case for other CSP technologies, the parabolic dish systems are also developed to rotate and track the sun's path throughout the day. The tracking system positions the system reflectors in the most optimal location to harness the maximum amount of solar energy and focus it on the heat absorber.

1.5.2 CSP with thermal energy storage

The main challenge associated with the use of renewable energy sources to produce power is their intermittent nature and irregularity. In fact, wind, PV, and CSP without storage are known as variable energy resources. This issue greatly affects the electric power system reliability. It also makes it inevitable to complement these clean sources with conventional ones to ensure continuous and dispatchable energy supply (i.e., available on demand) and meet the rapid spikes in energy demand [87].

The traditional grid stability and balance are also affected by the penetration of renewable energies, in the sense that the grid was first built and designed to integrate the conventionally produced electricity that relies on fossil fuels. In other words, the conventional grid is stable and well balanced as the electricity that is fed to it and that it transmits and distributes to end users is predictable, foreseeable, and controllable. As a matter of fact, when the energy demand is forecasted, the energy producers can determine exactly how much fuel should be fed to the system to meet that demand and vice versa, i.e., when the amount of available fuel is known, the energy output that could be generated from it is well known, as well. The margin of error is very minim in the traditional power systems [87]. On the other hand, power generation from solar energy, as an example of renewable energy sources, is not stable and hard to predict. Hence, it could lead to production forecast errors. As it is known, even during the sunrise-to-sunset period during which the sun is supposed to be shining, there may be some cloudy periods that could affect the energy output of the power station and the stability of the grid as a whole. The grid balance and the overall utilization of the CSP plant may also be affected during times of excess energy-generation that gets lost when no storage options are available. To overcome the previously stated challenges and performance limitations related to the CSP, the plants must be equipped with thermal energy storage (TES). The latter acts like a battery that stores some of the solar thermal energy captured when the sun is shining during the daylight hours and shifts or delays the energy production to when needed, for example, during cloudy periods of the day or beyond the sunrise-to-sunset period [87]. The integration of thermal energy storage in CSP plants provides flexibility in the power system operation, ensures long-term reliability, and helps control the power generation and output of the power system, making the technology a dispatchable source of electricity. Through the use of different methods to store excess solar energy, CSP plants can operate independently without the need for a backup traditional fossil-fueled thermal power plant and to provide not only peak power but also the baseload power. As a matter of fact, when 10h of storage are available, the CSP plant can produce a relatively stable energy output throughout the day and night, which puts the technology at a position to occupy a larger percentage of the energy mix [88]. All of these benefits are added to the environmental friendliness of the technology as both the generated and the stored thermal energies are absorbed from the sun.

Hence, it is certified as renewable energy that will contribute to a clean energy future. Moreover, as the energy required for heating the storage medium is harnessed from the free solar energy and not from the plants' electric power output, minimal cost is required for the storage of energy in CSP plants [87].

1.5.2.1 Thermal energy storage working principle and types

Thermal energy storage (TES) is defined as a system that enables the transfer and storage of heat energy. This thermal storage in CSP plants allows for storing surplus available energy during low energy demand periods to be used subsequently. The working principle of CSP technology explained previously remains the same, except that instead of directly producing energy from the absorbed sunlight heat, the superhot fluid in this case flows to an insulated thermal storage tank and is kept as standby. In the latter, the energy is stored as high-temperature molten salt, which acts as both the heat transfer fluid and the heat storage medium in several existing and planned CSP plants. The reason why salts are preferred as fluid as they are considered an ideal heat capture medium that remains stable at high temperatures (up to 600°C) and they maintain a wide operating temperature range in the liquid state, which makes them suitable for conventional Rankine steam power cycles used in CSP systems. Continuous research and experimentation are needed to determine alternative heat transfer and storage media that would offer superior heat-transfer and thermal storage capabilities and, thus, increased efficiency. An example of a material that is under consideration is molten glass, which can withstand even higher temperatures than salt. When solar-generated power is needed, the stored molten salt flows into a heat exchanger. Water is piped into the latter simultaneously. The high-temperature fluid transfers its heat to water flowing in the heat exchanger creating superheated steam. The latter is then used to drive the conventional power block and produce electric power. After the creation of steam using molten salt, the latter is piped back into a storage tank. The cool salt will later be reheated in the receiver, and the process will continue. Likewise, after the steam is used to turn the turbine, it is sent to the condenser to convert it back to water. The latter is then stored in the water holding reservoir.

1.5.2.2 Thermal energy storage options

There exist three main thermal energy storage (TES) options that could be integrated into CSP plants, which are

- Sensible heat storage
- Latent heat storage
- Thermochemical storage

The most widely used form of thermal energy storage in the CSP sector is the sensible heat storage, which is considered relatively less expensive and less complicated compared to the two other storage options. In general, sensible heat storage relies on a liquid or solid storage medium (molten salt, sand, or rock) that gets heated or cooled to store energy. In the case of CSP systems, the storage medium gets heated and not cooled. Most current commercial applications use molten salt as explained previously.

Latent heat storage is not commonly used even though it has been proved to be a promising technology in some designs of CSP systems, especially because it offers higher energy density when compared to the sensible heat storage. In fact, the latter requires large volumes of storage medium due to its low energy density. To store thermal energy, this technique relies on phase change materials as the storage medium to take advantage of the latent heat absorbed/released as the medium changes its state at constant temperatures.

The third option for thermal energy storage is the thermochemical storage, which, as its name implies, relies on chemical reactions to store heat. The technology offers even higher energy density, hence, higher storage capacity, than the latent heat storage technique. However, the cost of implementation and required equipment procurement are very high, which hinder the deployment of this storage technique.

1.5.3 Hybridization of CSP

Besides incorporating energy storage, the combination of solar thermal power with a backup power system is a promising method to provide dispatchable power from solar energy and to avoid grid reliability issues associated with renewable energies. The most common form of hybridization for CSP systems is the use of a fossil fuel-based system that would operate in the absence of solar energy [89]. The hybridization of CSP technology could also be done with some other types of renewable technologies, such as biomass, geothermal, PV, or wind.

Concentrated solar power is considered an ideal renewable energy technology to hybridize with other power production technologies as many benefits could be obtained from solar-aided power systems. No matter what types of CSP technology and hybrid energy sources are used, the main advantages of thermal solar hybrid systems include the following [89–90]:

- **Reduced capital cost through the use of similar equipment used by traditional systems**: As explained in the previous sections, CSP relies on the conventional power generation blocks, which are similar to the ones used in traditional fossil-fueled power plants. Therefore, in the case of a hybrid solar-fossil fuel system, no extra investments are needed for the procurement, installation, and integration of some equipment used in power production.
- **Increased dispatchability and reliability of the power system**: As stated earlier, the variability of solar energy affects the continuous supply of electricity. When combining solar thermal energy with some other types of more reliable power generation technologies, the on-demand electricity supply can be ensured. The second system would act as a backup that provides the necessary energy input when the sun is not shining. The two systems would, thus, complement each other to ensure reliable power supply.
- **Flexible operation**: Since different systems (at least two) are available for electricity generation in a hybrid system, the system operators will have the flexibility to switch and alternate between the two or more energy sources used or combine them, as well.
- **Operation optimization**: The possibility to choose allows for optimal use of the different energy sources that are available in the hybrid system based on their availability, prices, and whether there are any restrictions with regard to the allowed harmful emissions and so on.
- **Increased capacity utilization and improved efficiency of the power plant**: In some cases, the combination of two (or more) electric power generation systems creates a synergy that improves the whole system efficiency and capacity utilization. As a matter of fact, when the energy collected from sunlight is used at high temperatures, the conversion efficiency of the thermal cycle is significantly improved. However, as the solar radiation may not be all the time as high as necessary, the boosting aspect of an additional system would bring the steam to the required temperature and help in achieving the desired efficiency and making full use of the system capacity. Another example would be to increase the hot temperature T_H of the system even further and, hence, make the difference between T_H and T_C higher and improve the overall system efficiency. No need to mention that the maximum temperature

allowed inside the system should not be exceeded to avoid any equipment damage or safety issues.

- **Reduced greenhouse gas emissions**: The possibility to use both thermal energy from the clean source (the sun) and the dirty one through the integration of CSP in conventional power stations could help reduce the reliance on fossil fuels. Instead of getting all the necessary heat from fuel combustion, one can get a fraction of it, if not all of it (when solar energy is abundant), from the solar collected heat, and eventually reduce the polluting aspect of traditional power plants.

Depending on the energy sources used in the solar hybrid system, the CSP technology used, the power station configuration, and design, different advantages could be achieved. Some of the benefits are applicable to all hybrid CSP hybridized systems.

1.5.3.1 Hybridization of concentrating solar power (CSP) with coal-fired power plants

As coal is considered a cost-effective and abundant energy source, it provides various opportunities while being combined with solar thermal energy. This kind of hybrid system, which relies on solar and coal-fired steam production, is most suitable for countries that have significant coal resources (Australia and China [89]) or for countries that already have coal-fired power stations operating. The hybrid system combines the benefits of both using coal, which is a low-cost, dispatchable, and reliable source of energy, and the sun, which is abundant, free, and environmentally friendly [90]. Such systems are also considered flexible in hybridization as they allow for easily injecting the solar heat at different levels of the power cycle as will be explained below [91].

The main challenge that could be faced, however, when trying to hybridize existing coal-fired plants with CSP technology is the fact that the majority of these aging plants have been operating for so many years. Their equipment may not match the new equipment used in the new technologies; they may also be approaching their end of life, making it inevitable to make extra investments for replacing the old equipment.

Several benefits are achieved with the hybridization of coal with CSP. This includes reliability, low cost, and higher energy production. Hybridizing with coal guarantees reliability as coal is considered a dispatchable fuel source. Additionally, the total cost of constructing a solar thermal plant is reduced with the utilization of coal. The authors in Ref. [91] have shown that the overall cost of a CSP plant is 28% higher than the cost of a hybrid solar/coal plant. In addition, the latter would produce over 25% more energy than a stand-alone solar farm [92]. Solar heat can be injected at multiple points in a coal power plant such as preheating of boiler air, boiler preheating of feedwater, postcombustion for the regeneration of solvent, and generation of direct steam [91]. Furthermore, the use of CSP with coal plant reduces the total CO_2 emissions [93]. There exist different solar-aided system configurations in which the solar field works in parallel to the coal-fired conventional power station to generate electricity.

- *Direct steam generation*

One of the possibilities for integrating CSP technology into coal-based power stations consists of generating the steam that will drive the typical Rankine power cycle of the plant using solar thermal energy and then make use of the traditional boiler to superheat the water. The solar fraction, which refers to the percentage or fraction of energy provided to the power cycle from the sun, in the hybrid system designs explored in the literature so far does not exceed 10%. However, even though the solar fraction is relatively low, the conversion efficiency from solar energy to electric energy is higher than that of typical stand-alone CSP plants [90].

In his work [94], Zhang et al. found out that generating steam from a power tower solar system and then supplementing the necessary heat to the steam using the coal-fired boiler improve the solar-to-electric efficiency from around 15% to 20% (efficiency of a stand-alone system) to around 28%.

The hybridization of CSP with coal helps improve the efficiency of the system through reaching superhot temperatures. As a matter of fact, coal combustion in the coal-fired boiler helps raise the solar-generated steam temperature to as high as 500°C as opposed to less than 400°C, which is usually the maximum temperature that parabolic trough systems are limited to [90].

- *Preheating the boiler feedwater*

In a traditional coal-fired plant, a feedwater heater is used to preheat the water that is delivered to the boiler. The energy required to bring the feedwater to the desired temperature is usually derived from valuable steam that gets extracted from the turbine. This extraction affects the thermal efficiency and, hence, the power output of the system because part of the steam that should have been used to perform expansion work in the turbine and generate power gets redirected to mix with feedwater and heat it.

In a hybrid solar-coal fired system, the thermal energy captured from the sun in the CSP field could be used to preheat the feedwater to the required temperature (around 220°C) before it gets fed to the boiler. As a result, no steam needs to be extracted from the steam turbine as all the feedwater preheating requirement will be extracted from solar thermal energy. When the latter is not available, the system operates in a conventional way getting its required heat from the steam extracted from the turbine [89–90].

Among the benefits associated with these hybrid systems, we could distinguish the following:

- The power output of the plant after hybridization is improved as the steam that is normally redirected to the feedwater heater will be further converted into mechanical work in the turbine. Therefore, after hybridization, the same amount of input coal would provide more electric power output with no extra emissions and without the need to oversize the turbine.
- The solar heat is usually collected at lower temperatures of around 300°C; as a result, the solar field thermal losses, which are proportional to these temperatures, are reduced [90].
- Another advantage of using solar heat to preheat the feedwater is increased efficiency of the solar-to-electricity conversion. A study conducted by Zhao et al. [95] proved that this type of hybridized configuration results in around 21.2% solar-to-electricity conversion efficiency, as compared to 19.4% in a solar-only thermal power station [90].
- Another study conducted by Feng et al. and published in Ref. [96] explored a different system configuration that combines both the solar-generated steam and steam extracted from the turbine to be used for preheating feedwater. This design reduces the system extraction fraction (i.e., the percentage of the steam mass flow extracted from the turbine and used for the feedwater heater in traditional plants) and, hence, the power output losses. However, and as expected, the power output is not as high as when all the feedwater preheating is done from solar-generated steam. The author also tested injecting the solar thermal energy at different stages of the power generating station, and the maximum net solar thermal efficiency reached was around 28.5% [6]. Yan et al. also explored different designs for the solar-aided preheating of feedwater with different sizes for the CSP plant and found out that the highest reachable efficiency is around 40% [89,90,97].

1.5.3.2 Hybridization with natural gas

One of the most interesting hybridization of CSP is the incorporation of natural gas in the solar power plant. Power production from natural gas has gained popularity lately due to the different characteristics it offers; such as easy transportation through pipelines and high energy density. Natural gas has a lower emission than coal because of its molecular structure-low carbon-to-hydrogen ratio (mainly CH_4). Similar to coal fuel, natural gas power can be hybridized with CSP in many different ways. As the operation of gas turbines requires very high temperature compared to steam turbines, the collector end will be facing some technical challenges as the injection of heat through solar necessitates higher temperatures (usually above 1000°C) [90].

The concept of solar-gas turbines, however, has the potential to achieve high solar portions by replacing high-temperature heat typically provided by natural gas, while other systems, such as solar steam generation with combined cycle plant, depend on fossil fuel as the main heat source. Gas turbine generation uses the Brayton cycle and usually works as an open system. In the latter, the combustion of natural gas heats the compressed atmospheric air, which drives a steam turbine coupled to a generator. Gas turbines' operating flexibility enables the system to combine with an intermittent source, such as solar thermal energy. Flexible operation capability is achieved because of the adjustable flow of air and natural gas via vanes, which enables the operation of the turbine to regulate diverse solar conditions. Solar heat is usually incorporated in the after-compression and precombustion process, where the utilization of fuel combustion would maintain acceptable operating temperature, irrespective of the usable sunlight at any given time. The solar fraction of the power plant can also be significantly improved by the use of thermal energy storage.

For a Brayton cycle power plant, the direct use of solar energy typically necessitates the use of a pressurized receiver because the best way to add solar heat is usually postcompression. It typically makes use of significantly higher temperatures than the Rankine cycle. The concept has been demonstrated at a large scale (MWs). Through heat recovery, the performance of simple cycle gas turbines can also be improved if the waste heat from the flue gas is used to preheat the air before reaching the solar heater component. The overall efficiency of the system has been predicted to reach 35% [98]. Santos et al. have shown that fuel savings of about 11% can be obtained using the aforementioned recuperation approach in the summer months. However, in winter months, this savings is significantly reduced to 4% [99].

The world's first integrated solar combined cycle (ISCC) plant has been built by a Spanish company at Ain Beni Mathar, in Morocco. This power firm has a power capacity of 470 MW; it makes use of a parabolic trough system to integrate solar thermal technology to a conventional gas-fired power plant. The solar portion supplies 20 MWe with an output of approximately 450 MW from traditional thermal plants. The net annual production is estimated to be 3538 GWh per year. The plant occupies a solar area of 180.000 m^2, using 224 solar collector units in 56 loops. The heat transfer fluid inside the thermo-oil is injected to the power block at about 400°C. Although thermal storage capacity is not included in the current generation of ISCC installations, this technology enables CSP to extend its operating range.

The development of efficient high-temperature thermal energy storage systems will encourage the construction of solar CSP/natural gas power plant operating with solar thermal energy, natural gas, and energy storage. This approach would result in soundly cost-effective hybrid technologies with the flexibility to ultimately become almost environmentally friendly while benefiting from the dispatchability aspect offered by natural gas.

1.5.3.3 Hybridization with biofuels

Similar to fossil fuels, good synergies are achieved with the hybridization of solar thermal energy with biofuels. This includes flexible, dispatchable, and reliable operation. In addition, this hybridization has the ability to offer 100% renewable power source [100]. Replacing fossil fuel with biomethane or biogas removes almost all the negative environmental impacts caused by the burning of fossil fuel. Various biomass sources can be used in combination with CSP such as wood waste, bagasse, forestry residues, and stubble. Furthermore, the production of biofuels can also make use of CSP for the provision of additional heat source. This hybridization enables the production of liquid transportation fuels and syngas, which is considered a storage mean. The latter mechanism enables the storage of long term energy. The configuration of CSP with biofuel hybrid system is similar to the hybridization of fossil fuel with solar energy as biofuels are combustible fuel.

1.5.3.4 Hybridization with geothermal

Geothermal energy is another power source that could be combined with solar thermal to create a hybrid renewable system. Similar to CSP with biofuel, this hybrid system has the potential to provide 100% sustainable energy. One major advantage of this hybridization is that both geothermal and solar energies are carbon-free. Even though low-temperature heat is produced from geothermal energy, the efficiency of the hybridized system can be improved by the high temperatures collected from solar thermal energy. A low capacity factor is obtained from combining solar with geothermal in the absence of thermal energy storage. The lower grade heat generated from geothermal needs to be utilized in a certain manner when solar heat is not available to enhance it. Researchers have proposed multigeneration systems with multiple heat-injection points.

1.5.3.5 Hybridization with photovoltaics

Solar photovoltaic energy is another common system that is used for hybridization with CSP. Very good synergies are obtained from the combination of the aforementioned technologies even if they both make use of solar power as an energy source. Power output and efficiency of PV cells are reduced with high temperature, resulting in degraded performance. It is, therefore, necessary to either passively or actively cool the photovoltaic cells to enable them to perform optimally. Rather than cooling them, the wasted thermal energy is used by cogeneration to enhance the total efficiency of the cell. This has induced the introduction of photovoltaic thermal solar collectors (PVT). Michael et al. [101] and Makki et al. [102] performed a comprehensive review of this hybrid system. In this case, the solar thermal energy is used for low-temperature thermal applications, while PV cells are used for the production of energy. This PVT technology is different from other hybrid systems discussed in this section as they mostly make use of CSP mainly for power generation.

1.5.3.6 Hybridization with wind

Because wind power generation technology is so different from conventional thermal generation technology that CSP is founded upon, opportunities to hybridize CSP with wind are less prevalent in literature. This hybrid system would be considered weak as CSP and wind technologies are typically only coupled at the grid level. Still, the combined power generation profiles of wind and CSP can provide benefits, such as a better fit for consumer demand or improved grid stability [103]. A study by Vick and Moss showed that CSP was a match with wind farms to meet the electrical demand in the Texas Panhandle. They proposed a share of 67% from wind with 33% from CSP (6 h of TES) to meet the electrical demand of that area. Although the hybrid system was more expensive (on an LCOE basis) than a pure wind farm, it had better ability to meet peak demand than a wind farm [104]. Similarly, Reichling and Kulacki found that wind power had lower cost in Minnesota case study, but electrical load matching favored the hybrid combination of wind and CSP [105].

One key benefit of CSP with wind is its capability to store energy at a low cost using TES, which has the effect of helping to stabilize the grid by providing a semidispatchable power source. Thermal storage is such a benefit to CSP as a technology that is why its use has also been proposed for implementation with the wind to provide a stabilized power output. Although this configuration does not use solar, it does provide a rare example of the use of CSP technology to enhance wind power with hybridization at the component level, rather than electrically at the grid level only.

1.5.3.7 Hybridizing CSP with other sources

Table 1.2 presents the different advantages and drawbacks of the discussed CSP hybrid systems. Although this hybridization may offer a number of advantages, it sometimes may hinder a concentrated solar power system. Combining wind energy with CSP does not provide many synergies. Their hybridization means that only the use of their power outputs is scheduled together. Whereas each system has its own merit, it is usually irrational to combine them at the component level. On the other hand, the combination of geothermal with CSP is considered a dispatchable and all-renewable solution. Even though the combined system efficiency is limited by the low-temperature power cycles needed by geothermal, finding an alternative use for low-temperature heat can enhance the overall system efficiency. However, it is difficult to find regions with adequate geothermal and solar thermal resources with several low-temperature heat sinks.

PVT system has several advantages; It does not depend on a solar thermal system for energy production and might not be suitable for centralized energy generation. This could necessitate colocation of a bulk plant and a useful heat sink for the provision of low-temperature thermal energy. While this hybrid technology has multiple advantages, it does not intrinsically improve dispatchability as both solar thermal and photovoltaic systems still depend on variable solar source.

Hydrocarbon-CSP hybrid systems also have various advantages such as dispatchability as fuel can be combusted on demand. However, the use of fossil fuels by the hybrid system is not an environmentally friendly option. Similarly, hybrid CSP systems with biofuels offer dispatchability. The energy produced by the latter is purely renewable and carbon-free. An ideal hybridization of CSP system with other energy sources would result in the following:

- A decreased plant LCOE as compared to a stand-alone CSP plant
- An increased plant efficiency as compared to a stand-alone CSP plant

Table 1.2 Comparison of hybridizing CSP with other energy sources.

Source	Disadvantages	Advantages
Natural gas	• Integrated Thermo Solar Combined Cycle (ISCC) has rather a low share of solar energy. • More development is still needed in high-temperature solar systems	• Various configurations • Low cost • Low emissions • Saving in capital cost due to the share of equipment in the hybrid system. • Dispatchability • Efficient operation (ISCC). • Provide different possible solar injection spots at various temperature level.
Geothermal	• Low efficiency power cycles. • Corrosion due to geothermal brine. • Solar equipment scaling issue. • Colocation of geothermal, solar, and heat sinks is needed for multiple generation technologies.	• Dispatchabilty. • Make use of low cost and temperature solar collectors. • Saving in capital cost due to the share of equipment in the hybrid system. • Potential for improvement of efficiency due to multigeneration • Opportunity to produce 100% renewable energy.
Coal	• High emissions • Development in technology is needed for high-temperature gasification. • Retrofit issue. • Limited efficiency due to Rankine cycle.	• Dispatchability • Low cost • Provide different possible solar injection spots at various temperature level. • Abundant fuel source. • Saving in capital cost due to the share of equipment in the hybrid system.

Continued

Table 1.2 Comparison of hybridizing CSP with other energy sources—cont'd

Source	Disadvantages	Advantages
Photovoltaic	• Not dispatchable. • Heat sink is needed for thermal energy.	• PV system efficiency can be improved with thermal heat removal. • Utilization of wasted thermal energy.
Wind	• No saving in capital cost from the sharing of equipment in the hybrid system. • Low synergy at system component level (equipment) • Not dispatchable.	• Opportunity to produce 100% renewable energy. • Reliability of several energy availability profiles.
Biofuels	• Limited availability necessitates the use of other fuels. • Development in technology is needed for high temperature gasification.	• Saving in capital cost due to the share of equipment in the hybrid system. • Dispatchabilty • Provide different possible solar injection spots at various temperature level. • Various configurations • Opportunity to produce 100% renewable energy.

- Optimal plant performance
- Flexible operation
- Better capacity factor
- Improved plant reliability
- Lower CO_2 emissions as compared to conventional plants.

There are several obstacles facing the development of large-scale renewable energy plant such as cost-effectiveness. The implementation of large scale renewable generation systems requires the use of fossil fuel sources as backup in the plant to overcome the variability nature of renewable energy sources. It is important to investigate synergistic hybridization at the facility level that can improve the total efficiency of the plant and increase the utilization of intermittent energy sources.

1.6 Conclusion

This chapter presented an overview of different hybrid energy systems. The electricity production system can be optimized from both the economic and the technical point of view through the combination of renewable energy sources and conventional power systems. There exist a number of combinations of hybrid energy technologies such as photovoltaic-conventional source, wind-diesel and wind-diesel-battery, wind-photovoltaic-diesel, wind-photovoltaic, and PV-energy storage, and others. The same energy sources can be hybridized with other energy sources such as hydropower. In addition, hybridization of CSP can be achieved with the use of other energy sources including fossil fuel, biomass, geothermal, photovoltaic, and wind. The combination of these energy sources has a number of advantages and disadvantages. While each hybrid system has its own merit, it depends on the economic and the technical models of each system.

Cost competitiveness is one of the main difficulties faced in the implementation of large-scale renewable energy power plants. Further, to overcome the variability of the latter systems, the use of fossil fuel as a backup is required in order to enable an easier penetration of renewable generation technologies at large scale. Therefore, it is necessary to explore the synergies existing between the different hybrid systems because their configuration consists of a mix of several energy systems at the grid level. This could be performed to enhance system reliability, efficiency, and cost-effectiveness. In the immediate future, hybrid energy systems will allow a deeper and higher penetration of renewable energy systems into the power market.

References

[1] IEA. World Energy Statistics and Balances (database), 2019, www.iea.org/statistics. [Accessed on 15 April 2020].

[2] Shaahid SM, Elhadidy MA. Prospects of autonomous stand-alone hybrid (photovoltaic + diesel + battery) power systems in commercial applications in hot regions. Renew Energy 2004;29:165–77.

[3] Muñoz J, Narvarte L, Lorenzo E. First operating year of two village PV-diesel plants in the south of Morocco. In: 19th European photovoltaic solar energy conference, Paris, France, 7–11 June; 2004. p. 3462–5.

[4] van Sark WGJHM, Lysen EH, Cocard D, Beutin P, Merlo GF, Mohanty B, van den Akker J, Razzak Idris A, Firag A, Waheed A, Shaheed A, Latheef M, Wajeeh A. The first PV-diesel hybrid system in the Maldives installed at Mandhoo Island. In: 21st European photovoltaic solar energy conference, Dresden, Germany, 4–8 September; 2006. p. 3039–43.

[5] Muselli M, Notton G, Poggi P, Louche A. PV-hybrid power systems sizing incorporating battery storage: analysis via simulation calculations. Renew Energy 2000;20:1–7.

[6] Klinghammer W, Nörenberg K. Wide-scale village electrification with PV hybrid power systems in western China-experience gained. In: 21st European photovoltaic solar energy conference, Dresden, Germany, 4–8 September; 2006. p. 3023–6.

[7] Tina GM, Brunetto C, Gagliano S, Petino S, Guerra M, Schioppo R, Candio A. Monte Aquilone hybrid photovoltaic-diesel power generation system testing site-experimental tuning of subsystem models. In: 20th European photovoltaic solar energy conference, Barcelona, Spain, 6–10 June; 2005. p. 2319–22.

[8] Ramesh M, Saini RP. Dispatch strategies based performance analysis of a hybrid renewable energy system for a remote rural area in India. J Clean Prod 2020;259:120697. https://doi.org/10.1016/j.jclepro.2020.120697.

[9] Oldach R, Bates J, Derrick A, Gantulga D, Hasnie S, Enebish N. PV hybrid system for a remote village in Mongolia. In: 19th European photovoltaic solar energy conference, Paris, France, 7–11 June; 2004. p. 3540–3.

[10] Wickert B, Nayar CV, Lawrance WB. Photovoltaic-diesel hybrid energy systems for off-grid rural electrification. Int J Renew Energy Eng 1999;1(1):7–17.

[11] Mahian O, Javidmehr M, Kasaeian A, Mohasseb S, Panahi M. Optimal sizing and performance assessment of a hybrid combined heat and power system with energy storage for residential buildings. Energy Convers Manage 2020;211:112751. https://doi.org/10.1016/j.enconman.2020.112751.

[12] Shaahid SM, Elhadidy MA. Opportunities for utilization of stand-alone hybrid (photovoltaic + diesel + battery) power systems in hot climates. Renew Energy 2003;28:1741–53.

[13] Suponthana W, Ketjoy N, Rakwichian W. Transforming solar home system to village grid system by using PV-farmer diesel hybrid system. In: 21st European photovoltaic solar energy conference, Dresden, Germany, 4–8 September; 2006. p. 3011–5.

[14] Khawaja Y, Allahham A, Giaouris D, Patsios C, Walker S, Qiqieh I. An integrated framework for sizing and energy management of hybrid energy systems using finite automata. Appl Energy 2019;250:257–72. https://doi.org/10.1016/j.apenergy.2019.04.185.

[15] Monné C, Bravo Y, Moreno F, et al. Analysis of a solar dish–Stirling system with hybridization and thermal storage. Int J Energy Environ Eng 2014;5:80. https://doi.org/10.1007/s40095-014-0080-x.

[16] Wu SY, Xiao L, Cao Y, Li YR. A parabolic dish/AMTEC solar thermal power system and its performance evaluation. Appl Energy 2010;87(2):452–62.

[17] Kleih J. Dish-Stirling test facility. Sol Energy Mater 1991;24(1):231–7.

[18] Nepveu F, Ferriere A, Bataille F. Thermal model of a dish/Stirling systems. Sol Energy 2009;83(1):81–9.
[19] Balzar A, Stumpf P, Eckhoff S, Ackermann H, Grupp M. A solar cooker using vacuum-tube collectors with integrated heat pipes. Sol Energy 1996;58(1):63–8.
[20] Grupp M, Balmer M, Beall B, Bergler H, Cieslok J, Hancock D, et al. On-line recording of solar cooker use rate by a novel metering device: prototype description and experimental verification of output data. Sol Energy 2009;83(2):276–9.
[21] Muthusivagami RM, Velraj R, Sethumadhavan R. Solar cookers with and without thermal storage—a review. Renew Sustain Energy Rev 2010;14(2):691–701.
[22] Badran AA, Yousef IA, Joudeh NK, Al Hamad R, Halawa H, Hassouneh HK. Portable solar cooker and water heater. Energy Convers Manage 2010;51(8):1605–9.
[23] Mohammed IL. Design and development of a parabolic dish solar water heater. Int J Eng Res Appl 2012;2(1):822–30.
[24] Manukaji JU, Akinbode FO. Design and performance analysis of a solar cooker using Fresnel reflectors. Spect J 2002;9(1).
[25] Dafle VR, Shinde NN. Design, development & performance evaluation of concentrating monoaxial Scheffler technology for water heating and low temperature industrial steam application. Int J Eng Res Appl 2012;2(6):848–52.
[26] Lemmer V. Comparative market analysis and economic simulation for Morocco of the parabolic trough and dish CSP technologies [Doctoral dissertation]; 2014.
[27] Anon, http://www.solarpowerworldonline.com/2010/01/how-do-solar-parabolic-dishes-work/.
[28] Hafez AZ, Soliman A, El-Metwally KA, Ismail IM. Solar parabolic dish Stirling engine system design, simulation, and thermal analysis. Energy Convers Manage 2016.
[29] Stone KW, Lopez CW, McAlister R. Economic performance of the SCE Stirling dish. In: Proceedings of the IECEC, Atlanta, Georgia; 1993.
[30] Carta JA, González J, Gómez C. Operating results of a wind-diesel system which supplies the full energy needs of an isolated village community in the Canary Islands. Sol Energy 2003;74:53–63.
[31] Notton G, Cristofari C, Poggi P, Musseli M. Wind hybrid electrical supply system: behaviour simulation and sizing optimization. Wind Energy 2001;4:43–59.
[32] Bowen AJ, Cowie M, Zakay N. The performance of a remote wind-diesel power system. Renew Energy 2001;22:429–45.
[33] Ulleberg Ø, Pryor TL. Optimization of integrated renewable energy hydrogen systems in diesel engine mini-grids. In: 14th world hydrogen energy conference, Montreal, 9–14 June; 2002.
[34] Lund H, Clark WW. Management of fluctuations in wind power and CHP comparing two possible Danish strategies. Energy 2002;27:471–83.
[35] Elhadidy MA, Shaahid SM. Role of hybrid (wind + diesel) power systems in meeting commercial loads. Renew Energy 2004;29:109–18.
[36] Garcia RS, Weisser D. A wind-diesel system with hydrogen storage: joint optimisation of design and dispatch. Renew Energy 2006;31:2296–320.
[37] Elhadidy MA, Shaahid SM. Decentralized/stand-alone hybrid wind–diesel power systems to meet residential loads of hot coastal regions. Energy Convers Manage 2005;46:2501–13.
[38] Xu X, Hu W, Cao D, Huang Q, Liu W, Liu Z, Lund H. Designing a standalone wind-diesel-CAES hybrid energy system by using a scenario-based bi-level programming method. Energy Convers Manage 2020;211:112759. https://doi.org/10.1016/j.enconman.2020.112759.
[39] Lund H, Münster E. Modelling of energy systems with a high percentage of CHP and wind power. Renew Energy 2003;28:2179–93.

[40] Phuangpornpitak N, Kumar S. PV hybrid systems for rural electrification in Thailand. Renew Sustain Energy Rev 2007;11:1530–43.
[41] Baghdadchi J, Varmette DS. Wind-based hybrid power systems in rural Western New York. In: WINDPOWER conference proceedings, Portland, Oregon June 2–5; 2002.
[42] Elhadidy MA. Performance evaluation of hybrid (wind/solar/diesel) power systems. Renew Energy 2002;26:401–13.
[43] El-Tamaly HH, Mohammed AAE. Impact of interconnection photovoltaic/wind system with utility on their reliability using a fuzzy scheme. Renew Energy 2006;31:2475–91.
[44] Bernal-Agustín JL, Dufo-López R, Rivas-Ascaso DM. Design of isolated hybrid systems minimizing costs and pollutant emissions. Renew Energy 2006;31:2227–44.
[45] Dufo-López R, Bernal-Agustín JL. Multi-objective design of PV–wind–diesel–hydrogen–battery systems. Renew Energy 2008;33:2559–72.
[46] Steinhüser A, Kaiser R, Kosack F, Reich N, Adelmann P. Photovoltaics and fuel cells for a decentralised power supply at "Rappenecker Hof". In: 19th European photovoltaic solar energy conference, Paris, France, 7–11 June; 2004. p. 2278–80.
[47] Pinho JT, Araújo RG. Wind-PV-diesel hybrid system for the electrification of the village of Sao Tome-municipality of Maracana-Brazil. In: 19th European photovoltaic solar energy conference, Paris, France, 7–11 June; 2004. p. 3511–4.
[48] Suresh V, Muralidhar M, Kiranmayi R. Modelling and optimization of an off-grid hybrid renewable energy system for electrification in a rural areas. Energy Rep 2020;6:594–604. https://doi.org/10.1016/j.egyr.2020.01.013.
[49] Rothert M, Wollny M. Optimized operation management in hybrid systems. In: 21st European photovoltaic solar energy conference, Dresden, Germany, 4–8 September; 2006. p. 2335–7.
[50] Mulder G, De Ridder F, Six D. Electricity storage for grid-connected household dwellings with PV panels. Sol Energy 2010;84:1284–93.
[51] Vosseler I, Ramírez E, Vallvé X, Carreras JM. PV hybrid village electrification in Spain: 6 years experience with multi-user solar hybrid grids. In: 19th European photovoltaic solar energy conference, Paris, France, 7–11 June; 2004. p. 3361–4.
[52] Cao Y, Wang Q, Cheng W, Nojavan S, Jermsittiparsert K. Risk-constrained optimal operation of fuel cell/photovoltaic/battery/grid hybrid energy system using downside risk constraints method. Int J Hydrogen Energy 2020;45(27):14108–18. https://doi.org/10.1016/j.ijhydene.2020.03.090.
[53] Taghizadeh M, Bahramara S, Adabi F, Nojavan S. Optimal thermal and electrical operation of the hybrid energy system using interval optimization approach. Appl Therm Eng 2020;169:114993. https://doi.org/10.1016/j.applthermaleng.2020.114993.
[54] Berrada A, Loudiyi K. Gravity energy storage. Elsevier; 2019. p. 186. ISBN: 9780128167175.
[55] Berrada A, Loudiyi K, Garde R. Dynamic modeling of gravity energy storage coupled with a PV energy plant. Energy J 2017;134:323–35.
[56] Fragaki A, Markvart T. Stand-alone PV system design—results using a new sizing approach. Renew Energy 2008;33:162–7.
[57] Aktaş A, Kırçiçek Y. A novel optimal energy management strategy for offshore wind/marine current/battery/ultracapacitor hybrid renewable energy system. Energy 2020;199:117425. https://doi.org/10.1016/j.energy.2020.117425.
[58] El-Shatter TF, Eskandar MN, El-Hagry MT. Hybrid PV/fuel cell system design and simulation. Renew Energy 2002;27:479–85.
[59] McConnell RD, Lasich JB, Elam C. A hybrid solar concentrator PV system for the electrolytic production of hydrogen. In: 20th European photovoltaic solar energy conference, Barcelona, Spain, 6–10 June; 2005. p. 167–70.

[60] Conibeera GJ, Richards BS. Acomparison of PV-electrolyser and photoelectrolytic technologies for use in solar to hydrogen energy storage systems. Int J Hydrogen Energy 2007;32:2703–11.
[61] Busquet S, Leroux P, Albuisson M, Gatt P, Ménard L, Mayer D, Metkemeijer R. PV/ fuel cell hybrid systems perofrmance analysis and possible improvements. In: 19th European photovoltaic solar energy conference, Paris, France, 7–11 June; 2004. p. 2426–9.
[62] Uzunoglu M, Onar OC, Alam MS. Modeling, control and simulation of a PV/FC/ UC based hybrid power generation system for stand-alone applications. Renew Energy 2009;34:509–20.
[63] Thounthong P, Raël S, Davat B. Analysis of supercapacitor as second source based on fuel cell power generation. IEEE Trans Energy Convers 2009;24(1):247–55.
[64] Dong W, Li Y, Xiang J. Sizing of a stand-alone photovoltaic/wind energy system with hydrogen and battery storage based on improved ant colony algorithm. In: Proc. 28th Chinese control decis. conf. CCDC 2016; 2016. p. 4461–6.
[65] Nouni MR, Mullick SC, Kandpal TC. Techno-economics of small wind electric generator projects for decentralized power supply in India. Energy Policy 2007;35:2491–506.
[66] Zhou T. Commande et Supervision Energétique d'un Générateur Hybride Actif Eolien incluant du Stockage sous forme d'Hydrogène et des Super-Condensateurs pour l'Intégration dans le Système Electrique d'un Micro Réseau [PhD thesis]. Ecole Doctorale SPI 072, L2EP à l'Ecole Centrale de Lille; 2009.
[67] Zini G, Tartarini P. Wind-hydrogen energy stand-alone system with carbon storage: modeling and simulation. Renew Energy 2010;35:2461–7.
[68] Denholm P. Improving the technical, environmental and social performance of wind energy systems using biomass-based energy storage. Renew Energy 2006;31:1355–70.
[69] Kasseris E, Samaras Z, Zafeiris D. Optimization of a wind-power fuel-cell hybrid system in an autonomous electrical network environment. Renew Energy 2007;32:57–79.
[70] Bechrakis DA, McKeogh EJ, Gallagher PD. Simulation and operational assessment for a small autonomous wind-hydrogen energy system. Energy Convers Manage 2006;47:46–59.
[71] Shi J, Tang YJ, Ren L, Li JD, Chen SJ. Application of SMES in wind farm to improve voltage stability. Physica C 2008;468:2100–3.
[72] Pecen R, Salim MD, Zora A. A LabView based instrumentation system for a wind-solar hybrid power station. J Ind Technol 2004;20(3).
[73] Urli NB, Kamenski M. Hybrid photovoltaic/wind grid-connected power plants in Croatian renewable energy program. Renew Energy 1998;15:594–7.
[74] Tina G, Gagliano S, Raiti S. Hybrid solar/wind power system probabilistic modelling for long-term performance assessment. Sol Energy 2006;80:578–88.
[75] Max Boegl—wind-hydro hybrid, n.d., Retrieved from: https://www.ge.com/renewableenergy/stories/wind-hydro-hybrid-power.
[76] Pilot Project, n.d., Retrieved from: https://www.mbrenewables.com/en/pilot-project/.
[77] Anagnostopoulos JS. Study of hybrid wind-hydro power plants operation and performance in the autonomous electricity system of Crete Island; 2012.
[78] Canales FA, et al. A comparative study of a wind hydro hybrid system with water storage capacity: conventional reservoir or pumped storage plant? J Energy Storage 2015;4:96–105. https://doi.org/10.1016/j.est.2015.09.007.
[79] Hamann A, Hug G. Integrating variable wind power using a hydropower cascade. Energy Procedia 2016;87:108–15. https://doi.org/10.1016/j.egypro.2015.12.339.

[80] Hirth L. The benefits of flexibility: the value of wind energy with hydropower. Appl Energy 2016;181:210–23. https://doi.org/10.1016/j.apenergy.2016.07.039.

[81] Bakos GC. Feasibility study of a hybrid wind/hydro power-system for low-cost electricity production. Appl Energy 2002;72(3–4):599–608. https://doi.org/10.1016/s0306-2619(02)00045-4.

[82] Martínez-Lucas G, et al. Frequency regulation of a hybrid wind–hydro power plant in an isolated power system. Energies 2018;11(1):239. https://doi.org/10.3390/en11010239.

[83] Scaria R. Hybrid systems: wind solar and wind hydro hybrid systems with battery storage; 2014.

[84] Castronuovo ED, et al. An integrated approach for optimal coordination of wind power and hydro pumping storage. Wind Energy 2013;17(6):829–52. https://doi.org/10.1002/we.1600.

[85] Alliance CSP. The economic and reliability benefits of CSP with thermal energy storage: literature review and research needs. CSPA; 2014.

[86] Zafar S. Renewable energy in Morocco. [Online]. Available: https://www.ecomena.org/renewable-energy-in-morocco/; June 7, 2019. [Accessed November 2019].

[87] Forrester J. The value of CSP with thermal energy storage in providing grid stability. Energy Procedia 2014;49:1632–41.

[88] Shultz A. Where does CSP fit on the US power grid? Renew Energy World 2017.

[89] Jin H, Hong H. Hybridization of concentrating solar power. Woodhead Publishing Limited; 2012. p. 395–420.

[90] Powell KM, Rashid K. Hybrid concentrated solar thermal powersystems: a review. BYU ScholarsArchive; 2017.

[91] Popov D. An option for solar thermal repowering of fossil fuel fired power plants. Sol Energy 2011;85(2):344–9.

[92] Pierce W, Gauché P, von Backström T, Brent AC, Tadros A. A comparison of solar aided power generation (SAPG) and stand-alone concentrating solar power (CSP): a South African case study. Appl Therm Eng 2013;61:657–62. https://doi.org/10.1016/j.applthermaleng.2013.08.014.

[93] Zhai R, Li C, Chen Y, Yang Y, Patchigolla K, Oakey JE. Life cycle assessment of solar aided coal-fired power system with and without heat storage. Energy Convers Manage 2016;111:453–65.

[94] Zhang M, Du X, Pang L, Yang CXL. Performance of double source boiler with coal-fired and solar power tower heat for supercritical power generating unit. Energy 2016;104:64–75.

[95] Zhao Y, Hong H, Jin H. Mid and low-temperature solar–coal hybridization mechanism and validation. Energy 2014;74:78–87.

[96] Feng L, Chen H, Zhou Y, Zhang S, Yang T, An L. The development of a thermo-economic evaluation method for solar aided power generation. Energy Convers Manage 2016;116:112–9.

[97] Yan Q, Yang Y, Zhai R. Evaluation of solar aided thermal power generation with various power plants. Int J Energy Res July 2011.

[98] Fernández P, Miller F. Assessment of the overall efficiency of gas turbine-driven CSP plants using small particle solar receivers. Energy Procedia 2014;49:334–43. https://doi.org/10.1016/j.egypro.2014.03.036.

[99] Santos MJ, Merchán RP, Medina A, Calvo HA. Seasonal thermodynamic prediction of the performance of a hybrid solar gas-turbine power plant. Energy Convers Manage 2016;115:89–102. https://doi.org/10.1016/j.encon-man.2016.02.019.

[100] Domínguez R, Conejo AJ, Carrión M. Operation of a fully renewable electric energy system with CSP plants. Appl Energy 2014;119:417–30. https://doi.org/10.1016/j.apenergy.2014.01.014.

[101] Michael JJ, Goic SI, Flat R. plate solar photovoltaic–thermal (PV/T) systems: a reference guide. Renew Sustain Energy Rev 2015;51:62–88. https://doi.org/10.1016/j.rser.2015.06.022.
[102] Makki A, Omer S, Sabir H. Advancements in hybrid photovoltaic systems for enhanced solar cells performance. Renew Sustain Energy Rev 2015;41:658–84. https://doi.org/10.1016/j.rser.2014.08.069.
[103] Santos-Alamillos FJ, Pozo-Vázquez D, Ruiz-Arias JA, Von Bremen L, Tovar-Pescador J. Combining wind farms with concentrating solar plants to provide stable renewable power. Renew Energy 2015;76:539–50. https://doi.org/10.1016/j.renene.2014.11.055.
[104] Vick BD, Moss TA. Adding concentrated solar power plants to wind farms to achieve a good utility electrical load match. Sol Energy 2013;92:298–312. https://doi.org/10.1016/j.solener.2013.03.007.
[105] Reichling J, Kulacki F. Utility scale hybrid wind–solar thermal electrical generation: a case study for Minnesota. Energy 2008;33:626–38. https://doi.org/10.1016/j.energy.2007.11.001.

CHAPTER 2

Optimization modeling of hybrid DG systems

Asmae Berrada[a], Arechkik Ameur[b,c], Anas El Maakoul[a], and Rachid El Mrabet[d]
[a]College of Engineering, LERMA Lab, International University of Rabat, Sala El Jadida, Morocco
[b]Faculty of Science, Ibn Tofail University, Kenitra, Morocco
[c]School of Science and Engineering, Al Akhawayn University, Ifrane, Morocco
[d]Senior Expert and General Director of the Public Utility Development Company, Morocco

2.1 Introduction

In the last decades, power plants were used as small-scale electricity generation as they supplied power to customers who were close to the generation site. This supply of voltage was very limited because the electric grid was DC-based. Yet, the transmission distance was also short [1]. With the evolution of technology, the electricity sector has seen the emergence of AC grid, which allowed the transmission of electricity over large distances and an increase in the power generation output. Large central power plants were then constructed to produce convenient, inexpensive, and secure supply [1]. This security was achieved by the interconnection system in which power plants compensate others in the case of failure. The economy of scale has also become possible due to this interconnected system. However, this way of generating electricity causes several environmental and technical issues. Further concerns about climate change and the increase in customer demand for reliable power have contributed to the development of DG technologies. For this reason, there has been a huge interest in distributed generation [1].

Unlike centralized power plants, where power is produced and transmitted to the customer through long power lines, distributed systems are placed near customers' sites offering various potential benefits. These latter provide lower cost and reliable power with less environmental consequences. The power capacities of DG technologies can reach 100 MW [2]. Today, distributed energy is playing an essential role in the energy portfolio. They are used to meet peaking power, heating and cooling needs, baseload power, remote power, and backup power.

Hybrid Energy System Models
https://doi.org/10.1016/B978-0-12-821403-9.00005-6

This chapter is organized as follows. Section 2.2 defines and classifies distributed generation. Different renewable distributed generation technologies are presented, and the main features of each technology are also described. Section 2.3 discusses the evolution factors, the benefits, and the applications of DG. Section 2.4 analyses the different issues that are facing distributed generation. In Section 2.5, optimization modeling approaches for DG systems are presented.

2.2 Classification and types of distributed generation

Distributed generation systems are small-scale electricity generation. There exist several definitions of distributed energy generation, but there is no precise one. Each county and organization has its own definition. For example, IEEE defines distributed generation as "the generation of electricity by facilities that are sufficiently smaller than central generating plants so as to allow interconnection at nearly any point in a power system" [1]. Distributed generations are classified into four categories depending on the amount of energy they generate as presented in Table 2.1 [3].

Compared to large central power stations, distributed energy generators are small and encompass a variety of technologies. They include wind turbines, internal combustion reciprocating engines, fuel cells, Stirling engines, photovoltaic systems, and others. Each technology has its own characteristics and application though it is best suited to meet a specific need.

A fuel cell is an electrochemical system that converts chemical energy of a fuel directly into electrical energy. It consists of two electrodes and an

Table 2.1 DG classification [3].

DG classification	Characteristics
Micro-DG	Generation ≤10 kW. It is for customer own use.
Small DG	Generation ≤500 kW connected on distribution system voltage <15 kV.
	Generation ≤1 MW connected on distribution system voltage ≥15 kV.
Mid-sized DG	Generation between 500 kW and 10 MW connected on distribution system voltage <15 kV.
	Generation between 1 and 10 MW connected on distribution system voltage ≥15 kV.
Large DG	Generation that is greater than 10 MW.

electrolyte. Fuel cell operation is similar to that of a battery except that in fuel cells, the reactants are continuously fed and are not stored. The efficiency of fuel cells in DG applications ranges from 30% to 80%. These units work with a group of other technologies to provide all the electrical requirements and heating of a home or a small assembly of houses [4]. Barriers facing fuel cells used in DG applications are basically similar to those facing other DG systems. These include grid interconnection challenges and feed-in tariff. For residential use, the expensive discount rate being considered for purchase decisions and the lack of awareness about DGs' potential are crucial barriers. The next upcoming years could experience a potential change in demand patterns for distributed generation technologies, with fuel cells expected to take a market share of this.

The off-grid sector is likely to see high interest in fuel cells under 10kW in regions with high grid vulnerability, such as in hurricane regions and areas with no grid connections. Further, fuel cells operating as CHP units could see significant adoption by facilities with important heat demands such as swimming pools, schools, and hospitals. Apart from combined heat and power, other fields which will contribute considerably to the increasing demand of DG-based fuel cell are UPS, mainly data centers and telecoms. This application is already applied in the United States and could experience a huge demand within the next decade. Finally, fuel cells will become a major player in the distributed generation application. Utilized in combination with other distributed generation systems, the next years could be the time of electricity generation becoming more localized and smaller [5].

Photovoltaic solar panels are broadly available for both domestic use and commercial use. They are environmentally friendly and thus produce no emissions. They also require minimal maintenance. However, PV systems are expensive. To overcome this economic barrier, progress in the manufacturing process needs to be done to produce less expensive components. Currently, photovoltaic systems are used to produce green power and are commonly being used in off-grid remote areas. Photovoltaic systems can be either connected to the grid or used in off-grid application. They react quickly to dispatching commands because they use power storage systems [6]. Today, there is a significant interest in the installation of large PV technologies integrated into the utility grid. Therefore, it is essential to ensure the compatibility of higher levels of distributed generation. Further, interconnection issues need to be addressed to guarantee safe integration.

Small wind turbines are viable devices in DG application. Generally, the size of small-scale turbines used in residential applications ranges from 1 to

10 kW. To make sure that the turbine will operate efficiently throughout the year, the average wind speed where the turbine is sited should be above 10 mph. Further, the location topography of the site is also crucial [7]. Small turbines require a tail fin to orient the direction of the rotor toward the wind. On the other hand, large turbines use computer control and wind indicators to direct the rotor. Wind turbines are typically coupled directly to the user's home and to the utility grid through a power meter and an inverter. The role of this latter is to convert DC voltage provided by the wind turbine to AC-based used by the appliances and the grid. Generally, the user buys electricity from the grid if the wind speed is low. However, if the user has excess power, the power meter runs backward. This directly balances the power bill or creates credit depending on the quantity of energy produced [7]. Wind power provides various applications for distributed generation. They can be either small- or large-scale size power generation. For small applications, they are used, for example, to charge batteries or to pump water. For large-scale applications, wind turbines are used for generating power for the utility grid. Wind turbines cause several impacts on the surroundings although they have no harmful emissions. Typically, they are noisy, affect wildlife, and damage the aesthetic of the environment.

Biogas is an alternative renewable energy source for distributed generation. It is produced from raw materials such as green waste, crops, municipal waste, and plant material. Further, it is an environmentally friendly generation system. Biogas is generally used to produce both heat and electricity.

Other forms of renewable distributed generation include solar thermal, small hydro, geothermal, and ocean. The generating capacity of these small-scale generators is up to 10 MW. It generally serves a small community or industrial plant. On the other hand, nonrenewable DG generation system includes, for example, gas turbines, internal combustion engines, and Stirling engines.

2.3 Overview of hybrid distributed generation

2.3.1 Evolution factors

The evolution of distributed generation was provoked by many triggers. According to the International Energy Agency (IEA), there are some factors which have contributed to this high interest in distributed generation. This includes greater demand for reliable electricity, environmental concerns, increased development of distributed generation technologies, and restrictions on the construction of new transmission lines [1].

In the last decades, centralized power plants provided electricity directly to end users in the region. Utility companies started forming interconnections between transmission networks to supply electricity outside their regions. In case, a local plant experiences failure, the end user would receive power from other power plants. This way, the reliability of the electricity supply is increased. However, with this solution, the electric grid still faces technical problems during peak hours. An increase in the electricity demand causes technical problems. Thus, inefficacies in the power grid result in various power failures. Power shortage will then become a common issue. To overcome this problem, new transmission lines should be constructed. However, the cost of this upgrade is very high. Therefore, a better alternative is the implementation of distributed electricity generation. Depending on the demand, distributed generators can be turned on and off when necessary. Establishing a group of decentralized generators close to consumers prevents power shortage and grid overload used in conjunction with centralized power plants. Distributed electricity generation will then ensure a reliable electricity supply to consumers [8].

A reliable power supply is considered very essential for some industries such as telecommunication. Therefore, to increase the reliability of the grid supply, industries are interested to invest in hybrid distributed energy generation to ensure their electricity supply. In addition to reliability, distributed generation has an effect on power quality. It has both positive and negative impacts. For the positive ones, distributed generation has a remedial effect on power quality problems. It has a positive impact on power factor corrections and voltage support. Moreover, connecting DG to the grid may cause a rise in voltage which is an issue encountered in regions where voltage support is difficult.

One last important trigger is environmental concerns. This factor is considered a driving force for the increased development of hybrid distributed energy generation. Electricity market players are forced to generate clean energy by worldwide environmental regulations. Many of the systems utilized in distributed energy generation are known as environmentally friendly. Renewable technologies have only indirect emissions. However, for the nonrenewable ones, CO_2 emissions are still an issue; therefore, they do not significantly lower greenhouse emissions.

2.3.2 Benefits of hybrid distributed energy generation

One important potential benefit of hybrid distributed energy generation is the electrification of remote areas. There are some customers who are located in areas that are far away from the grid. Expanding the grid is not always possible. Therefore, to energize them, one of the best solutions is the use of distributed generation technologies such as photovoltaic systems [9].

Another potential advantage of a hybrid distributed generation is the ability to use wasted energy resources that are not feasible to convert or transport. Such an energy resource is biomass gas from landfills, wastewater treatment, and farms. Many DG technologies can use this energy resource such as microturbines and gas turbines [9]. Hybrid DG has also the capability to increase the energy supply security. Renewable energy technologies use diversified energy resources. However, for nonrenewable energy technologies, they have a negative effect on energy security.

As electricity travels through transmission lines, a percentage of this electricity is lost. Since DG produces energy close to the demand load being served, it has potential benefits in minimizing the transmission and the distribution losses. Electricity produced by DG does not flow through large transmission lines to reach the customers' houses. Therefore, this reduces the transmission and distribution cost by almost 30%. It was reported by the World Bank that the loss percentage for Morocco was 7% in 2012 [10].

2.3.3 Applications of hybrid distributed generation

– Emergency generation

Some loads do not tolerate interruption in the supply of electricity. Hybrid DG units can be used by these latter to generate electricity in case the grid experiences a failure. Therefore, this emergency application would allow the facility to operate adequately. Examples of such loads are police stations, airports, and hospitals. In case of power failure, a backup generated is automatically started by an automatic system which monitors the electric supply. The load is then connected to the backup generator by automatic switches. Typically, distributed generation technologies that provide immediate start-up are mostly needed such as reciprocating engines [11].

– Peaking/load shaving

This hybrid DG application is used to reduce costs during peak periods. Further, it is used to prevent grid power failure and therefore increase the reliability of the power supplied to customers. Demands for peaking applications have increased due to the increasing power quality problems and sensitive loads. Those problems mainly result from the inadequate electricity generation or from the distribution and transmission constraints. Peak applications are different from emergency backup use. The operation duration of peak applications is higher than that of emergency applications. Moreover, they do not require a specific type of DG because immediate start-up is not necessary in this case. Therefore, various types of technologies can be used in this application [11].

– Continuous power

As its name implies, continuous power application operates on a continuous basis at least 6000 h a year to generate some or all the power required by a facility. It is mostly used by industrial applications such as plastics and metals manufacturing. The characteristics of this application include low emissions, high efficiency, and low maintenance cost.

– Green power

This application is used by facilities to reduce greenhouse emissions which result from generating energy. It is mainly offered to end users who request clean energy with low emissions. The characteristics of this application include low emissions, high efficiency, and low maintenance cost.

– Combined heat and power (CHP)

Distributed energy generation could be utilized as a combined heat and power (CHP) application. It is also known as cogeneration, heating, and cooling. Similar to continuous power, it operates at more than 6000 h per year. Further, it makes use of wasted exhaust heat to heat water and space or generate steam. The characteristics of this application include low emission, high thermal output, and low maintenance cost. Because of the aforementioned characteristics are almost the same as those of continuous power applications, they both target the same types of customers [11].

– True premium power

This hybrid distributed generation application provides uninterrupted power that is free from all the challenges related to the electric grid concerning power quality, such as changes in voltage and frequency. This power quality is not offered by the utility grid. Therefore, this system is used as the normal source while the grid is utilized as a backup. Further, this application is offered to critical clients such as hospitals and airlines. The characteristics of this type of hybrid DG application include low maintenance costs and a quick start-up.

2.4 Challenges facing hybrid distributed generation

As discussed in the previous section, distributed generation provides various potential benefits. However, it still faces many challenges such as financial cost, impact on energy security, and connection issues. Various economic, technical, and regulatory concerns arise while integrating these resources into the exciting network.

2.4.1 Financial cost

The major challenge facing hybrid distributed generation systems is financial cost. Even though many industrial combined heat and power (CHP) systems are economic, most renewable distributed generation technologies are not. Compared to large central power plants, DG experiences a high capital cost. There is also a large difference in capital cost between distributed energy generation systems. Fig. 2.1 and Table 2.2 present the different prices of commercially available systems in the market [12]. These prices change with the development of technologies and the electricity market.

2.4.2 Energy security

Energy security has two meanings. It is first interpreted as the diversification of primary energy supplies. As this diversification increases, energy security improves. Distributed energy generation systems that are based on renewables use diversified energy resources such as sunlight for photovoltaic systems and wind for turbines. However, for nonrenewable DG technologies, they directly and indirectly use natural gas. Many commercialized distributed generation technologies depend on oil or natural gas, and most of the investment in research has been focused on fuel cells as well as gas turbines which make use of natural gas. The second interpretation of energy security concerns the reliability of the power system. As discussed before, DG helps increase the reliability of the electricity supplied to customers. However, it is not always the case. DG can also have a negative effect on

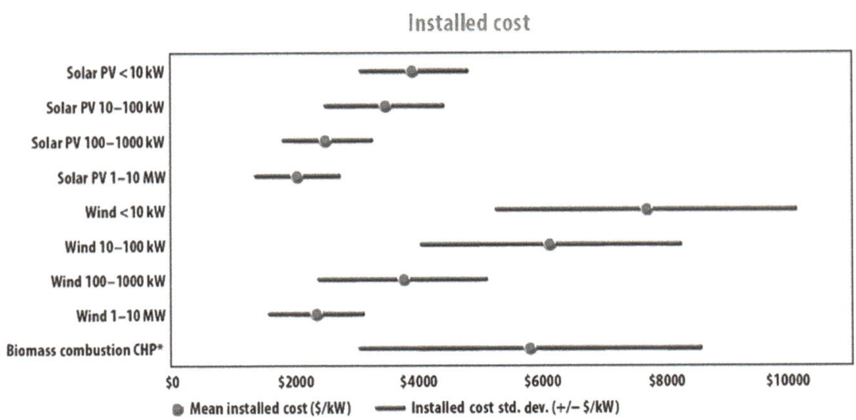

Fig. 2.1 Capital cost of distributed energy generation technologies.

Table 2.2 Thermal technologies cost.

Energy system	Mean installed cost ($)	Standard deviation of installed cost (+/−) in ($)
Biomass wood heat ($/kW)	575	252
Solar water heat (with plastic collector) ($/ft^2)	59	15
Heat pump (ground source) ($/ton)	7765	4632
Solar water heat (evacuated tube, flat plate) ($/ft^2)	162	61
Solar vent preheat ($/ft^2)	31	14

reliability. This occurs when it is difficult to control the fluctuating output of nondispatchable energy generation systems. Therefore, regulation of power is necessary. This issue has a negative impact on energy security [13, 14].

2.4.3 Environmental concerns

The production of electricity using distributed generation results in some environmental issues. Generating electricity close to customer has a negative effect on regional and local air quality. This concern is mainly associated with the emission of greenhouse gas especially CO_2. For renewable distributed generation, only indirect emissions are taken into consideration. Indirect emissions refer to greenhouse gas emissions associated with the purchase of electricity. Eqs. (2.1)–(2.3) are used for the calculation of indirect emissions from electricity use:

$$\begin{aligned} CO_2\,\text{Emission (metric tons)} = {} & \text{Electricity use (MWh)} \\ & \times \text{Emission factor (ILS } CO_2/\text{MWh)} \\ & \div 2204.625 \left(\frac{\text{Lbs}}{\text{metric ton}}\right) \end{aligned} \tag{2.1}$$

$$\begin{aligned} CH_4\,\text{Emission (metric tons)} = {} & \text{Electricity use (MWh)} \\ & \times \text{Emission factor (ILS } CH_4/\text{MWh)} \\ & \div 2204.625 \left(\frac{\text{Lbs}}{\text{metric ton}}\right) \end{aligned} \tag{2.2}$$

$$\begin{aligned} N_2O\,\text{Emission (metric tons)} = {} & \text{Electricity use (MWh)} \\ & \times \text{Emission factor (ILS } N_2O/\text{MWh)} \\ & \div 2204.625 \left(\frac{\text{Lbs}}{\text{metric ton}}\right) \end{aligned} \tag{2.3}$$

With metric ton is 2204.62 pounds

The wide range of applications for decentralized generation, the potential benefits of DG, and the trend to produce clean energy using renewables sources will pave the way to the development of decentralized generation that may have a large contribution in power generation in the upcoming years. To obtain more benefits from distributed generation, it is important to consider the issues that DG is facing and the impacts of integrating it into the distribution system. To overcome these barriers, more research has to be done for successful integration of DG.

2.5 Hybrid distributed generation system models

Due to the variable nature of renewable energy resources, it is essential to effectively choose a diversified configuration of multisource generation technologies to guarantee the reliability and efficiency of the hybrid distributed generation systems. The use of several sources to harness energy has been proved to be challenging as the system necessitates a required coordination level. Typically, complex control models are required to maximize the overall system in a sustainable manner. In addition, appropriate sizing and selection of hybrid technologies are the key needs for an effective design of such systems. To achieve high efficiency and reliability, optimal operation modeling of the entire system is necessary using robust control techniques and schemes. This section discusses the most common approaches and techniques used to analyze and model hybrid DG systems.

2.5.1 Optimization approaches for off-grid hybrid distributed energy generation systems

A number of techniques are used to model off-grid distribution energy systems such as hybrid optimization techniques, artificial intelligence-based techniques, and analytical optimization approaches.

2.5.1.1 Hybrid techniques

An optimization technique using fuzzy programming is utilized to examine an energy management system for a combined heat and power (CHP) technology [15]. Linear programming algorithms along with PSO are employed in the hybrid optimization method to investigate the optimal capacity of CHP and the boiler optimal range. A case study analyzing the feasibility of constructing a combined heat and power for a hospital has been performed to examine the viability of the proposed model.

Authors in Ref. [16] modeled the optimal operation of a hybrid PV/fuel cell system with battery energy storage using the technique of weighted sum with a fuzzy satisfying method. The main aim of this optimization model is to operate the hybrid system in such a manner that the load is properly managed within the tolerable limits. The objective function of this model is based on the minimization of the hybrid system cost and CO_2 emissions.

A stand-alone wind PV hybrid system with battery storage is optimized using hybrid simulated annealing and flower pollination (FPA) algorithms (FPA/SA) in Ref. [17]. This latter hybrid method is utilized to maximize the reliability of the aforementioned system while reducing its costs. In addition, computational fluid dynamics (CFD) simulation is employed to investigate the effect of PV tilt angle on wind speed. The model objective function takes into consideration power loss probability and payback period. On the other hand, the decision variables consist of the number of batteries and the PV panels used as well as their tilt angle.

A study in Ref. [18] identifies the optimal design configuration of a stand-alone hybrid desalination system. A novel optimization model using weather forecast is developed to optimize the hybrid wind solar system. Harmony search (HS) and a combination of HS with chaotic search algorithm have been employed in this analysis with an aim to minimize the life cycle cost and to respect the tolerable loss of power supply probability. The implementation of HS algorithm is simple that is why it is used in this study [18].

The optimal sizing of a hydrogen-based stand-alone PV/wind system has been modeled in a comparative study using artificial intelligence (AI) techniques [19]. The performance of various AI methods used to optimally size the aforementioned hybrid system has been assessed in this study. The aim of these models is to meet the energy demand of the load with minimum cost. The investigated hybrid system has been compared to a PV/wind/fuel cell system. Different heuristic algorithms are employed in this analysis including HS, TS, simulated annealing (SA), and particle swarm optimization (PSO). The obtained results demonstrate that the most attractive and robust technique is PSO.

Table 2.3 presents the different hybrid AI optimization methodologies. It is shown that the employed approaches have been used to optimize the different characteristics of hybrid DG technologies. It can be observed that one of the most prevalent approaches used in this classification is HS algorithm.

2.5.1.2 Single artificial intelligence-based approach

A hybrid DEG system composed of PV/Diesel with battery storage has been modeled using particle swarm optimization (PSO) with an aim to minimize

Table 2.3 Hybrid approaches used to model the optimization of off-grid hybrid DG systems.

Hybrid DG system	Objective	Approach used	Reference
CHP system Boiler Combined heat	Optimization of CHP capacity Boiler optimal range	Linear programming algorithm Particle swarm optimization	[15]
Solar PV Fuel cell Battery storage	Minimization of system cost Reduction of CO_2 emissions	Fuzzy satisfying approach Weighted Sum method	[16]
Wind plant Energy storage	Minimization of investment cost Minimization of operational cost	Fischer-Burmeister algorithm Self-adapted evolutionary strategy	[20]
PV Wind Battery	Maximization of reliability Minimization of cost	Simulated annealing Flower pollination algorithm	[17]
Solar Wind Battery	Minimization of the life cycle cost	Hybrid harmony search-based chaotic search	[18]
PV Wind Battery	Minimization of cost.	Particle swarm optimization Simulated annealing Harmony search TS	[19]

the total cost, the probability loss of the load, and the greenhouse gas emissions [21]. The obtained results demonstrated the ability of the PV/storage system to decrease the energy cost by reducing the amount of diesel needed and the cost of the hybrid system. To identify the annual fuel of DG fuel consumption, CO_2 emissions, and the annual unmet load, simulation targets are specified.

To optimally size a wind/PV hybrid system, discrete harmony search (HS) algorithm has been developed by Askarzadeh [22]. The implementation of the algorithm is simple and is used to effectively solve the discrete problem to identify the optimum solution. The optimization sizing of the hybrid system is found using the proposed discrete harmony search algorithm by the following steps:

Step 1: N harmony solutions are produced with each one consisting of two integers. The first integer is about the number of wind turbines while the second one indicates the number of solar PV panels. The harmony memory (HM) is used to memorize the N harmonies.
Step 2: The annual cost is determined for each harmony.
Step 3: Setting of adjustable parameters used in the algorithm.
Step 4: The generation of a new harmony solution using the following pseudocode:
Step 5: If the generated harmony is considered a feasible option with superior quality than the worst harmony solution saved in the harmony memory, it is added to this latter with the extension of the existing worst harmony.
Step 6: Repetition of Steps 3–5 until reaching the preventing criterion.
Step 7: The optimal number of solar PV panels and wind turbines are returned by the model as the HM best harmony.

Similarly, an optimization model has been developed using PSO to identify the optimal sizing of a stand-alone hybrid RE system for providing energy to a remote region [23]. The model decision variables include the number of batteries, the PV panels occupied area, and the swept area of the turbines' blades. Variants of particle swarm optimization are also applied such as PSO with repulsion factor, PSO with adaptive inertia weight, and PSO with constriction factor. These latter are applied to boost a particle to alter the direction of its search, to improve the convergence factor and the stability of PSO, and to avoid premature convergence issue.

A genetic algorithm technique is applied to optimize a stand-alone hybrid system consisting of a diesel/PV/wind with different battery systems [24]. The objective of the model is to minimize the system total cost with the use of diverse model settings and component sizes. This analysis has shown that it is more economical and ecological to used battery energy storage in combination with renewable energy systems.

Another work investigated an interactive operation management methodology using a single objective optimization known as uniform water cycle algorithm for modeling a microgrid with multiple DG systems [25]. This latter algorithm is inspired from the hydrologic cycle. The system operation cost and emissions are minimized by the objective function of the model.

To identify the optimal configuration of DG for an autonomous hybrid power system, authors in Ref. [26] developed a systematic particle swarm optimization-based technique. The model decision variable is the levelized cost of the electricity (LCOE). This latter is used to decide the optimum sizing of the system components using different DG configurations. Likewise, design optimization of a microgrid hybrid PV/CSP/LPG system is studied using particle swarm optimization. The aim of this model is to minimize the computational cost [27].

Management strategy and design optimization have been investigated in Ref. [28] for a PV plant with a diesel CHP engine, a reversible heat pump, and a boiler. A particle swarm optimization has been applied for optimizing both the operation strategies and the sizing of the aforementioned hybrid system. The aim of this analysis is to minimize the total cost of the system with an objective to meet the customer's needs in terms of electricity, heat, drinking water, and cooling.

Authors in Ref. [29] applied an optimization technique using multiobjective for CHP technology incorporated with low-energy buildings. The main of the optimization model is to improve the efficiency of the system while maximizing its power output. It is demonstrated that the small enhancement of exergetic efficiency has resulted in an important rise in the overall cost rate.

A small autonomous hybrid system has been studied using biogeography-based optimization (BBO) algorithm with the prediction of wind speed and solar radiation [30]. This analysis has revealed the usefulness of forecasting in this optimization study.

A summary of the presented artificial intelligence techniques is presented in Table 2.4. It is shown that the most used method to model off-grid hybrid distribution energy systems is PSO technique and its variants. These latter are employed to reach diverse objectives on various hybrid DG configurations.

2.5.1.3 Analytical approaches

A linear programming model represented as a decision support tool is proposed for optimizing the operation and investment of the hybrid residential system. This latter consists of a PV, heat pump, solar thermal collector, fuel cell, and battery energy storage [31]. The achievable reduction of cost can go up to 60% under specific conditions, while the emissions reduction is between 45% and 90%. Meteorological data including temperature and solar radiation data are obtained from satellite measurements, while it is difficult to have multiple years' data of hot water and electricity consumption for residential application.

A method that employs a hybrid approach consisting of Fischer-Burmeister algorithm and a self-adapted evolutionary strategy is presented

Table 2.4 Single artificial intelligence-based approach used to model the optimization of off-grid hybrid DG systems.

Hybrid DG system	Objective	Approach used	Reference
Wind PV	Optimal sizing of hybrid system	Discrete harmony search approach	[22]
PV Battery Diesel	Minimization of system cost Minimization of CO_2 emissions Minimization of load loss	Particle swarm optimization	[21]
PV Wind Diesel	Minimization of the overall cost	Generic algorithm	[24]
PV Wind Battery	Sizing of an autonomous hybrid system	PSO with repulsion factor PSO with adaptive inertia weight PSO with constriction factor	[23]
Fuel cell PV Microwater turbine Wind Battery	Minimization of greenhouse gas emissions Minimization of operation cost	Uniform water cycle algorithm	[25]
PV Wind Battery Diesel system	Minimization of hybrid system LCOE Minimization of system units' costs	Systematic approach based on particle swarm optimization	[26]
Solar Wind	Forecasting of solar radiation and wind speed	biogeography-based optimization	[30]
CHP technology incorporated with low-energy buildings	Maximization of the exergetic efficiency Maximization of the net power output	Generic algorithm	[29]
LPG CSP PV	Minimization of propane consumption	Particle swarm optimization-based algorithm	[27]

to determine the optimal sizing of a hybrid DG system composed of wind and energy storage system [20]. The aim of the model is to minimize the system investment and operation costs.

The design of a hybrid system with PV, diesel generator, and battery system has been studied based on environmental and economic biobjectives [32]. The aim of this model is to determine the optimal technical configuration, the battery capacity, and the PV rated power which result in the lowest carbon footprint of energy and levelized cost of energy. The results show that the best trade-off between economic and environmental performances leads to a reduction in the carbon footprint of energy of 48% and a small decrease in the economic performance (about 2%).

Another study investigated the operation of energy storage in a DG hybrid system with an objective to guarantee the operation stability of the system [33]. General algebraic modeling system has been used to address issues associated with system performance. This has been carried out through the development of a mixed-integer linear programming (MILP). Based on system cost, the daily and weekly operation models have been compared to investigate the optimal operation. The obtained results reveal that it is more economically interesting to operate energy storage on a daily mode because of the high investment cost it requires.

A model is used to optimize the sizing of a stand-alone DG with PV/Wind hybrid energy system using a statistical approach [34]. This hybrid model makes use of fast and slow dynamics and aims to remove the peak power resulting from load consumption. The results of this study show that the incorporation of supercapacitors takes benefits from the characteristics of batteries, extends their lifetime, enhances power flow exchange, and impacts the sizing of energy storage by accommodating the fast fluctuations of power.

Power management for an autonomous hybrid battery/diesel/PV system is analyzed. Frequency modulation is used in this assessment in such a way that the grid frequency of the battery inverter increases where there is a need for power reduction. To avoid battery overcharge, an automatic strategy is set to identify the battery operation and perform the necessary actions [35]. Both the diesel operating life and the hybrid energy system efficiency are optimized in this study.

An optimization method named ontology is employed to study the optimal DG system compromising multirenewable energy sources [36]. A combination of ontology and optimization approaches are used in this analysis. It should be noted that the utilization of the ontology domain in renewable energy enables the update of knowledge conceptualization without altering the objectives of the system. Furthermore, the structure optimization of a hybrid DG system is modeled using mathematical programming approaches [37].

Another study has also covered the sizing of hybrid renewable DG systems and their optimal operation in a microgrid application [38]. The study also investigates the optimal type of DG using a distributed energy resources customer adoption model. Based on the hybrid DG annual energy costs and CO_2 emissions, the model outputs the optimal operation schedule, the sizing, and the optimal DG type.

A methodology for integrating renewable energy technologies to the grid is proposed in Ref. [39] studying the economic and practical approaches for hybridizing an autonomous hybrid off-grid system compromised of a small-scale hydropower plant, PV, and wind farm. A grid tie inverter and a charger of battery storage are used by the two hybridization approaches. The hybrid charge controller is employed for the hybridization of wind and solar PV. The output power of the hybrid charge controller is used to charge the batteries, while the grid tie inverter power is fed to the electric grid. The obtained results have shown that the hybridization of the aforementioned system could be performed with the utilization of the hybrid charge controller and the grid tie inverter with an appropriate electronic load controller.

Another work presented a hybrid model composed of Fourier series and autoregressive moving average. The aim of this study is to assess the probabilistic and the dynamic behavior of hybrid systems. The aforementioned techniques are used to create synthetic weather data with the characterization of historical measurements [40]. A probabilistic study of a hybrid system is conducted using the produced synthetic wind speed data. The hybrid energy system consists of a wind plant, natural gas, nuclear plant, battery energy storage, and chemical power plant. The objectives of this analysis are to assess the environmental and economic impacts of a hybrid distribution system and to evaluate the impact of using batteries of different sizes for the purpose of smoothing the intermittency of renewable energy resources.

A summary of the analytical method employed in the modeling of off-grid hybrid systems is presented in Table 2.5. It is shown that the analytical approaches make use of statistical and computer-based deterministic techniques, as well as nonlinear and linear programming methods.

Table 2.5 Analytical approaches used to model the optimization of off-grid hybrid DG systems.

Hybrid DG system	Objective	Approach used	Reference
PV Fuel cell Solar thermal collectors Heat pump Boiler Battery	Operation optimization of hybrid DG system Minimization of cost and emissions	Linear programming	[20]
PV Diesel Battery	Minimization of carbon footprint of energy Minimization of LCOE	Bi-objective design algorithm	[20]
Wind Energy storage	Sizing optimization	Stochastically model and sequential Monte Carlo simulation	[32]
Hybrid DG with battery	Optimization of DG operation mode	Linear programming model	[33]
PV Wind Battery Supercapacitor	Sizing of storage system	Statistical approach based on capacity distributions	[34]
PV Diesel system Battery	Maximization of system efficiency Maximization of diesel operating life	Strategy based on grid frequency modulation	[35]
Multisource Hybrid renewable DG	Optimize size of DEG resource	Ontology method	[36]
PV Microgrid diesel generator battery	Sizing and type optimization of DG system Minimize the costs, and the CO_2 emissions	Distributed energy resources customer adoption model	[38]
PV Microhydro wind	Optimization of system reliability	Grid tie inverter hybrid charge controller-based technique	[39]
Hybrid DG with wind	Analyzing economic and environmental impact on hybrid DEG system	Hybrid of autoregressive moving average (ARMA) and Fourier series	[40]

2.5.2 Optimization approaches for hybrid DG connected to the grid

Significant work has been carried out to identify the optimal planning, location, and configuration of a hybrid DG system connected to the electric grid, as well as to reduce the cost and emission related to the latter. In this section, only models dealing with system sizing and location optimization along with emissions reduction are analyzed. It mainly discusses the optimization technique used, the system performance, and the models' objective functions.

2.5.2.1 Hybrid techniques

A study investigating the optimal placement and sizing of hybrid DG systems has been performed in Ref. [41] using a combined technique consisting of artificial bee colony (ABC) algorithm and ant colony optimization (ACO). The aim of this model is to minimize the energy cost, the power losses, as well as the CO_2 emissions. In addition, the model aims to enhance the stability of the voltage. The obtained results demonstrate the validity and the potential of the proposed algorithm as compared to other optimization approaches. A hybrid BB-BC algorithm has been applied for the optimization of DG power allocation and network reconfiguration using a multiobjective model [42]. This algorithm integrates a fuzzy-based mutation operator to enhance exploration competency. The aim of the proposed model is to improve voltage stability, decrease the system cost, and reduce CO_2 emissions as well as power losses. The objective function of the model is subject to operational and technical constraints of the system.

Optimal sizing of a hybrid PV/wind DG system with an aim to minimize the design cost of the system has been performed in Ref. [43] using an ant colony optimization technique based on integer programming. A comparison of the obtained results with other artificial intelligence methods such as artificial bee colony, genetic algorithms, and conventional optimization B&B approach is done in this study.

It is important to consider the operation reliability of the electric network when DG systems are integrated to this latter. A study done in Ref. [44] covered the reliability aspect related to the connection of DG units using a multistage framework based on the fuzzy satisfying method and a hybrid self-adaptive global-based harmony search algorithm (SGHSA). Customer interruption cost, as well as investment, operation, and maintenance costs are the main optimization parameters used by the objective functions of this model.

An approach that utilizes modified teaching-learning-based optimization (MTLBO) algorithm is proposed to identify the optimal sizing and placement of hybrid DG in a distribution network system [45]. Mixed-integer nonlinear programming is employed in this model due to the independence of this approach from the constraints and objective function types. The employed MTLBO approach has been compared to the brute force method to investigate the model performance. The obtained results indicate that the studied algorithm has a better performance and effectiveness than several other methods including brute force.

Table 2.6 summarizes the different hybrid optimization methodologies employed to model grid-connected hybrid systems.

2.5.2.2 Single AI-based optimization technique

Pareto optimal technique for DG system allocation in a radial distribution network is studied employing an improved differential search algorithm [46]. The model objective is to identify the optimal sizing and location of DG in radial distribution system with an aim to diminish the system losses and operating costs, as well as enhance the voltage profile.

Another work assessed the optimal location and size of multiple DG systems in a distribution power grid applying particle swarm optimization with constriction factor approach [47]. The obtained results demonstrated that the integration of multiple DG units in the distribution power network leads to an important improvement in various parameters within the distribution system such as voltage deviation, active and reactive power loss, purchased energy cost, and the system overall cost.

A cuckoo search (CS) algorithm has been applied by authors in Ref. [48] to optimally size a hybrid system composed of PV and wind connected to the electric grid. The proposed algorithm offers a number of benefits as compared to particle swarm optimization method. This includes better accuracy, less computation time, and faster convergence. Similarly, the optimum sizing and location of various DG systems with several load models has been analyzed using Shuffled Bat algorithm with multiobjective [49]. The objective functions of this algorithm are the improvement of voltage profile, the minimization of DG cost, and the reduction of system losses.

Another study proposed a Big Bang-Big Crunch (BB-BC) algorithm analyzing the optimal reconfiguration and distributed generation power allocation approach [50]. Load uncertainty is considered by the proposed

Table 2.6 Hybrid approaches used to model the optimization of off-grid hybrid DG systems.

Hybrid DG system	Objective	Approach used	Reference
Fuel cell Wind Gas turbine	Sizing optimization Location optimization Minimization of power losses Minimization of energy cost Minimization of emissions	Artificial bee colony (ABC) algorithm Ant colony optimization (ACO)	[41]
PV Wind Fuel cell	Sizing optimization Optimal location Minimization of costs Minimization of power losses Minimization of GHG emissions Voltage stability	BB-BC algorithm with fuzzy-based mutation operator	[42]
PV Wind	Minimization of total design cost	ACO method for continuous domain-based integer programming	[43]
Hybrid DG systems	Minimization of investment cost, minimization of operation and maintenance costs, minimization of expected customer interruption cost	Self-adaptive global-based harmony search algorithm with fuzzy satisfying method	[44]
PV Fuel cells Minihydro turbines Diesel generator Wind Microturbines	Sizing optimization Optimal placement	Modified teaching-learning-based optimization algorithm based on mixed-integer nonlinear programming	[45]

algorithm. This latter has been developed by authors in Ref. [51] based on the evolution of the universe theory. The decision variables include maximizing the index of voltage stability, reducing power losses and emissions, as well as minimizing the operation cost. It has been revealed that the applied approach offers more varied Pareto solutions, showing that an improved exploration ability is achieved even with better suitability. In addition, the optimization of the DG unit within the distribution network with an aim to minimize the power losses has been done in Ref. [52]. A GA approach is used in this study. Other objectives of the model include voltage profile enhancement and short circuit current reduction.

Table 2.7 presents a summary of artificial intelligence optimization techniques used to model hybrid systems connected to the electric grid.

Table 2.7 Single artificial intelligence-based approaches used to model the optimization of on-grid hybrid DG systems.

Hybrid DG system	Objective	Approach used	Reference
DEGs in radial distribution systems	Optimization sizing and location Minimization of operation cost Reduction of losses Improvement of voltage profile	Improved differential search algorithm	[46]
Multiple wind turbines	Optimum size and location	PSO with constriction factor method	[47]
PV/wind	Optimal sizing	Cuckoo search (CS) algorithm	[48]
PV Wind plant Microhydro	Optimization sizing and location Minimization of cost Improvement of voltage profile	Shuffled Bat algorithm	[49]
PV Fuel cells Wind plant	Minimization of operation cost Reduction of power losses and emissions Maximization of voltage stability index	Big Bang-Big Crunch	[51]
PV Fuel cells Diesel generator Microhydro Wind plant Biogas	Minimization of real and reactive power losses	GA technique	[52]

2.5.2.3 Analytical optimization technique

Optimization of hybrid renewable system's operation and demand response is done in Ref. [53] with the use of a receding horizon optimization technique. The objective of this analysis is to identify the optimum power of several energy generation systems by minimizing the cost function. The aforementioned strategy is used as a starting layer of model predictive control, which is considered a well-established technique that is commonly used in process engineering. Minimization of the cost function is performed at each time step of the receding horizon optimization technique, and over a finite time horizon. The objective function is subject to some constraints.

The optimal operation of a small-scale natural gas grid fueled by hybrid DG systems is performed using an optimization algorithm based on centralized decision-making scheme [54]. The objective is to reduce peak energy demand. This analysis has shown that the investigated region encounters difficulties related to the control of the natural gas grid due to the considerable increase in natural gas consumption near end users for all the studied scenarios. It is though recommended that it is vital to implement a centralized decision-making scheme where numerous hybrid DG units are available; it is also important to have an optimal sizing of these latter.

A dynamic programming search technique and a separately optimized algorithm are being applied to study power system losses and stability when DG units are installed. The aim of this analysis is to optimally size and locate hybrid DG in a power system [55]. To retain the voltage profile of the system within the tolerable security limits, several DG systems are installed. The obtained results have shown that the optimization of DG number, size, and location is affected by the consideration of DG units' reactive limits as it has an impact on voltage stability.

Stochastic optimization is performed by authors in Ref. [56] using a sampling average method to analyze hybrid renewable energy systems. This study makes use of uncertainties linked to the load and to the renewable energy resources for optimal sizing of the hybrid DG unit. The objectives of this model are to maximize the share of renewable energy and to minimize the CO_2 emissions as well as the system net present cost while maintaining a required level of load probability loss. The obtained Pareto front of the stochastic optimization is compared to a deterministic one. The results have shown that the stochastic case has a higher net present cost. However, the Pareto front generated from the deterministic optimization has a better quality.

In another development, a mixed-integer linear programming model is applied for an eco-industrial park on a hybrid system connected to the eclectic grid [57]. The model aims to determine the optimal cost planning of the aforementioned hybrid system with an objective to minimize the energy cost net present value (NPV). The model also enables the identification of the most optimal energy storage technology to be used in the studied case. The results demonstrate that the lead-acid battery is the optimal storage system because of its low investment cost.

Another work comprising the optimal system allocation for DG units is presented in Ref. [58]. The study has been performed to identify the model objective, which includes environmental, technical, and economic considerations. This is done using a two-stage multiobjective system-policy framework. The first one is used by customers to assess the system performance according to three objectives based on production frontier estimation modeling. The second stage deals with bottleneck multicriteria decision modeling. As input, this latter makes use of the outputs of the first stage such as capacity limitations, efficiencies, demand, renewable penetration, and dispatchability. The aim of this model is to provide end users with the optimal allocation of energy resources. It can also be used by stakeholders such as policymakers, users, and distributed generation developers to decide on regulatory policy.

The development of analytical expressions based on the derivation of a multiobjective index is done in Ref. [59] to determine the hybrid DG optimal power factor with an aim to improve voltage stability and reduce power losses. The multiobjective index is expressed as a combination of power loss indices. The optimal location, size, and DG number are provided as outputs of the cost-benefit analysis.

To assess a hybrid system consisting of renewable electricity technologies and natural gas in a distributed network, a real option and a discounted cash flow analysis are performed [60]. It has been observed that it is interesting to invest in the aforementioned hybrid system in the studied distributed application as compared to a single system especially when renewable energy incentives are offered.

Table 2.8 presents a summary of the analytical optimization methods for hybrid distributed systems connected to the electric grid.

2.6 Conclusions

This chapter presented an overview of hybrid distributed energy generation systems and their optimization approaches. These latter have been briefly discussed for both off-grid and grid-connected hybrid technologies. The aim of this work is to provide the basics for analyzing the optimization techniques of hybrid DG technologies. This study has demonstrated that the most dominating approach for optimizing hybrid DG systems is artificial intelligence techniques, while the most commonly used AI method is the particle swarm optimization approach. Further, the most popular hybrid technologies used in the presented studies are PV and wind turbines. It has been noticed that the aim of the presented studies shifted from the reduction of costs and emissions to maximization of reality and optimization of the system operation. Such studies aim to enhance the operational efficiency of the hybrid DG units as well as to improve their implementation into the grid.

Table 2.8 Single artificial intelligence-based approaches used to model the optimization of on-grid hybrid DG systems.

Hybrid DG system	Objective	Approach used	Reference
PV Wind	Optimal sizing Minimization of emissions Minimization of operating costs	A receding horizon optimization approach	[53]
Natural gas fueled DG	Sizing optimization	Optimization framework based on centralized decision-making scheme	[54]
Hybrid DEG with PV	Optimization sizing Optimal location	Separately optimized algorithm and dynamic programming search technique	[55]
PV Biomass Wind Natural gas Solar collector	Maximization of RE ratio Minimization of the NPC Minimization of emissions	Stochastic sampling average method	[56]
PV Wind Battery	Minimization of net present value Optimization of energy storage	Mixed-integer linear programming	[57]

Continued

Table 2.8 Single artificial intelligence-based approaches used to model the optimization of on-grid hybrid DG systems—cont'd

Hybrid DG system	Objective	Approach used	Reference
PV Fuel cell Wind Gas microturbine CHP Geothermal Biodiesel Solar thermal collector	Optimization of DG allocation	A two-stage multiobjective strategic technology-policy framework	[58]
Gas turbine engine-based DEG units	Optimization of location Sizing optimization Optimal power factor	Analytical expressions based on derivation of a multiobjective index	[59]
PV Natural gas microturbines	Optimization of costs Optimization of incentives	Discounted cash flow analysis and real options analysis	[60]

References

[1] Pepermansa G, Driesen J, Haeseldonckx D, Belmans R, D'haeseleer W. Distributed generation: definition, benefits and issues. Energy Policy 2005;779–98.

[2] Rugthaicharoencheep N, Auchariyamet S. Technical and economic impacts of distributed generation on distribution system. World Acad Sci Eng Technol 2012;288–92.

[3] Guan FH, Zhao DM, Zhang X, Shan BT, Liu Z. Research on distributed generation technologies and its impacts on power system, In: IEEE conference on sustainable power generation and supply; February 2009. p. 1–7.

[4] Fuel cells for distributed generation, a technology and marketing summary. Energy Center; 2001 Rep. 21.

[5] Adamson K. Are fuel cells the future for distributed generation, [Online]. Available FTP: http://www.cleantechinvestor.com/portal/fuel-cells/1368-are-fuel-cells-the-future-for-distributed-generation.html; 2008.

[6] Whitaker C, Newmiller J, Ropp M, Norris B. Distributed photovoltaic systems design and technology. USA, Rep. 52.

[7] Schaefer L. Distributed generation of wind and solar, [Online]. Available: http://me1065.wikidot.com/distributed-generation-of-renewable-sources.

[8] Tress GM. Ensuring reliable electricity supplies. Dept. Elect. Eng.; August 2008 Washington, Rep. 25.

[9] Zareipour H, Bhattacharya K, Canizares CA. Distributed generation: current status and challenges. World Acad Sci Eng Technol 2004;1–8.
[10] Electric power transmission and distribution losses (% of output) [Online]. Available: http://www.worldbank.org/.
[11] Bluestein J. Environmental benefits of distributed generation. Energy and Environmental Analysis, Inc.; December 20001–16
[12] NREL. Distributed generation energy technology capital costs. [online]. Available: https://www.nrel.gov/.
[13] Sarabia AF. Impact of distributed generation on distribution system. [PhD dissertation] Denmark: Dept. Energy Technology, Aalborg Univ.; 2011.
[14] Khan UN. Distributed generation and power quality. IEEE conference 2009;1–4.
[15] Moradi MH, Hajinazari M, Jamasb S, Paripour M. An energy management system (EMS) strategy for combined heat and power (CHP) systems based on a hybrid optimization method employing fuzzy programming. Energy 2013;49(1):86–101.
[16] Majidi M, Nojavan S, Nourani Esfetanaj N, Najafi-Ghalelou A, Zare K. A multi-objective model for optimal operation of a battery/PV/fuel cell/grid hybrid energy system using weighted sum technique and fuzzy satisfying approach considering responsible load management. Sol Energy 2017;144:79–89.
[17] Tahani M, Babayan N, Pouyaei A. Optimization of PV/wind/battery stand-alone system, using hybrid FPA/SA algorithm and CFD simulation, case study: Tehran. Energy Convers Manage 2015;106:644–59.
[18] Maleki A, Khajeh MG, Rosen MA. Weather forecasting for optimization of a hybrid solar-wind-powered reverse osmosis water desalination system using a novel optimizer approach. Energy 2016;114:1120–34.
[19] Maleki A, Askarzadeh A. Comparative study of artificial intelligence techniques for sizing of a hydrogen-based stand-alone photovoltaic/wind hybrid system. Int J Hydrogen Energy 2014;39(19):9973–84.
[20] El Motaleb AMA, Bekdache SK, Barrios LA. Optimal sizing for a hybrid power system with wind/energy storage based in stochastic environment. Renew Sustain Energy Rev 2016;59:1149–58.
[21] Yahiaoui A, Benmansour K, Tadjine M. Control, analysis and optimization of hybrid PV-diesel-battery systems for isolated rural city in Algeria. Sol Energy 2016; 137:1–10.
[22] Askarzadeh A. Developing a discrete harmony search algorithm for size optimization of wind-photovoltaic hybrid energy system. Sol Energy 2013;98:190–5.
[23] Askarzadeh A, dos Santos Coelho L. A novel framework for optimization of a grid independent hybrid renewable energy system: a case study of Iran. Sol Energy 2015;112:383–96.
[24] Merei G, Berger C, Sauer DU. Optimization of an off-grid hybrid PV-wind- diesel system with different battery technologies using genetic algorithm. Sol Energy 2013;97:460–73.
[25] Deihimi A, Keshavarz Zahed B, Iravani R. An interactive operation management of a micro-grid with multiple distributed generations using multi-objective uniform water cycle algorithm. Energy 2016;106:482–509.
[26] Paliwal P, Patidar NP, Nema RK. Determination of reliability constrained optimal resource mix for an autonomous hybrid power system using particle swarm optimization. Renew Energy 2014;63:194–204.
[27] Ghaem Sigarchian S, Orosz MS, Hemond HF, Malmquist A. Optimum design of a hybrid PV-CSP-LPG microgrid with particle swarm optimization technique. Appl Therm Eng 2016;109:1031–6.
[28] Destro N, Benato A, Stoppato A, Mirandola A. Components design and daily operation optimization of a hybrid system with energy storages. Energy 2016;117:569–77.

[29] Pirkandi J, Jokar MA, Sameti M, Kasaeian A, Kasaeian F. Simulation and multi-objective optimization of a combined heat and power (CHP) system integrated with low-energy buildings. J Build Eng 2016;5:13–23.
[30] Gupta RA, Kumar R, Bansal AK. BBO-based small autonomous hybrid power system optimization incorporating wind speed and solar radiation forecasting. Renew Sustain Energy Rev 2015;41:1366–75.
[31] Lauinger D, Caliandro P, Van Herle J, Kuhn D. A linear programming approach to the optimization of residential energy systems. J Energy Storage 2016;7:24–37.
[32] Bortolini M, Gamberi M, Graziani A, Pilati F. Economic and environmental bi-objective design of an off-grid photovoltaic-battery-diesel generator hybrid energy system. Energy Convers Manage 2015;106:1024–38.
[33] Ho WS, Macchietto S, Lim JS, Hashim H, Muis ZA, Liu WH. Optimal scheduling of energy storage for renewable energy distributed energy generation system. Renew Sustain Energy Rev 2016;58:1100–7.
[34] Abbassi A, Dami MA, Jemli M. A statistical approach for hybrid energy storage system sizing based on capacity distributions in an autonomous PV/wind power generation system. Renew Energy 2017;103:81–93.
[35] Urtasun A, Sanchis P, Barricarte D, Marroyo L. Energy management strategy for a battery-diesel stand-alone system with distributed PV generation based on grid frequency modulation. Renew Energy 2014;66:325–36.
[36] Saba D, Laallam FZ, Hadidi AE, Berbaoui B. Optimization of a multi-source system with renewable energy based on ontology. Energy Procedia 2015;74:608–15.
[37] Sakakibara K, Taniguchi I, Matsumoto T, Tamaki H. Structural optimization of a self-sustainable decentralized energy system by mathematical programming techniques. Procedia Technol 2014;18:68–71.
[38] Jung J, Villaran M. Optimal planning and design of hybrid renewable energy systems for microgrids. Renew Sustain Energy Rev 2016;2015:1–12.
[39] Bhandari B, Lee KT, Lee CS, Song CK, Maskey RK, Ahn SH. A novel off-grid hybrid power system comprised of solar photovoltaic, wind, and hydro energy sources. Appl Energy 2014;133:236–42.
[40] Chen J, Rabiti C. Synthetic wind speed scenarios generation for probabilistic analysis of hybrid energy systems. Energy 2016;1–11.
[41] Kefayat M, Lashkar Ara A, Nabavi Niaki SA. A hybrid of ant colony optimization and artificial bee colony algorithm for probabilistic optimal placement and sizing of distributed energy resources. Energy Convers Manage 2015;92:149–61.
[42] Sedighizadeh M, Esmaili M, Esmaeili M. Application of the hybrid Big Bang-Big Crunch algorithm to optimal reconfiguration and distributed generation power allocation in distribution systems. Energy 2014;76:920–30.
[43] Fetanat A, Khorasaninejad E. Size optimization for hybrid photovoltaic–wind energy system using ant colony optimization for continuous domains based integer programming. Appl Soft Comput 2015;31:196–209.
[44] Shivaie M, Ameli MT, Sepasian MS, Weinsier PD, Vahidinasab V. A multistage framework for reliability-based distribution expansion planning considering distributed generations by a self-adaptive global-based harmony search algorithm. Reliab Eng Syst Saf 2015;139:68–81.
[45] Martín García JA, Gil Mena AJ. Optimal distributed generation location and size using a modified teaching–learning based optimization algorithm. Int J Electr Power Energy Syst 2013;50:65–75.
[46] Injeti SK. A Pareto optimal approach for allocation of distributed generators in radial distribution systems using improved differential search algorithm. J Electr Syst Inf Technol 2017.

[47] Mistry KD, Roy R. Enhancement of loading capacity of distribution system through distributed generator placement considering techno-economic benefits with load growth. Int J Electr Power Energy Syst 2014;54:505–15.
[48] Nadjemi O, Nacer T, Hamidat A, Salhi H. Optimal hybrid PV/wind energy system sizing: application of cuckoo search algorithm for Algerian dairy farms. Renew Sustain Energy Rev 2016;0–1.
[49] Yammani C, Maheswarapu S, Matam SK. A multi-objective Shuffled Bat algorithm for optimal placement and sizing of multi distributed generations with different load models. Int J Electr Power Energy Syst 2016;79:120–31.
[50] Esmaeili M, Sedighizadeh M, Esmaili M. Multi-objective optimal reconfiguration and DG (Distributed Generation) power allocation in distribution networks using Big Bang-Big Crunch algorithm considering load uncertainty. Energy 2016;103:86–99.
[51] Erol OK, Eksin I. A new optimization method: Big Bang-Big Crunch. Adv Eng Softw 2006;37(2):106–11.
[52] Singh B, Mukherjee V, Tiwari P. GA-based multi-objective optimization for distributed generations planning with DLMs in distribution power systems. J Electr Syst Inf Technol 2016.
[53] Wang X, Palazoglu A, El-Farra NH. Operational optimization and demand response of hybrid renewable energy systems. Appl Energy 2015;143:324–35.
[54] Touretzky CR, McGuffin DL, Ziesmer JC, Baldea M. The effect of distributed electricity generation using natural gas on the electric and natural gas grids. Appl Energy 2016;177:500–14.
[55] Esmaili M, Firozjaee EC, Shayanfar HA. Optimal placement of distributed generations considering voltage stability and power losses with observing voltage-related constraints. Appl Energy 2014;113:1252–60.
[56] Sharafi M, ElMekkawy TY. Stochastic optimization of hybrid renewable energy systems using sampling average method. Renew Sustain Energy Rev 2015;52:1668–79.
[57] Theo WL, Lim JS, Alwi SRW, Rozali NEM, Ho WS, Abdul-Manan Z. An MILP model for cost-optimal planning of an on-grid hybrid power system for an eco-industrial park. Energy 2016;116:1423–41.
[58] Mallikarjun S, Lewis HF. Energy technology allocation for distributed energy resources: a strategic technology-policy framework. Energy 2014;72:783–99.
[59] Hung DQ, Mithulananthan N, Bansal RC. An optimal investment planning framework for multiple distributed generation units in industrial distribution systems. Appl Energy 2014;124:62–72.
[60] Pless J, Arent DJ, Logan J, Cochran J, Zinaman O. Quantifying the value of investing in distributed natural gas and renewable electricity systems as complements: applications of discounted cash flow and real options analysis with stochastic inputs. Energy Policy 2016;97:378–90.

CHAPTER 3

Off-grid hybrid renewable energy systems and their contribution to sustainable development goals

Binayak Bhandari[a] and Sung-Hoon Ahn[b]
[a]Department of Railroad Engineering and Transport Management, Woosong University, Daejeon, South Korea
[b]Department of Mechanical and Aerospace Engineering, Seoul National University, Seoul, South Korea

3.1 Introduction

According to the Energy Information Administration (EIA) 2019 report, global energy consumption will increase by 28% between 2015 and 2040; a large portion of this energy will be provided by fossil fuels. Fossil fuels such as coal and oil contribute significantly to carbon dioxide emissions, global warming, and air pollution. As a part of the 2015 Paris Agreement, governments around the world are engaged to ramp down greenhouse gas emissions to prevent the adverse effect of climate change. One alternative is to shift the generation of energy from fossil fuels to renewable energy sources. A sharp decrease in fossil fuels reserve worldwide is also a reason to reduce dependency on these nonrenewable energy resources.

Global access to electricity has been steadily rising in recent decades. In 2017, the total number of the population without electricity fell below one billion for the first time in decades. The data show that global access to electricity surpassed 87% in 2016 but still far from meeting the universal access goal by 2030. Thus, off-grid power systems can help meet the energy needs of the population in remote places. In addition, remote systems, such as repeater stations and radio telecommunication stations, depend on the off-grid power systems. Many factors must be carefully analyzed before deciding on a full-phase renewable energy-based off-grid power generation system.

The majority of the population without access to electricity live in rural areas. Fig. 3.1 shows the percentage of electricity access in Asia, Africa, and Latin American countries. To provide electricity access in an environmentally sustainable manner to these remote population, the off-grid power

Hybrid Energy System Models
https://doi.org/10.1016/B978-0-12-821403-9.00007-X

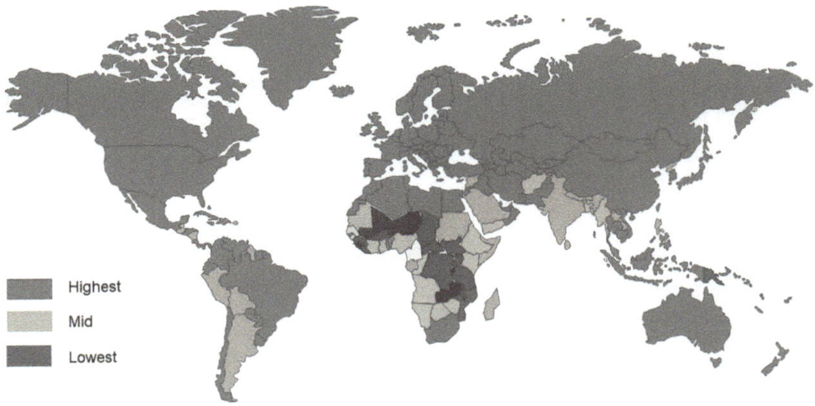

Fig. 3.1 Share of the population with access to electricity, 2018. *(Redrawn from IEA. SDG7: data and projections; 2019. Available: https://www.iea.org/reports/sdg7-data-and-projections.)*

system with minigird can be a solution. Access to electricity in the developing countries has improved over the years, and the share of renewable energy in the electricity sector has also increased [1]. Access to reliable and affordable electricity affects the quality of life, access to basic services, and livelihoods [2]. Table 3.1 compares the electricity access and related data for the years 2010 and 2017.

Over the past decade, the deployment of off-grid power and minigrid systems has seen huge progress because technology cost has decreased and efficiency increased [2]. However, the systems depend on weather conditions which may hinder stable power supply and therefore the use of multiple renewable energy resources to produce power is required. One of the most challenging tasks of the off-grid power system is to balance the load demand and electrical load anomaly detection. Wang and Ahn [3] proposed an electrical load anomaly detection framework, which includes a hybrid one-step ahead load predictor (OSA-LP), and a rule-

Table 3.1 Primary indicators of the global process toward SDG7 targets.

	2010	2017
People without electricity access	1.2 billion	840 million
People without clean cooking	2.96 billion	2.90 billion
Total % energy from renewables	16.6	17.5
Primary energy intensity	5.9 MJ/USD	5.1 MJ/USD

IEA, IRENA, UNSD, WBG, and WHO, 2019 Tracking SDG7: the energy progress report; 2019.

engine-based load anomaly detector (RE-AD) to detect the load of the off-grid solar plant in Ngurdoto, Tanzania.

Despite the challenges, according to the 2018 IRENA report [2], nearly 133 million people in 2016 were served globally by off-grid renewable energy, a sixfold increase from 2011. This is evident from the number of recent publications highlighting the implementation of the off-grid renewable system in various countries and its socioeconomic impacts [4–7]. The attraction toward the off-grid system is also because of the engagement of diverse stakeholders such as local entrepreneurs, international investors, and financial institutions with customized financing models, such as pay-as-you-go (PAYG) and microfinancing [2].

This chapter reviews the background of renewable energy systems and the impact of off-grid systems in providing affordable and clean energy to un-electrified areas and finally discusses the role of the off-grid power system in achieving the seventh objective of the SDGs.

3.2 Types of power systems (based on the connection)

A large portion of the global supply of electricity is generated by the fossil-fuel-based power system. However, because of the adverse effect on the environment, restrictions are implied to reduce the use of fossil fuels for energy generation. On the other hand, there is a global trend toward the use of renewable energy-based power systems for expanding, upgrading. and modernizing power systems. The decentralized power generation with renewable energies helps in reducing the dependence on fossil fuels, lower the investment for grid extension, and check various losses incurred during grid extensions. The main challenge in developing a renewable energy-based power system is that the power production fluctuates depending on the availability of energy resources, weather, and climatic conditions.

The majority of the household without electricity live in remote and isolated rural areas, mostly in developing countries sparsely scattered over a vast geographic location. Because of economic and technical reasons, the governments often have difficulty in providing electric power to their entire citizen through the centralized grid. The off-grid power system is an alternative to grid extension in rural and isolated areas.

3.2.1 Off-grid or stand-alone power systems

The stand-alone power system is also known as an off-the-grid power system that is not fitted with an electricity distribution system and is independent of the electric grid. Some of the popular electrical power generation systems are photovoltaic systems using PV panel, wind turbine, micro-hydro, and diesel or biofuel generator. An energy storage device such as a battery bank is typically implemented in a stand-alone power system which makes the system expensive.

Remote and isolated power systems depend entirely on renewable energy systems such as repeater tower stations, radio telecommunication stations, etc. Depending on the resource availability at a particular location, two or more renewable energy sources with complementary characters such as Photovoltaic (PV) and Wind hybrid systems can be used to increase the reliability [8]. Such a system can range from a small home system to a large system that can power a village or an island.

A minigrid is an independent distribution grid that operates in isolation from a centralized national grid. Minigrid serves a limited number of consumers with a small-scale electricity generation system.

3.2.1.1 A case of Lama Hotel village in Nepal

Lama Hotel village is located at an altitude of 2478 m above sea level and is within the conservation area of Langtang National Park. It is surrounded by high mountains on the east, west, and the north, and it takes 8–10 h of trekking along the Langtang River to reach this place from the nearest motor-able road in Syafrubeshi. It has only seven houses, all in all serving as a temporary staying in for trekkers and a camping site for many dwellers as seen in Fig. 3.2.

For decades, the community of Lama Hotel lived without electricity and their life was centered on sun and firewood. Because of the difficult geographical location, there is a feeble chance of national grid extension to this area. Such is the place for an off-grid power system. A team of volunteers including authors designed 2.2 kW off-grid PV system in 2011 which is serving the villagers and thousands of tourists every year [9]. Fig. 3.3 shows the installed PV system with the battery house.

3.2.2 Grid-tied power systems

On-grid or grid-tied systems allow the excess electricity generated by the system to be added onto the national-grid instead of storing it in batteries. Batteries are inefficient in storing energy and often degrades with time. Grid-tie inverter (GTI) is used in this type of power system to synchronize the phase and frequency of the current to fit the utility grid (50 Hz or 60 Hz). The output voltage is adjusted in such a way that the excess electricity flows outward to the grid. In the case of insufficient electricity production,

Fig. 3.2 An isolated Lama Hotel village in Rasuwa, Nepal.

Fig. 3.3 Solar panel with battery bank house.

electricity can be drawn in from the grid. Consumers only need to pay for what they consume after deducting the power they put into the grid. This scheme is called net-metering. Net metering is a national policy to promote private investment in renewable energy-based power production. Net metering uses a bidirectional energy meter (or net meter) which can measure incoming and outgoing current. The other scheme is called Net Billing where prosumers (producers + consumers of electricity) are charged/compensated based on the actual market value of electricity taking into consideration of what they consume against what they supply into the grid.

3.2.2.1 A case of New South Wales, Australia

Most of the houses in Australia are connected to the electricity grid that is powered by large power plants using fossil fuels such as coal. Australian government's plan to tackle climate change and harness clean energy from the abundant solar renewable energy introduced solar home programs. That is why the majority of newly built Australian houses have solar panels on their roof and are grid connected without the need for the batteries. The house owners are attracted by the Australian Government subsidy scheme. According to the scheme up to Australian Dollar (AUD) 5000 or 60% rebate is provided by the Australian Government for grid-tied solar power systems. Majority of houses have 5kw - 10 kW PV installation. For example, the cost of 6.5 kW PV system installation is around AUD 7000, of which 60% is refunded as rebate by government needing only about AUD 3000 by the house owner. Fig. 3.4 shows a typical house with a grid-tied PV system in New South Wales, Australia.

3.2.3 Hybrid power systems

Hybrid power systems combine the best of grid-tied and off-grid power systems. Hybrid power systems are less expensive than off-grid power systems because this eliminates the necessity of large battery storage capacity; however, it is expensive than the grid-tied power systems.

Although the grid-tied system can generate power during the outage, the inverter will be shut off and the power produced can neither be sent to the grid nor be used by household because of inverter's anti-islanding protection safety function. The anti-islanding protection feature is an inverter function that senses the problem with power grid, such as power outage, and shuts itself to stop feeding power to the electric grid. The feature assumes that the workers are diapatched for fixing outage; thus, there should be no electricity flow in the grid for the workers' safety. Unlike the grid-connected

Fig. 3.4 A typical Australian home with grid-tied PV system in New South Wales, Australia.

system, hybrid power system comprising battery backup can secure power even during grid outages. Hybrid power systems with energy storage can isolate from the grid (called islanding) and can supply power during blackout also.

Fig. 3.5 shows the schematics of various types of the renewable energy-based power generation system.

3.2.3.1 Case study of California, United States

Currently, in the USA, 37 states and Washington, DC offer net metering. Net metering offers the distributed energy producer to put excess power in the electric grid to be used by other consumers. The energy meter can rotate in both directions when the power is sent onto the grid, the meter "spins" backward and vice versa. Thus, net-metering means the total electricity used by consumer at the end of the month after deducting the power put to the grid from the power consumed from the grid.

Modified version of net-metering is called "energy storage net energy metering" and is colloquially called "NEM-paired storage" which has battery storage. NEM-paired storage was codified into law in California in

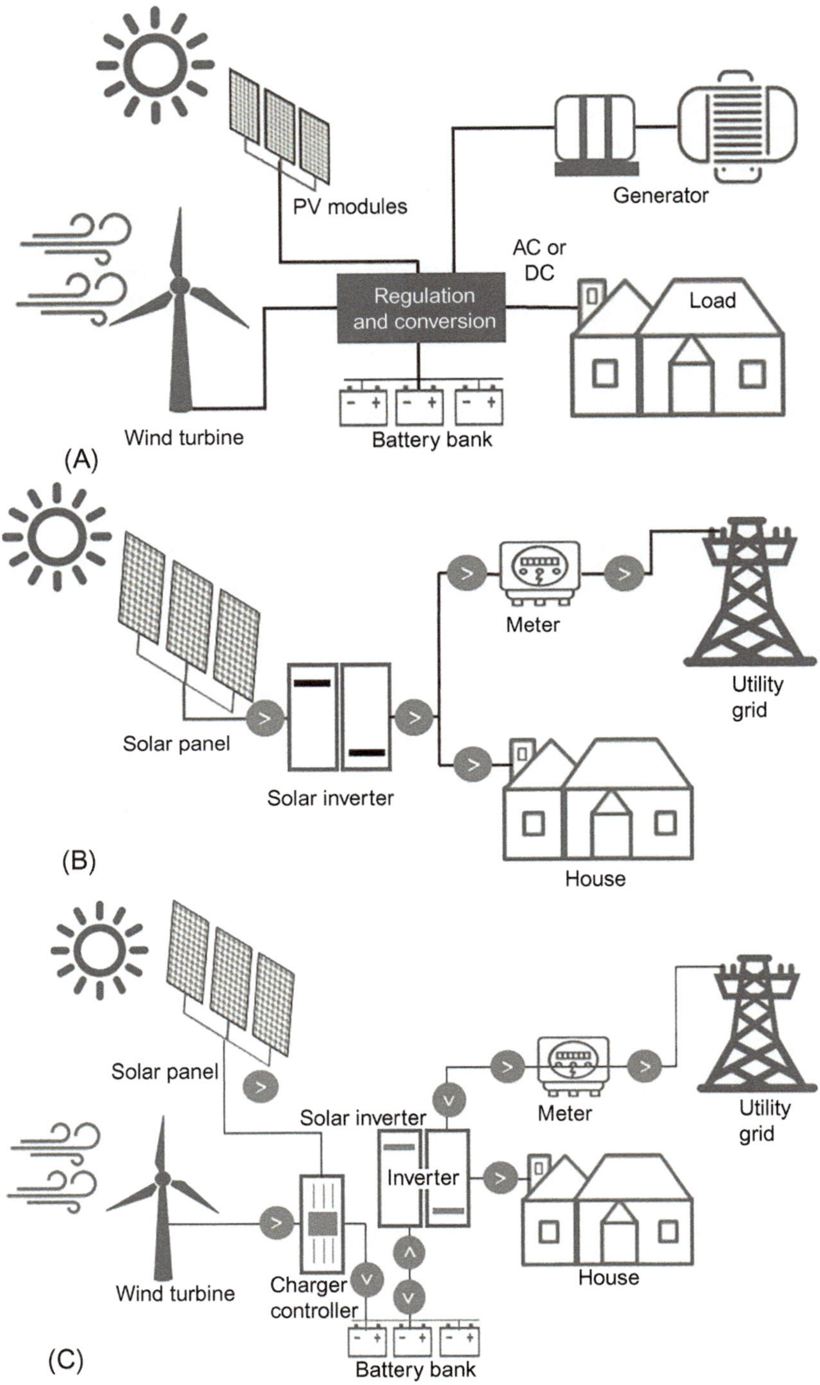

Fig. 3.5 Various types of power systems (A) off-grid power system, (B) grid-tied power system, and (C) hybrid power system.

February of 2019 permitting customers to receive credits for stored solar energy sent back to the grid. [10] California has lower rates for solar energy sent to the grid in the midday while it is higher in the evening. Therefore, grid-tied system (without battery storage) can only receive reduced payback who can export during the day. Contrary to that, hybrid power systems with energy storage can store midday generation and sell it to the grid at the peak time (in the evening). This can have a double benefit: first, it addresses the challenge of "duck curve," and second, the distributed generator makes more money for the same energy pumped to the grid.

3.3 Off-grid power systems based on renewable energies

The three end-uses of renewable energy are electricity, heat, and transport. The use of renewables for electricity has grown fastest, because of the rapid expansion of wind, solar, and hydropower technologies. As of 2016, the share of renewables in electricity generation was 24% [11].

Hydro, wind, and solar energy are the top renewable energy sources in terms of globally installed capacity. Because of the periodic nature of renewable energy sources such as wind and solar, often hybrid renewable energy sources (HRES) are employed for better reliability and the need for smaller energy storage devices. Commercially viable off-grid hybrid renewable energy systems for electricity generation include PV-battery, PV-diesel, wind-battery, wind-diesel, PV-wind-battery, and PV-wind-diesel-battery systems. Bhandari et al. [12] built the first off-grid hybrid power system comprised of solar photovoltaic, wind, and hydro energy sources in the remote mountainous village in Nepal. Development of off-grid renewables for other productive uses such as cooking, heating is also possible but it is beyond the scope of the present review.

According to the IRENA report, the total off-grid capacity in 2017 was 6.6 GW. Solar PV contributes to the majority of use in commercial and residential use while bioenergy has maximum share in industrial use. The detailed end-uses of off-grid renewable power are shown in Table 3.2.

3.3.1 Solar power system

The solar power system or a PV power system converts sunlight directly into electricity. The power output of a PV array is based on solar irradiance and ambient temperature and is calculated as [13]

$$P_{pv} = \eta_{pvg} \cdot A_{pvg} \cdot G_t, \tag{3.1}$$

Table 3.2 Off-grid electricity end-uses in 2017 in MW [2].

End-use sector	Hydropower	Solar PV	Wind	Bioenergy	Geothermal
Industry	70	10	10	2780	50
Commercial	10	430	–	10	–
Residential	60	280	–	10	–
Agriculture	10	210	–	20	–
Mixed-use (minigrid)	360	310	480	–	–
Not known	–	1500	–	–	–

where η_{pvg} is PV generation efficiency, A_{pvg} is PV generator area (m^2), and G_t is solar irradiation in tilted module plane (W/m^2). Fig. 3.6. shows the voltage and current characteristics for various irradiation levels.

In 2016, more than 124 million people accessed lighting and other electrical services in Africa using solar lights and solar home systems. In Asia, more than 70 million people benefited from solar light and solar home systems in the same year. Bangladesh alone deployed 4.1 million solar systems in 2017 to provide electricity to more than 18 million people in 2017.

3.3.2 Wind power system

The wind turbine is designed to produce power at a wide range of wind speeds. Location plays an important role in the performance and efficiency of a wind turbine. The governing equation for calculating the mechanical power of the wind turbine is given by

$$P_w = \frac{1}{2} C_p(\lambda, \beta) \rho A V^3 \tag{3.2}$$

where ρ is air density (kg/m^3), C_p is the power coefficient, A is intercepting area of the rotor blades (m^2), V is average wind speed (m/s), and λ is tip speed ratio. The theoretical maximum value of the power coefficient C_p is 0.593, known as Betz's coefficient [14]. Fig. 3.7 shows the wind turbine power curve for a range of wind velocity.

3.3.3 Hydropower system

Modern-day hydro turbines originated from ancient water wheels. Hydropower provides about 20% of all electricity worldwide. The mechanical power generated by the turbine is given by

$$P = \eta_{total} \rho g Q H \tag{3.3}$$

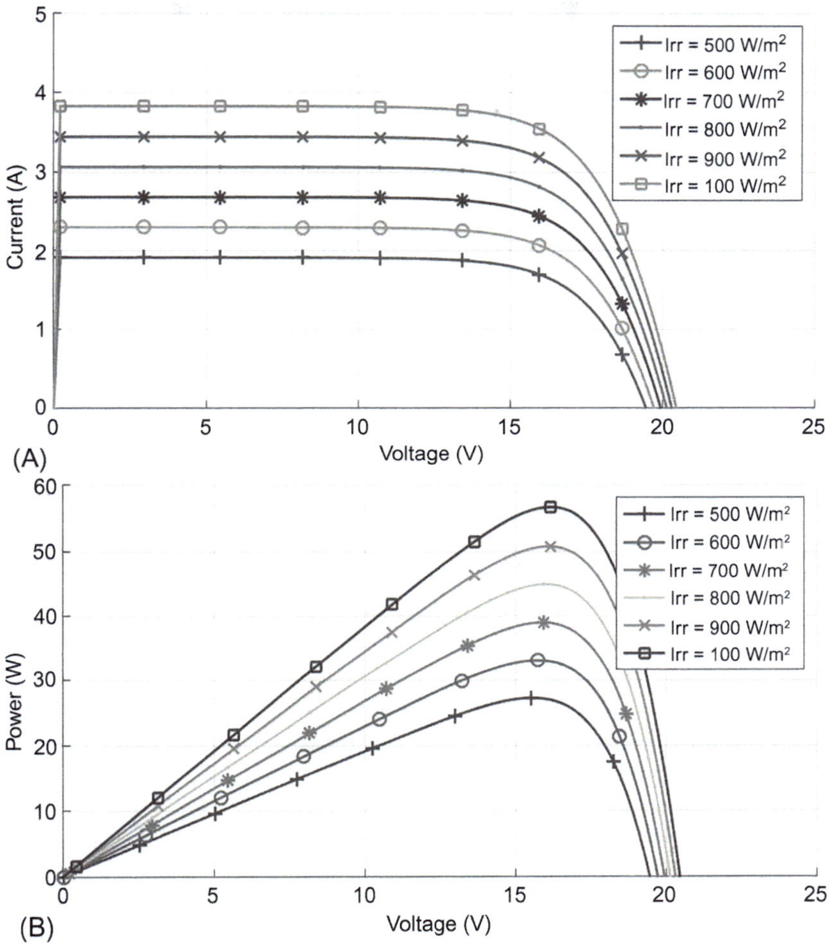

Fig. 3.6 Solar cell characteristics (A) voltage-current characteristics and (B) voltage-power characteristics.

where P is mechanical power output produced at the turbine, η_{total} is hydraulic efficiency of the turbine, ρ is the density of water (1000 kg/m^3), g is the acceleration due to gravity (9.81 m/s^2), and H is effective pressure head (m).

More than 6.4 million people were served by hydropower-based mini-grid in 2016 as reported by IRENA. In Nepal alone, the installed capacity of off-grid micro-hydro is 50 MW from over 2500 systems in 2017 [2].

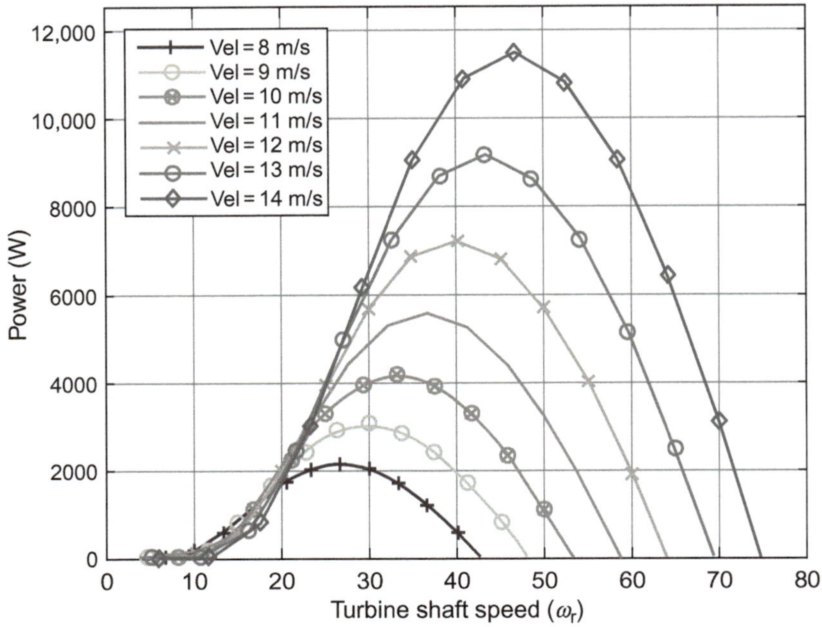

Fig. 3.7 Wind turbine power curve.

3.4 Role of off-grid power systems in attaining sustainable development goals

All United Nations (UN) member states in 2015 adopted 17 sustainable development goals (SDGs) for peace and prosperity for people and the planet to attain by the year 2030 [15]. SDGs are intertwined together, that is, one goal contributes to achieving other goals. 7th SDG is to ensure global access to affordable, reliable, sustainable, and modern energy for all. This highlights the importance of energy that contributes to other goals such as ending poverty and improving health, and to the development of education and gender equality.

To achieve the goal by 2030, the current pace of electricity expansion must be accelerated. It is estimated that rural electrification using the off-grid systems will supply nearly 60% of the additional generation needed to achieve universal electricity access [16]. Easy access to electricity will stimulate economic and social development and will lead to an improvement in the quality of life by helping reduce extreme poverty and providing universal primary education. Bhandari et al. [4] reported on the impact of the off-grid power system in the remote mountainous village in Nepal. The impact of electrification was measured on the household's lighting usage, education, social structure, environment, income, and level of satisfaction.

Since 2010, 80% of the 800 million people in Asia gained access to electricity. According to the World Bank report [17], Afghanistan, Nepal, and Bhutan are the three countries with the greatest increase in national electricity access rates from 2006 to 2016 where off-grid renewables have played a pivotal role in electrifying rural remote areas. The African continent has sharply increased the deployment of off-grid renewable energy in recent years too, and the population served by such systems grew rapidly from just over 2 million people in 2011 to over 53 million in 2016 [2]. However, African countries have uneven access to electricity, and the study suggests about 600 million people are still without access to electricity in sub-Saharan Africa [18].

Accelerating off-grid installations toward achieving the SDGs requires collaborative actions across multiple elements such as financing models, policies and regulations, technology, institutional frameworks, capacity building, and cross-sector linkages [19] as shown in Fig. 3.8. Off-grid renewable energy supports basic services, such as health, water, education, and

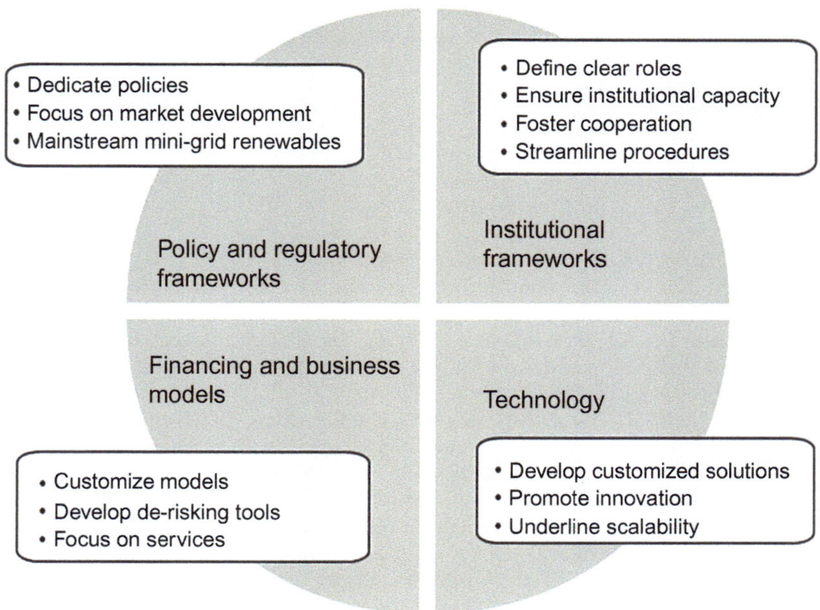

Fig. 3.8 Factors affecting off-grid system expansion. *(Redrawn from IRENA, International O¢-grid renewable energy conference. Key findings and recommendations; 2013.)*

livelihoods through productive end-uses. For larger population coverage, power from the off-grid systems can be integrated into a single distribution grid known as minigrid that is independent of a national grid. These minigrids are considered to be the most economical long-term solution for electricity access as they are flexible in terms of sizing and resource utilization. Off-grid and minigrid solutions powered by renewables already provide electricity to nearly 90 million people [20] which is evident that the off-grid power system has a bigger role to play in meeting SDGs objectives.

3.5 Conclusions

As of 2019, globally 840 million people are without access to electricity, the majority are concentrated in Sub-Saharan Africa. Although a significant portion of those will have access to reliable electricity by 2030, SDC 7 is likely to be reached partially with the current trend. It has been projected that around 650 million people are likely to remain without access to electricity by 2030 [9]. A large majority of such people will be in Sub-Saharan Africa.

The off-grid power system can play a significant role in providing reliable electricity without adverse environmental impact to the vast population living far from the national grid reach contributing to sustainability. The various options reviewed here offer opportunities to reach the 7th objective of SDGs.

References

[1] Progress of goal 7 in 2019. Available, https://sustainabledevelopment.un.org/sdg7; 2020.
[2] I. R. E. Agency. Off-grid renewable energy solutions.
[3] Wang X, Ahn SH. Real-time prediction and anomaly detection of electrical load in a residential community. Appl Energy 2020;259:1–10.
[4] Bhandari B, Lee KT, Chu WS, Lee CS, Song CK, Bhandari P, et al. Socio-economic impact of renewable energy-based power system in mountainous villages of Nepal. Int J Pr Eng Man-GT 2017;4:37–44.
[5] Ahn SH, Lee KT, Bhandari B, Lee GY, Lee CS, Song CK. Formation strategy of renewable energy sources for high mountain off-grid system considering sustainability. J Korean Soc Precis Eng 2012;29:958–63.
[6] Addin Z, Merida W. Hybrid energy systems for off-grid power supply and hydrogen production based on renewable energy: a techno-economic analysis. Energy Convers Manag 2019;196:1068–79.
[7] Rahman MM, Khan MMUH, Ullah MA, Zhang X, Kumar A. A hybrid renewable energy system for a north American off-grid community. Energy 2016;97:151–60.
[8] Bhandari B, Ahn S-H, Ahn T-B. Optimization of hybrid renewable energy power system for remote installations: case studies for mountain and island. Int J Precis Eng Manuf 2016;17:815–22.

[9] Bhandari B, Lee GY, Lee KT, Yoon HS, Kim DH, Song SH, Moon JS, Maskey RK, Ahn SH. Off-grid sustainable small-scale photovoltaic (PV) project in high altitude region of Nepal, In: Addressing climate change for sustainable development through up-scaling renewable energy technologies, October 12–14, Kathmandu, Nepal; 2011.
[10] Energy storage net metering: An Illustration of why it's so valuable. from, https://www.energystorageconsultants.com/energy-storage-net-metering-an-illustration-of-why-its-so-valuable/.
[11] IEA, IRENA, UNSD, WBG, WHO. 2019 Tracking SDG7: the energy progress report.
[12] Bhandari B, Lee KT, Lee CS, Song CK, Maskey RK, Ahn SH. A novel off-grid hybrid power system comprised of solar photovoltaic, wind, and hydro energy sources. Appl Energy 2014;133:236–42.
[13] Diaf S, Notton G, Belhamel M, Haddadi M, Louche A. Design and techno-economical optimization for hybrid PV/wind system under various meteorological conditions. Appl Energy 2008;85:968–87.
[14] Betz A. Introduction to the theory of flow machines. Oxford: Pergamon Press; 1966.
[15] UN. (13 March). Available, https://sustainabledevelopment.un.org/?menu=1300.
[16] IEA, UNDP, INIDO. Energy poverty how to make modern energy access universal? International Energy Agency; 2010.
[17] UN. Access to electricity (% of population), Available:https://data.worldbank.org/indicator/EG.ELC.ACCS.ZS; 2019.
[18] IEA. SDG7: data and projections, Available:https://www.iea.org/reports/sdg7-data-and-projections; 2019.
[19] IRENA. REthinking Energy 2017.
[20] B. a. L. Global. Off-Grid Solar Market Trends Report 2016. World Bank Group; 2016.

CHAPTER 4

Numerical simulation of hybrid systems based on solid oxide fuel cells

Haoran Xu[a] and Meng Ni[b,c]
[a]Department of Chemical Engineering, Loughborough University, Loughborough, United Kingdom
[b]Building Energy Research Group, Department of Building and Real Estate, The Hong Kong Polytechnic University, Kowloon, Hong Kong, China
[c]Environmental Energy Research Group, Research Institute for Sustainable Urban Development (RISUD), The Hong Kong Polytechnic University, Kowloon, Hong Kong, China

Nomenclature

Abbreviation

BR	Boudouard reaction
CHP	combined heat and power
DCFC	direct carbon fuel cell
DGM	dusty-gas model
DRM	dry reforming of methane
F-T	Fischer-Tropsch
SMR	steam methane reforming
N-S	Navier-Stokes
SCCM	standard cubic centime per minute
SGR	steam gasification reaction
SHE	Stirling heat engine
SOFC	solid oxide fuel cell
TiO_2	titanium dioxide
TPB	triple phase boundary
UV	ultraviolet
WGSR	water-gas shift reforming

Roman

B_0	permeability coefficient (m^2)
C_i	the number of C atoms contained in the fuel
C_p	fluid heat capacity at constant pressure
C_{total}	total number of C atoms in all fuels
D_{ik}^{eff}	Knudsen diffusion coefficient of i ($m^2\ s^{-1}$)
D_{ij}^{eff}	molecular diffusion coefficient of i ($m^2\ s^{-1}$)
E_{act}	activation energy ($J\ mol^{-1}$)
E_{CO}	equilibrium potential for carbon monoxide oxidization (V)
E_{CO}^0	standard equilibrium potential for carbon monoxide oxidization (V)

Hybrid Energy System Models
https://doi.org/10.1016/B978-0-12-821403-9.00003-2

E_{H_2}	equilibrium potential for hydrogen oxidization (V)
$E^0_{H_2}$	standard equilibrium potential for hydrogen oxidization (V)
E_{eq}	equilibrium potential (V)
F	Faraday constant (96,485 C mol^{-1})
ΔG	Gibbs free energy change (J mol^{-1})
i	current density (A m^{-2})
i_0	exchange current density (A m^{-2})
n	number of electrons transferred per electrochemical reaction
N_i	flux of mass transport (kg m^{-3} s^{-1})
P	pressure (Pa)
P_c	proportions of carbon fuels
P^L_{CO}	local partial pressure of carbon monoxide (Pa)
$P^L_{CO_2}$	local partial pressure of carbon dioxide (Pa)
$P^L_{H_2}$	local partial pressure of hydrogen (Pa)
$P^L_{H_2O}$	local partial pressure of steam (Pa)
$P^L_{O_2}$	local partial pressure of oxygen (Pa)
Q	heat (J s^{-1})
R	gas constant (8.314 J mol^{-1} K^{-1})
R_i	reaction rate of equation i (mol m^{-3} s^{-1})
R_j	mass source term (mol m^{-3} s^{-1})
t	time (s)
T	temperature (K)
u	velocity field (m^3 s^{-1})
V	voltage (V)
V_l	volume fractions of ionic phase
V_s	volume fractions of electronic phase
y_i	molar fraction of component i

Greek letters

α	charge transfer coefficient
γ	preexponential factor
ε	porosity
η	efficiency
η_{act}	activation overpotential (V)
η_{ohmic}	ohmic overpotential (V)
κ	permeability (m^2)
λ	thermal conductivity (W m^{-1} K^{-1})
μ	dynamic viscosity of fluid (Pa s)
ρ	fluid density (kg m^{-3})
σ	conductivity (S/m)
$\varnothing$	potential (V)

Subscripts

an	anode

ca	cathode
H_2O/H_2	reaction pairs of H_2O/H_2
CO_2/CO	reaction pairs of CO_2/CO
g	gas phase
l	ionic phase
s	electronic phase

Superscripts

0	parameter at equilibrium conditions
eff	effective
L	local

4.1 Introduction

Solid oxide fuel cells (SOFCs) are high-temperature electrochemical devices for energy conversion and storage with whole-solid structures [1–3]. Usually, SOFCs have dense electrolytes sandwiched between the porous anodes and cathodes. Applied with proper potentials, O_2 can be reduced at the cathode to generate the O^{2-} ions, which moves through the electrolyte to the anode and reacts with the fuels to release the electrons. The electrons move in the external circuit, and electricity is generated. Benefited from the simple structure, SOFCs are easily scalable based on different power demands. Therefore, they are suitable for both centralized and distributed applications [4–6].

As a fuel cell, SOFCs are eco-friendly with high efficiency (>60%). The high working temperature brings SOFCs high reaction activities and enables the use of nonnoble catalysts, e.g., nickel [7–9]. SOFCs can utilize a variety of fuels including H_2, CO, hydrocarbons, solid carbons, and NH_3. These characteristics decrease the cost of SOFC devices and broaden their applications [10–12].

SOFCs can work at the reversed mode for electrolysis purpose, solid oxide electrolysis cell (SOEC) [13–15], as shown in Fig. 4.1. An SOEC can effectively reduce oxidants (e.g., H_2O) into fuels (e.g., H_2), by consuming electricity and heat. Compared with low-temperature electrolyzers, SOECs consume less electricity as a portion of energy for electrolysis is provided by heat [16–18].

With the development of SOFC technologies, their applications in hybrid systems have been widely studied, as SOFCs can effectively convert energies among electricity, fuel, and heat. They can principally cooperate

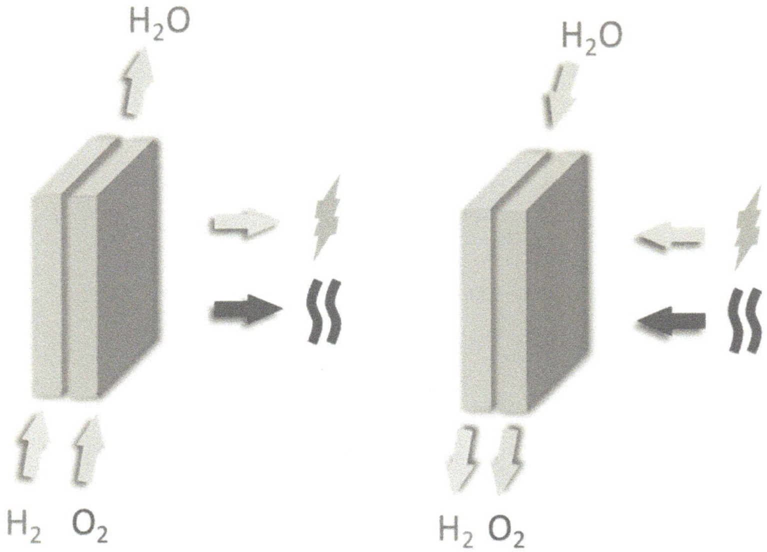

Fig. 4.1 Working process of an SOFC *(left)* and an SOEC *(right)*.

with the devices that can generate or utilize electricity, fuel, or heat. At present, these hybrid systems mainly focus on three aspects: (1) improving the performance (power density, efficiency, etc.) of SOFCs; (2) using the fuels generated from SOEC for the generation of value-added chemicals; and (3) improving the performance (efficiency, stability, etc.) of other devices [19–21].

In this chapter, hybrid systems using the SOFC as the core component will be introduced. These hybrid systems include the combined SOFC and heat engine system to raise the power density and fuel efficiency; the combined SOFC and photoreactor system for the utilization of solar fuels and online solar-fuel-to-power generation; and the combined SOFC and Fischer-Tropsch (F-T) process for low-carbon fuel generation from captured CO_2.

4.2 Model description

4.2.1 Electrochemical reaction model in the SOFC

H_2 and CO are typical fuels that can directly participate in the electrochemical reactions at the anode of SOFCs. The electrochemical oxidation reactions of H_2 and CO with O^{2-} ions are given in Eqs. (4.1), (4.2), respectively [22]:

$$H_2 + O^{2-} = H_2O + 2e^- \quad (4.1)$$

$$CO + O^{2-} = CO_2 + 2e^- \quad (4.2)$$

Air (oxygen) is the typical oxidant which provides the O^{2-} ions at the cathode. The electrochemical reduction reaction of O_2 to O^{2-} is given in Eq. (4.3):

$$O_2 + 4e^- = 2O^{2-} \quad (4.3)$$

The overall electrochemical reactions of Eqs. (4.1)–(4.3) can be written as shown in Eqs. (4.4), (4.5).

$$H_2 + 0.5O_2 = H_2O \quad (4.4)$$

$$CO + 0.5O_2 = CO_2 \quad (4.5)$$

Driven by electrochemical potentials, the cell's operating voltage can be calculated by deducting the activation overpotential (η_{act}) and ohmic overpotential (η_{ohmic}) from the equilibrium potential (E_{eq}), as shown in Eq. (4.6) [23]:

$$V = E_{eq} - \eta_{act} - \eta_{ohmic}. \quad (4.6)$$

4.2.1.1 Equilibrium potential

The equilibrium potentials (Nernst potential) are related to the thermodynamics of the reactions. For reactions in Eqs. (4.4), (4.5), the equilibrium potentials ($E_{H_2}$and E_{CO}) can be calculated by Eqs. (4.7), (4.8) [24]:

$$E_{H_2} = E_{H_2}^0 + \frac{RT}{2F} \ln \left[\frac{P_{H_2}^L \left(P_{O_2}^L \right)^{1/2}}{P_{H_2O}^L} \right], \quad (4.7)$$

$$E_{CO} = E_{CO}^0 + \frac{RT}{2F} \ln \left[\frac{P_{CO}^L \left(P_{O_2}^L \right)^{1/2}}{P_{CO_2}^L} \right], \quad (4.8)$$

where $E_{H_2}^0$ and E_{CO}^0 are standard potentials for the electrochemical oxidation of H_2 and CO, respectively. Their values can be calculated by Eqs. (4.9), (4.10) at high operating temperature [25]. P^L is the local partial pressure of gas species. R is the universal gas constant (8.3145 J mol^{-1} K^{-1}). T is the local temperature (K). F is the Faraday constant (96485 C mol^{-1}).

$$E_{H_2}^0 = 1.253 - 0.00024516T \text{ (V)}, \quad (4.9)$$

$$E_{CO}^0 = 1.46713 - 0.0004527T \text{ (V)}. \quad (4.10)$$

For more general situation, the equilibrium potential can be calculated based on oxygen partial pressure [26] as shown in Eq. (4.11):

$$E_{eq} = \frac{RT}{nF}\ln\left(\frac{\sum P^{L}_{O_2,ca}}{\sum P^{L}_{O_2,an}}\right). \tag{4.11}$$

For the pairs of H_2O/H_2 and CO_2/CO, their oxygen partial pressures can be expressed as shown in Eqs. (4.12), (4.13).

$$P^{L}_{O_2,\ (H_2O/H_2)} = \left(\frac{P^{L}_{H_2O}}{P^{L}_{H_2}}\cdot e^{\frac{\Delta G_{H_2O/H_2}}{RT}}\right)^2, \tag{4.12}$$

$$P^{L}_{O_2,\ (CO_2/CO)} = \left(\frac{P^{L}_{CO_2}}{P^{L}_{CO}}\cdot e^{\frac{\Delta G_{CO_2/CO}}{RT}}\right)^2. \tag{4.13}$$

Here, $\Delta G_{H_2O/H2}$ and $\Delta G_{CO_2/CO}$ are the Gibbs free energy changes in reactions (4.4), (4.5), respectively.

4.2.1.2 Activation overpotential

The activation overpotential reflects the activation barrier in the electrochemical reactions. It is affected by the reaction type and material properties. Butler-Volmer equation is used to describe the relationship between current density and activation overpotential as shown in Eq. (4.14) [27]:

$$i = i_0\left\{\exp\left(\frac{\alpha nF\eta_{act}}{RT}\right) - \exp\left(-\frac{(1-\alpha)nF\eta_{act}}{RT}\right)\right\}. \tag{4.14}$$

Here, i is the current density at applied voltages; α is the electron transfer coefficient, whose value is between 0 and 1; and n is the number of electrons transferred in each electrochemical reaction. i_0 is the exchange current density, which can be further calculated by Eq. (4.15):

$$i_0 = \gamma \exp\left(-\frac{E_{act}}{RT}\right). \tag{4.15}$$

Here, γ is a preexponential factor and E_{act} is activation energy determined by the reaction.

4.2.1.3 Ohmic overpotential

Ohmic overpotential is caused by resistances in conducting the ions and electrons. They can be described by Ohm's law, as given in Eqs. (4.16), (4.17):

$$i_l = -\sigma_l^{\text{eff}} \nabla(\varnothing_l), \tag{4.16}$$

$$i_s = -\sigma_s^{\text{eff}} \nabla(\varnothing_s). \tag{4.17}$$

Here, $\varnothing_l$ and $\varnothing_s$ are the electric potentials of ionic conduction and electronic conduction of the electrodes and electrolyte. σ_l^{eff} and σ_s^{eff} are the effective conductivities of ionic phase and electronic phase in porous electrodes, which are calculated as shown in Eqs. (4.18), (4.19):

$$\sigma_l^{\text{eff}} = \sigma_l \cdot \frac{V_l}{\tau_l}, \tag{4.18}$$

$$\sigma_s^{\text{eff}} = \sigma_s \cdot \frac{V_s}{\tau_s}. \tag{4.19}$$

Here, σ_l and σ_s are the intrinsic conductivity of ionic phase and electronic phase. V_l, V_s, τ_l, and τ_s are the volume fractions and tortuosity of ionic and electronic phases. The ionic/electronic conductivities of typical materials used in SOFCs are listed in Table 4.1.

Table 4.1 Properties of typical SOFC materials.

Parameters	Value or expression	Unit
Ionic conductivity		
CGO (dense)	$1.22 \times e^{\frac{16,054}{RT}}$ [28]	$\mathrm{S\,m^{-1}}$
GDC	$\frac{100}{T} \times 10^{\left(6.66071 - \frac{5322.92}{T}\right)}$ [29]	$\mathrm{S\,m^{-1}}$
LSCF	$\frac{100}{T} \times 10^{2.51289 - \frac{3036.75}{T}}$ [30]	$\mathrm{S\,m^{-1}}$
ScSZ	$6.92 \times 10^4 \exp(-9681/T)$ [16]	$\mathrm{S\,m^{-1}}$
SDC	$\frac{100}{T} \times 10^{5.48077 - \frac{3792.53}{T}}$ [31]	$\mathrm{S\,m^{-1}}$
YSZ	$3.34 \times 10^4\, e^{\frac{-10,300}{T}}$ [16]	$\mathrm{S\,m^{-1}}$
Electronic conductivity		
Ag	$\frac{1.59e^8}{(0.0038T - 0.1134)}$	$\mathrm{S\,m^{-1}}$
BSCF	$-24,455 + 82.321T - \frac{8.28T^2}{1000} + \frac{2.7778T^3}{10,000}$ [28]	$\mathrm{S\,m^{-1}}$
LSCF	$\frac{100}{T} \times 10^{4.32576 + \frac{1204.26}{T}}$ [32]	$\mathrm{S\,m^{-1}}$
LSM	$4.2 \times 10^7 \exp(-1150/T)$ [16]	$\mathrm{S\,m^{-1}}$
Ni	$3.27 \times 10^6 - 1065.3T$ [16]	$\mathrm{S\,m^{-1}}$

4.2.2 Chemical reaction model

Chemical reactions are key processes in the SOFC and other reactors. In the SOFC, the existence of anode catalyst (usual nickel) and high operating temperature favors many reactions to happen, including steam methane reforming (SMR), dry reforming of methane (DRM), Boudouard reaction (BR), steam gasification reaction (SGR), and water-gas shift reaction (WGSR). These reactions are very important in describing the SOFC using methane and solid carbon as fuels. In the photoreactor, the photosynthesis reaction converts H_2O and CO_2 into gaseous fuels, such as CO and CH_4. In the F-T process, the synthesis reactions will convert CO and H_2 into different kinds of hydrocarbons.

4.2.2.1 *Chemical reactions in the SOFC*

The electrochemical reaction rate of methane oxidation is much smaller than H_2 electrochemical oxidation and CO electrochemical oxidation. When methane is used as the fuel, H_2 and CO are the fuels considered to directly participate in electrochemical reactions instead of methane. CO and H_2 come from SMR and DMR as shown in Eqs. (4.20), (4.21), respectively [33]:

$$CH_4 + H_2O = CO + 3H_2 \tag{4.20}$$

$$CH_4 + CO_2 = 2CO + 2H_2 \tag{4.21}$$

As the reaction rate of SMR is usually much faster than DRM, usually DRM is not considered for model simplification when both two reactions occur in the anode [34].

When solid carbon is used as the fuel, gaseous fuels such as CO and H_2 are the fuels directly participate in the electrochemical reactions instead of solid carbon. These gaseous fuels are generated through carbon gasification reactions including the BR and SGR as shown in Eqs. (4.22), (4.23), respectively [35]:

$$C + CO_2 = 2CO \tag{4.22}$$

$$C + H_2O = CO + H_2 \tag{4.23}$$

WGSR catalyzed by nickel is considered when CO and H_2 both occur at the anode as shown in Eq. (4.24) [36]:

$$CO + H_2O \leftrightarrow CO_2 + H_2 \tag{4.24}$$

These widely used chemical reaction kinetics and related parameters are listed in Table 4.2.

Table 4.2 Kinetics of typical SOFC reactions.

Reaction/parameter	Unit/value
***Methane oxidization reaction rate* [37]**	
$5 \times 10^4 \times e^{\frac{-166{,}000}{RT}} \times p_{CH_4} \times p_{O_2}$(anode)	mol m^3 s^{-1}
$1.1 \times e^{\frac{-166{,}000}{RT}} \times p_{CH_4} \times p_{O_2}$(other area)	mol m^3 s^{-1}
***Hydrogen oxidization reaction rate* [37]**	
$5 \times 10^4 \times e^{\frac{-48{,}484}{RT}} \times p_{CH_4} \times p_{O_2}^{0.5}$	mol m^3 s^{-1}
***Carbon monoxide oxidization reaction rate* [37]**	
$5 \times 10^4 \times e^{\frac{-47{,}773}{RT}} \times p_{CO} \times p_{O_2}^{0.5}$	mol m^3 s^{-1}
***Methane steam reforming reaction rate* [38]**	
$2.395 \times 10^7 \times e^{\frac{-231{,}266}{RT}} \times \left(p_{CH_4} \times p_{H_2O} - \frac{p_{CH_4}^3 \times p_{CO}}{K_{pr}}\right)$	mol m^3 s^{-1}
$K_{pr} = 1.0267 \times 10^{10} \times e^{-0.2513 \times Z^4 + 0.3665 \times Z^3 + 0.5810 \times Z^2 - 27.134 \times Z + 3.277}$	
$Z = \frac{T}{1000}$	
***Water-gas shift reaction rate* [18]**	
$0.0171 \times e^{\frac{-103{,}191}{RT}} \times \left(p_{CO} \times p_{H_2O} - \frac{p_{H_2} \times p_{CO_2}}{k_{ps}}\right)$	mol m^3 s^{-1}
$k_{ps} = e^{-0.2935 \times Z^3 + 0.6351 \times Z^2 + 4.1788 \times Z + 0.0169}$	
$Z = \frac{T}{1000}$	
***Carbon gasification reactions* [39]**	
$R_{CCO_2} = \frac{K_1 p_{CO_2}}{1 + K_2 p_{CO} + K_3 p_{CO_2}}$	mol m^3 s^{-1}
$R_{CH_2O} = \frac{K_4 p_{H_2O}}{1 + K_5 p_{H_2} + K_6 p_{H_2O}}$	mol m^3 s^{-1}
K_1	9.32×10^{-4}
K_2	1.25×10^{-3}
K_3	3.82×10^{-5}
K_4	2.19×10^{-3}
K_5	9.88×10^{-4}
K_6	8.13×10^{-5}

4.2.2.2 Chemical reactions in the photoreactor

The chemical reactions in the photoreactor are simplified to calculate the photoreduction rates of CO_2 and H_2O under UV flux. According to Tahir and Amin [40], the reaction can be described by Eq. (4.25):

$$3CO_2 + 2H_2O = CH_4 + 2CO + 3O_2 \tag{4.25}$$

The kinetic parameters of this reaction are given in Table 4.3.

Table 4.3 Kinetics for the photoreaction [40].

Parameter	Value	Unit
$R_{\text{photo}} = kI^{\alpha}\left(\frac{K_{H_2O}P^{L}_{H_2O}K_{CO_2}P^{L}_{CO_2}}{\left(1+K_{H_2O}P^{L}_{H_2O}+K_{CO_2}P^{L}_{CO_2}\right)^2}\right)$		
k	5000	μmole g catal.$^{-1}$
I	150	mW cm^{-2}
β	0.60	
K_{H_2O}	0.75	bar^{-1}
K_{CO_2}	30	bar^{-1}

4.2.2.3 Chemical reactions in F-T reactor

The product from the F-T synthesis process is affected by the catalysts and operating parameters such as the feedstock, temperature, and pressure. For model simplification, the reaction kinetics is expressed as given in Eqs. (4.26)–(4.33) [41]:

$$CO + 3H_2 = CH_4 + H_2O \tag{4.26}$$

$$2CO + 4H_2 = C_2H_4 + 2H_2O \tag{4.27}$$

$$2CO + 5H_2 = C_2H_6 + 2H_2O \tag{4.28}$$

$$3CO + 7H_2 = C_3H_8 + 3H_2O \tag{4.29}$$

$$4CO + 9H_2 = n - C_4H_{10} + 4H_2O \tag{4.30}$$

$$4CO + 9H_2 = i - C_4H_{10} + 4H_2O \tag{4.31}$$

$$6.05CO + 12.23H_2 = C_{6.05}H_{12.36} + 6.05H_2O \tag{4.32}$$

$$CO + H_2O = CO_2 + H_2 \tag{4.33}$$

The reaction rates for these reactions are calculated in the form of Eq. (4.34) [42]:

$$R_i = 0.278\, k_i \exp\left(-\frac{E_i}{RT}\right) P^{m}_{CO} \cdot P^{n}_{H_2}. \tag{4.34}$$

Related parameters can be found in Table 4.4.

4.2.3 Mass transport model

The rate of mass transport is generally affected by operating conditions and material properties. In large spaces, it is dominated by free diffusion of molecules. In the porous area, where the pore size is smaller than the molecular

Table 4.4 Reaction parameters for F-T reactions [42].

Reaction number	*m*	*n*	*k*	*E*
26	−1.0889	1.5662	1.43×10^{5}	83423.9
27	0.7622	0.0728	5.16×10^{1}	65018.0
28	−0.5645	1.3155	2.47×10^{1}	49782.0
29	0.4051	0.6635	4.63×10^{-1}	34855.5
30	0.4728	1.1389	4.74×10^{-3}	27728.9
31	0.8204	0.5026	8.32×10^{-3}	25730.1
32	0.5850	0.5982	2.32×10^{-2}	23564.3
33	0.5742	0.7100	4.11×10^{2}	58826.3

mean-free path, Knudsen diffusion becomes significant. The dusty-gas model (DGM) is used to calculate mass transport of gas species (N), as shown in Eq. (4.35) [43]:

$$\frac{N_i}{D_{ik}^{\text{eff}}} + \sum_{j=1, j\neq i}^{n} \frac{y_j N_i - y_i N_j}{D_{ij}^{\text{eff}}} = -\frac{1}{RT}\left(\nabla(y_j P) + \frac{B_0 y_j P}{\mu D_{ik}^{\text{eff}}} \nabla P\right), \qquad (4.35)$$

where D_{ik}^{eff} is the Knudsen diffusion coefficient, D_{ij}^{eff} is the molecular diffusion coefficient, y is the mole fraction of species, B_0 is the permeability, and μ is the gas viscosity.

In steady state, the mass conservation can be described by Eq. (4.36):

$$\nabla \cdot N_j = R_j, \qquad (4.36)$$

where R_j is the mass source term.

4.2.4 Momentum transport model

The Navier-Stokes (N-S) equation is used to describe the behavior of fluid flow, as shown in Eq. (4.37):

$$\rho\frac{\partial u}{\partial t} + \rho u \nabla u = -\nabla p + \nabla\left[\mu\left(\nabla u + (\nabla u)^T\right) - \frac{2}{3}\mu \nabla u\right] - \frac{\varepsilon \mu u}{k} \qquad (4.37)$$

Here, ρ and u are the gas density and the velocity vector, respectively.

4.2.5 Heat transfer model

The heat balance equation is used to describe the heat transfer processes, as shown in Eq. (4.38):

$$\rho C_p u \cdot \nabla T + \nabla \cdot (-\lambda_{\text{eff}} \nabla T) = Q. \qquad (4.38)$$

Here, C_p, λ_{eff}, and Q are, respectively, the constant-pressure heat capacity, the effective thermal conductivity, and the heat source terms. In the porous area, λ_{eff} can be modified by the porosity as shown in Eq. (4.39):

$$\lambda_{\text{eff}} = (1 - \varepsilon)\lambda_s + \varepsilon\lambda_g. \tag{4.39}$$

Here, λ_s and λ_g are the heat conductivities of the solid phase and gaseous phases.

4.3 Heat management of direct carbon fuel cells

Solid carbon such as coal is one of the main energy sources in modern society. Coal source is abundant and cheap and contains high energy density in volume. However, the way they are used in the conventional thermal power plant causes air pollution, global warming, ocean acidification, and many other problems. More efficient and environmentally friendly technologies are urgently needed, such as direct carbon fuel cells (DCFCs) [44–46].

DCFCs are the SOFCs directly consume the solid carbon as fuel to generate electricity. The large size of the solid carbon limits its contact with the triple phase boundary (TPB) at anode, restricting its electrochemical oxidation. Gasification of solid carbons to CO and other gaseous molecules becomes the first step before these gaseous fuels can diffuse to the TPB area for electrochemical reactions [47]. Previous studies have demonstrated the importance of the in situ carbon gasification rate on the cell's performance, where a high-power density is usually related to a fast carbon gasification rate [48]. Moreover, the carbon gasification process was found to be an important heat sink in the DCFC. The heat released in the electrochemical reactions could be effectively absorbed by the carbon gasification reaction [49]. At different operating conditions (e.g., current density), the net heat generated by the DCFC could be negative, zero, or positive. It is thus important to combine the DCFC with other devices for its heat management. Such a hybrid system mainly includes a DCFC, an external heat source, and a thermal device as shown in Fig. 4.2. The external heat source is used at low current densities to sustain the operating temperature of the DCFC. The thermal device is used at high current densities to utilize the waste heat from the DCFC and improve the energy efficiency and output power density. A variety of thermal devices can be used in utilizing the waste heat from DCFC, including heat engines, gas turbines, and thermoelectric generators [50–52].

Stirling engines have received much attention recently because they can effectively convert the waste heat into mechanical work with low emission,

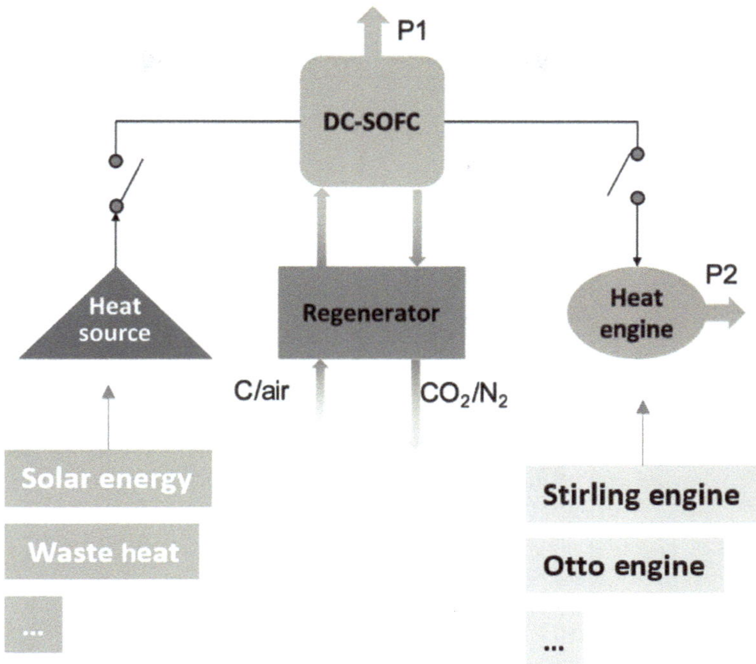

Fig. 4.2 Schematic of a hybrid system for the heat management of DCFC.

high reliability, and low cost for maintenance. In this section, a Stirling engine is used in the hybrid system to show the performance improvement of the DCFC by combining the thermal device.

4.3.1 Model description

In the hybrid system, the DCFC utilizes solid carbon to generate electricity, P_{DCFC}. External heat source is used to maintain the operation of the DCFC when working at the endothermic mode. The Stirling heat engine is used to utilize the waste heat from the DCFC and produce additional power, P_{SHE}. The regenerator is used to preheat the feedstock using the high-temperature products.

The parameters used in the Stirling heat engine and heat balance process are given in Tables 4.5 and 4.6.

4.3.2 Results and discussion

For the model validation, the *I*-*V* characteristics between the simulation results and experimental data of SOFCs using H_2 and solid carbon fuels

Table 4.5 Parameters used in the Stirling heat engine [53, 54].

Parameter	Value
Mole heat capacity of the working substance C (J mol^{-1} K^{-1})	15
Working substance volume ratio, V_2/V_1	2
Thermal conductance, k_1 (W K^{-1})	200
Thermal conductance, k_2 (W K^{-1})	200
Fractional deviation from ideal regeneration, x	20%
Heat transfer time ratio, b	1
Environment temperature, T_0 (K)	298

Table 4.6 Isobaric molar heat capacities for the involved species [55, 56].

Species	Isobaric molar heat capacity $C_{p,i}$(J/mol K)
CO_2	$4.3669+0.2046T-4.7133\times10^{-4}T^2+6.5788\times10^{-7}T^3$ $-5.199\times10^{-10}T^4+2.1458\times10^{-13}T^5-3.5992\times10^{-17}T^6$
O_2	$34.85-5.7975\times10^{-2}T+2.0368\times10^{-4}T^2-3.0037\times10^{-7}T^3$ $+2.3172\times10^{-10}T^4-9.1821\times10^{-14}T^5+1.4776\times10^{-17}T^6$
CO	$30.429-8.1781\times10^{-3}T-5.2062\times10^{-6}T^2+4.1974\times10^{-8}T^3$ $-6.6346\times10^{-11}T^4+3.7756\times10^{-14}T^5-7.6538\times10^{-18}T^6$
Solid carbon	$27.21876-1.21838\times10^{-4}T-4.5324\times10^{4}T^{-1}$ $-2.1815\times10^{6}T^{-2}+7.99859\times10^{8}T^{-3}-7.21429\times10^{10}T^{-4}$

are compared as shown in Fig. 4.3. In the following numerical studies, the same structural and tuning parameters are used.

As shown in Fig. 4.4, the power density of the hybrid system (P_{total}) is contributed by two devices, the DCFC (P_{DCFC}) and the Stirling heat engine (P_{SHE}). With the increase in operating current density, P_{DCFC} increases in the beginning and then decreases after 23,000 A m^{-2}. For the Stirling heat engine, its power density keeps growing after 8000 A m^{-2}. As a result, the total power density increases with the rising current density in the whole range. At a large operating current density (45,000 A m^{-2}), the power density improvement brought by the Stirling heat engine reaches 7000 W m^{-2}, which is about 160% of the power density provided by DCFC.

The improvement of system efficiency brought by the Stirling heat engine is further given in Fig. 4.5. At low operating current densities, the DCFC shows a very high efficiency, which even reaches 100% at heat balance situation. With the further increase in current density, its efficiency decreases quickly to less than 30% at 45,000 A m^{-2}. By utilizing the waste

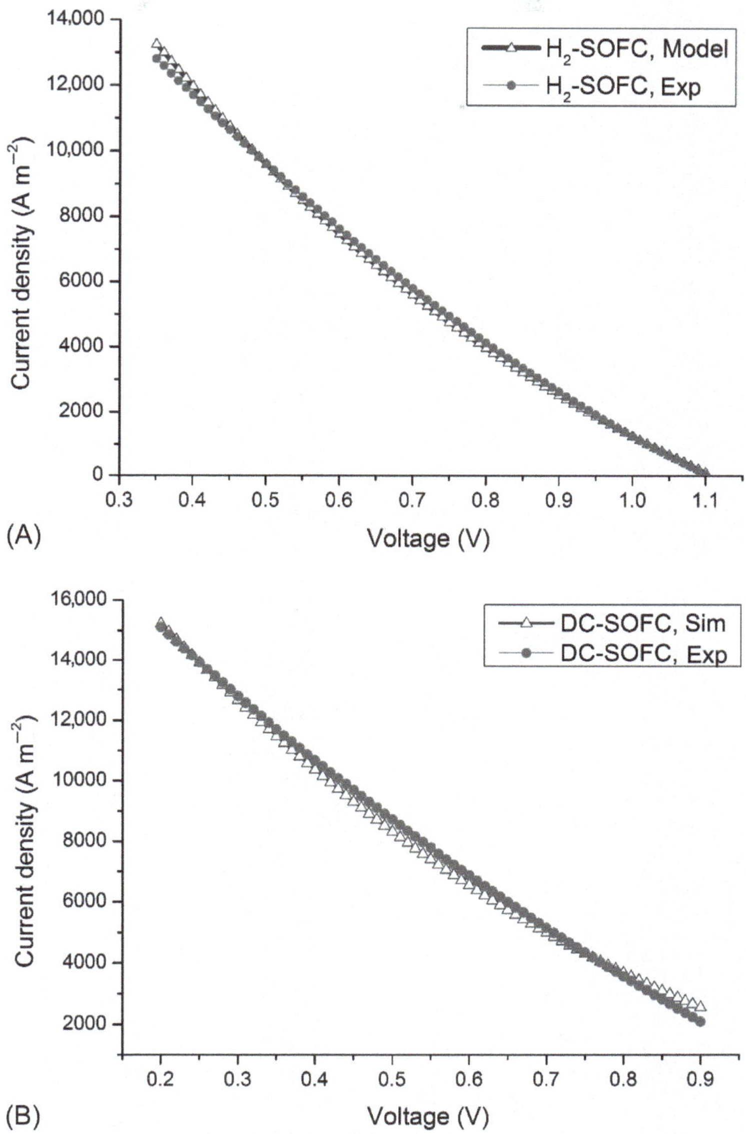

Fig. 4.3 Model validation of the electrochemical kinetics used in the SOFC model using H2 (A) and solid carbon (B) as the fuel [23].

heat from DCFC, its energy efficiency is much improved at large current densities. With about 54% of the waste heat reused for electricity generation, the total energy efficiency reaches 65%, which is about 40% higher than the single DCFC.

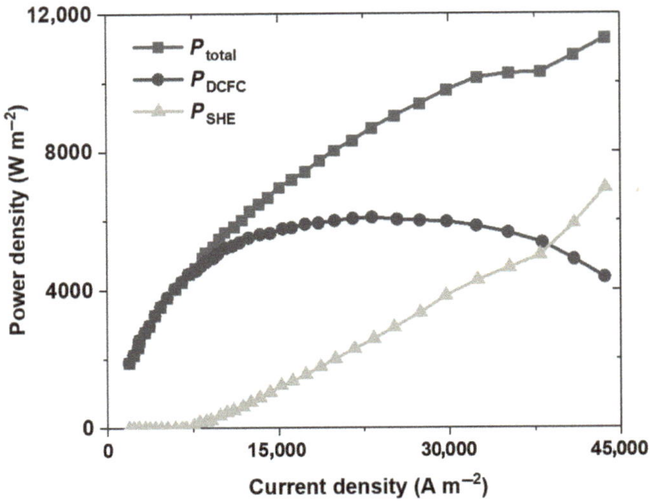

Fig. 4.4 Equivalent power densities of the DCFC, Stirling heat engine, and hybrid system at 1173 K [57].

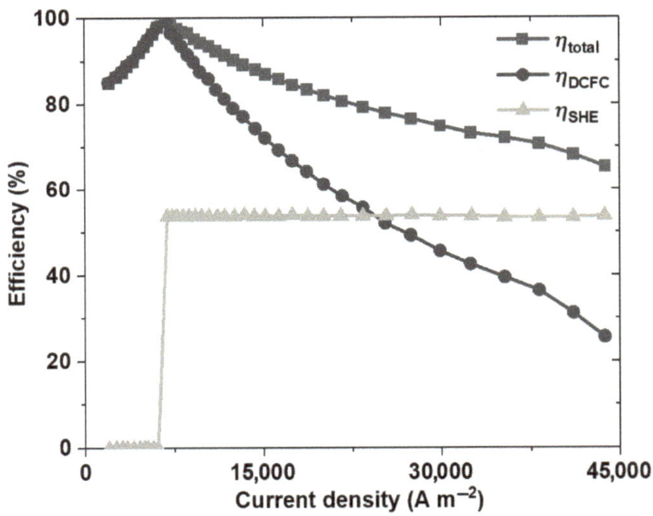

Fig. 4.5 Equivalent efficiencies of the DCFC, Stirling heat engine, and hybrid system at 1173 K [57].

4.4 On-site solar-fuel-to-electricity generation

Renewable energies such as wind power and solar radiation are clean and sustainable; thus, they have raised wide interest all over the world. However, these energies are intermittent and discrete and demand effective energy storage technologies for reliable energy supply [58–60]. Alternatively, photosynthesis of solar fuels is a promising method toward the utilization of solar energy [61–63].

To generate solar fuels, solar radiation is harvested by semiconductors (e.g., titanium dioxide, TiO_2) as the first step. Then, photosynthesis reactions occur when electron-hole pairs are generated on the catalyst [64]. Through the photoreduction of CO_2 and H_2O, fuels such as CO, methane and other hydrocarbons will be generated. Despite the significant progress achieved, utilization of the solar fuel mixture remains a problem; Because the product contains various species and the concentration of fuels in the mixture is low. To solve this problem, a hybrid system was proposed to combine the photoreactor with an SOFC [65]. As the SOFC can effectively utilize various fuels, we can hopefully convert the solar fuel mixture into electricity and realize the continuous power generation from solar radiation.

4.4.1 Model description

The proposed hybrid system consists of a photoreactor, an SOFC, and an oxygen separator, as shown in Fig. 4.6. CO_2 and H_2O are reduced in the photoreactor, where CH_4, CO, and O_2 are generated. After separation, the mixed solar fuel is introduced into the SOFC for power generation.

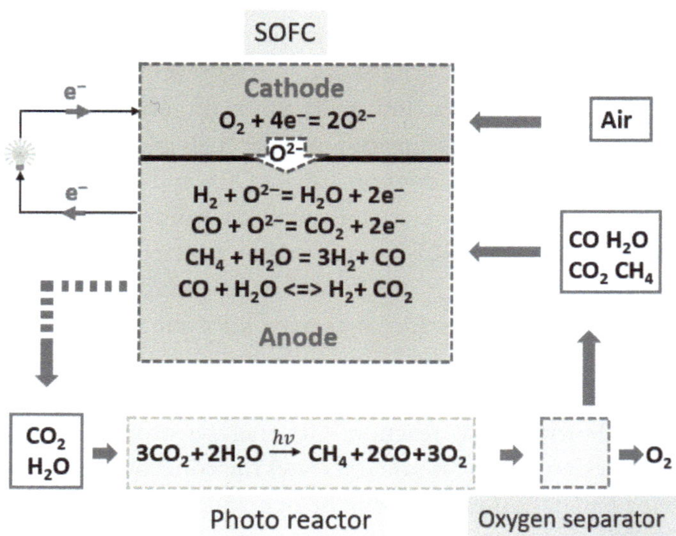

Fig. 4.6 Schematic of a hybrid system combining SOFC and the photosynthesis process [66].

4.4.2 Results and discussion

For the SOFC submodel, the reaction kinetics reported by Xu et al. [25] was adopted. This model is validated with a small difference in *I*-*V* characteristics as shown in Fig. 4.7A. For the model development of the photoreactor, the same operating temperature and pressure were adopted in accordance with the work conducted by Tahir and Amin [40]. The yield rates of methane under various partial pressures of CO_2 were compared to validate its reaction kinetics as shown in Fig. 4.7B.

Power generation of this hybrid system is affected by the operating parameters in both the photoreactor and the SOFC. The mole fraction of CO_2 in the feedstock and the applied light intensity are the key factors for the generation of solar fuels in the photoreactor, while the operating temperature is the key factors in the subsequent power generation as shown in Figs. 4.8–4.10.

In the SOFC, the operating temperature affects its kinetics of both chemical and electrochemical reactions. The higher temperature brings a faster reaction rate and improves the cell performance. As shown in Fig. 4.8, with the temperature increasing from 973 to 1173 K, the current density is raised from 1194 to 4728 $A\,m^{-2}$, which is a triple improvement. In the meanwhile, the concentration of CH_4 quickly decreases which CO and H_2 keep quite stable as shown in Fig. 4.8B. This is caused by the change of reaction rates of internal SMR and WGS reactions as shown in Fig. 4.8C. The SMR rate increases quickly with the increasing temperature, thus maintains the H_2 and CO concentration.

Due to the kinetics of reaction as shown in Eq. (4.25), a suitable CO_2 concentration is vital in the generation of solar fuels and the utilization rate of CO_2. In the studied case, adopting 40% of CO_2 and 60% of H_2O as inlet gas will achieve the highest fuel concentration from the photoreactor as shown in Fig. 4.9. A low CO_2 concentration at reactor inlet will reduce the generated solar fuel but it will raise the CO_2 utilization rate. In these considerations, the CO_2 fraction at reactor inlet should be controlled at 0.4 or smaller.

Solar radiation is the energy source of this system. Its intensity is vital to the speed of photosynthesis reaction. Generally, a stronger light intensity raises the synthesis reaction rate and increases the concentration of solar fuels at the reactor outlet. In this way, both the operating current density and output power of the hybrid system can be improved. Therefore, strong intensity of solar radiation is key to improving the performance of this hybrid system

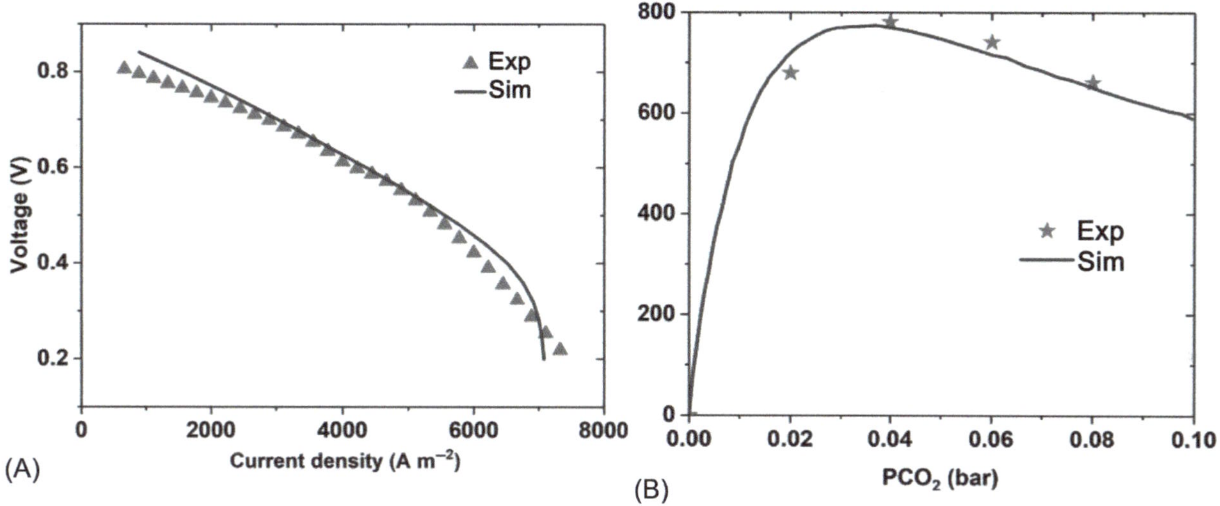

Fig. 4.7 Model validation of the kinetics used in the SOFC model (A) and the photoreactor model (B) [66].

and adopting methods such as concentrating the sunlight to the photoreactor will be helpful in practical application.

The effects of applied voltage on the output power at two different operating temperature [973 (A, C) and 1073K (B, D)] are further shown in Fig. 4.11. The peak power density reaches $1100\,W\,m^{-2}$ at 973K and $2200\,W\,m^{-2}$ at 1073K, respectively. The difference is not only caused by the electrochemical activity at a different temperature, but also caused by the SMR rate as can be observed in Fig. 4.11C and D. At the higher temperature, more methane is consumed to generate H_2 and CO and sustain a larger current density. This also explains the much larger limitation current density shown in Fig. 4.8B compared with that in Fig. 4.11A.

Despite the promising preliminary results, challenges such as the photosynthesis rate and O_2 separation efficiency remain to be overcome. With more studies to develop highly active photosynthesis catalyst and separator membranes, this hybrid system will be more competitive in continuously providing electricity from solar radiation.

4.5 Low carbon fuel generation from captured CO_2

To reverse global warming and achieve a carbon-neutral society, effective CO_2 capture storage and utilization (CCUS) methods are needed [67].

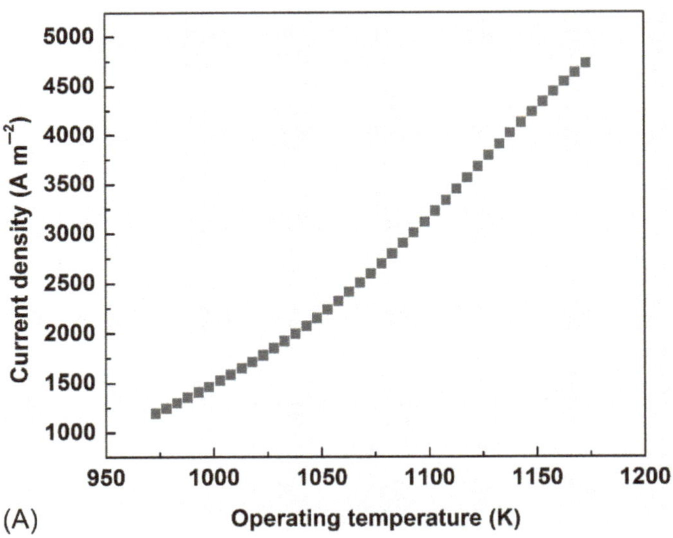

Fig. 4.8 The effects of temperature on (A) current densities.

(Continued)

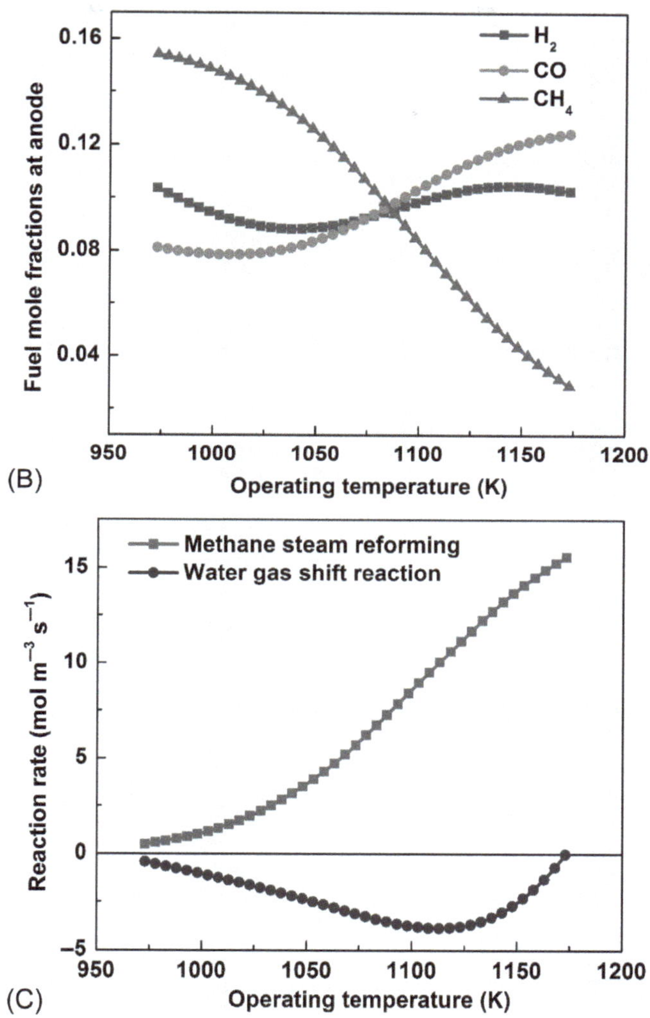

Fig. 4.8, Cont'd (B) mole fractions of fuels in the anode, and (C) the SMR and WGSR reaction rates of the SOFC.

SOECs are promising technologies for CO_2 reduction to generate CO. Moreover, SOECs can co-electrolyze CO_2 and H_2O to generate syngas (the mixture of CO and H_2), which is the feedstock to generate more valuable hydrocarbons through Fischer-Tropsch (F-T) process [68]. The generated hydrocarbons vary from C_1 (methane) to C_{5+} in distribution. Usually, methane is less wanted for industry [69], and they can be used to assist the

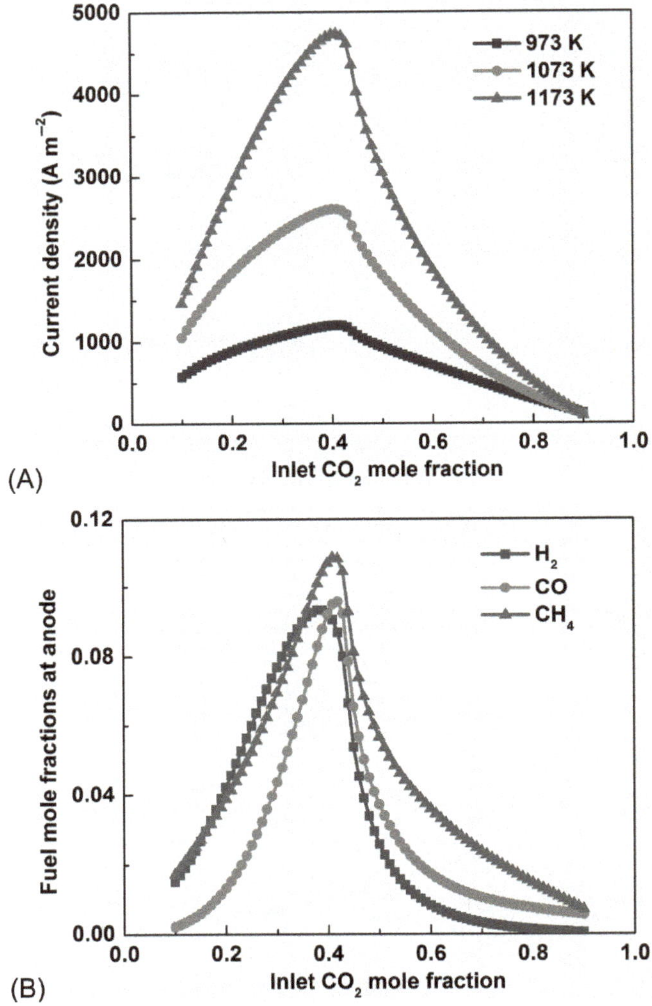

Fig. 4.9 The effects of the mole fraction of CO_2 in the feedstock on (A) current densities, (B) mole fractions of fuels in the anode,

(Continued)

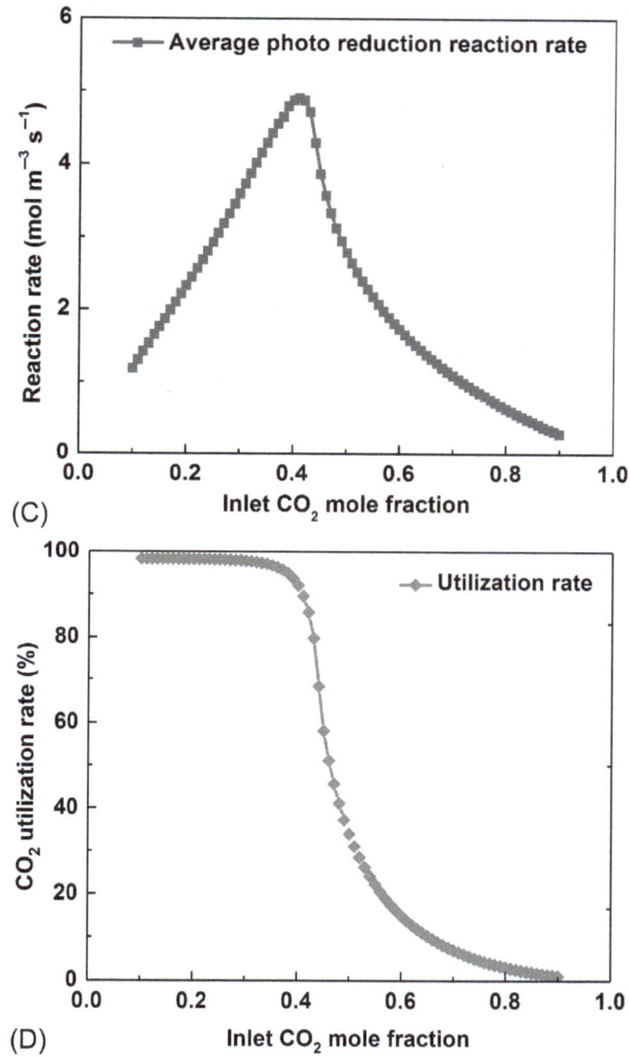

Fig. 4.9, Cont'd (C) average reaction rates of photoreduction, and (D) the utilization rate of CO_2 [66].

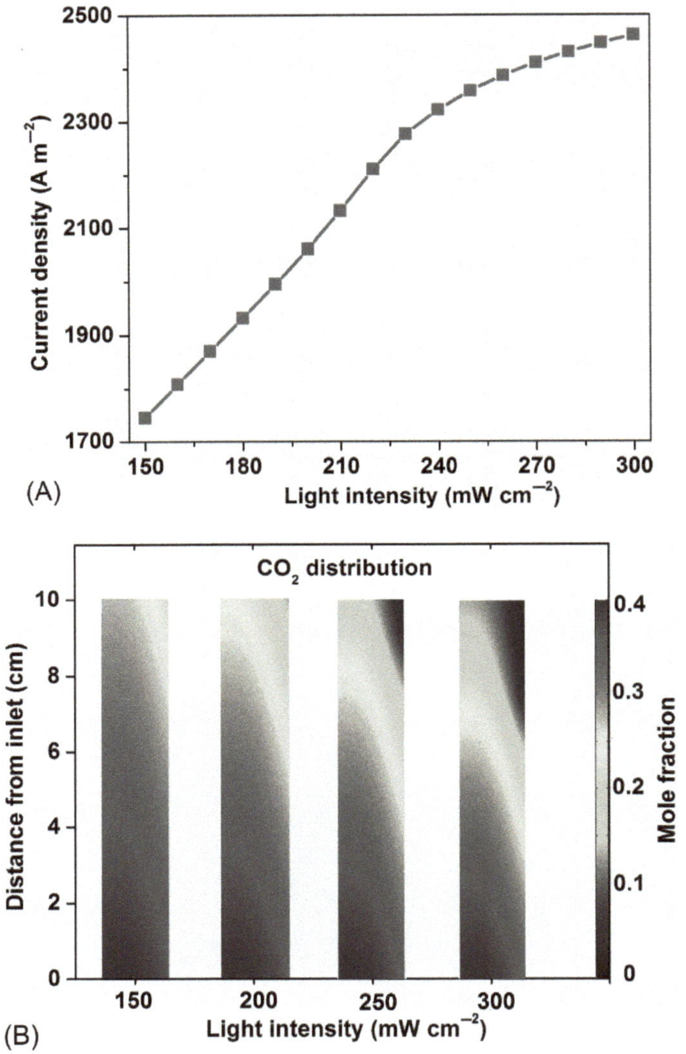

Fig. 4.10 The effects of light intensity on (A) current densities, (B) distribution of the mole fraction of CO_2.

(Continued)

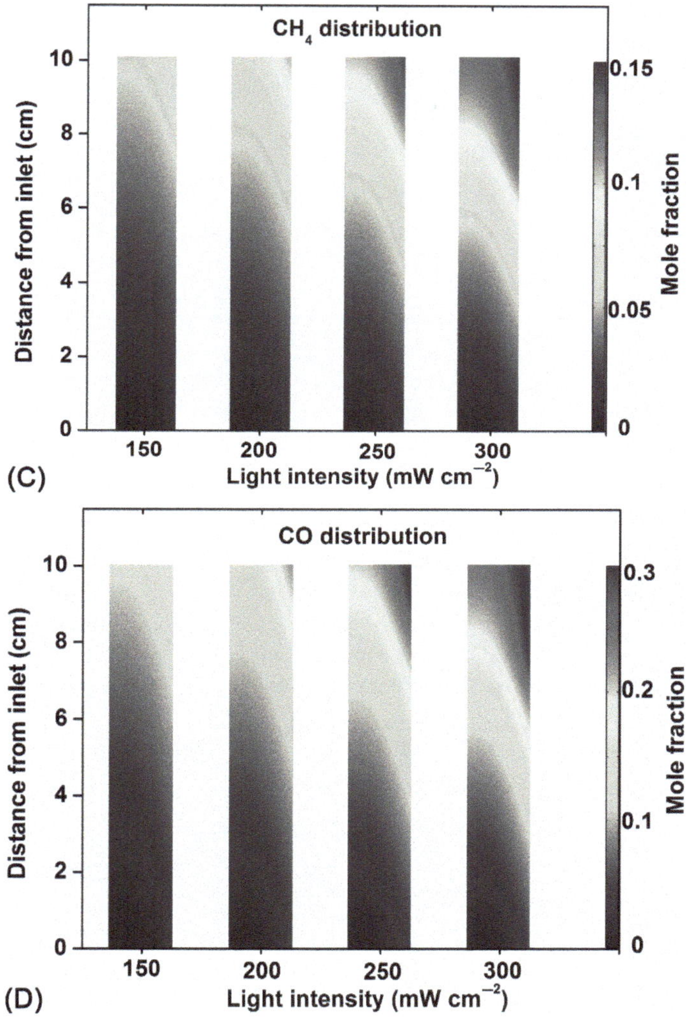

Fig. 4.10, Cont'd (C) distribution of the mole fraction of CH_4, and (D) distribution of the mole fraction of CO in the photoreactor [66].

SOEC to reduce the electricity cost [70]. In this section, a hybrid system combining the CH_4-assisted SOEC and the F-T reactor is proposed to generate hydrocarbons from CO_2.

4.5.1 Model description

The hybrid system combines an SOEC and an F-T reactor, as shown in Fig. 4.12. In the SOEC, CO_2 and H_2O are introduced to the cathode and

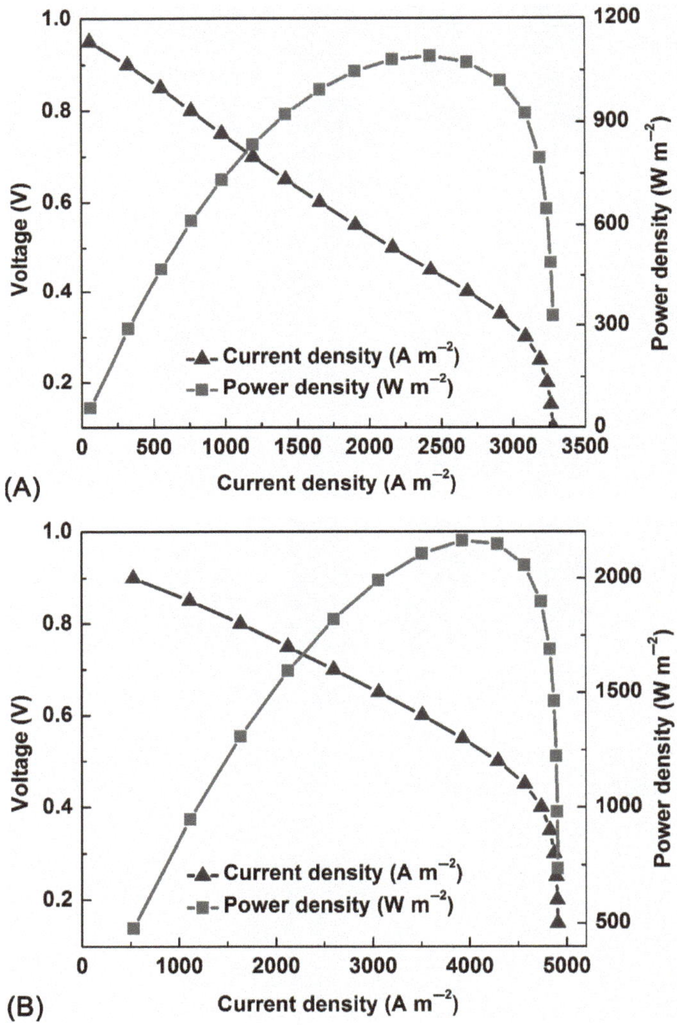

Fig. 4.11 The effects of voltage on the power density and mole fractions of the fuels at temperatures of (A, C) 973 K and (B, D) 1073 K at the anode [66].

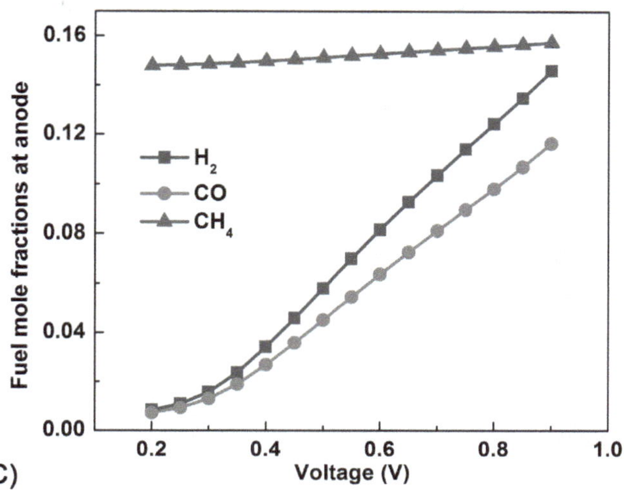

(C)

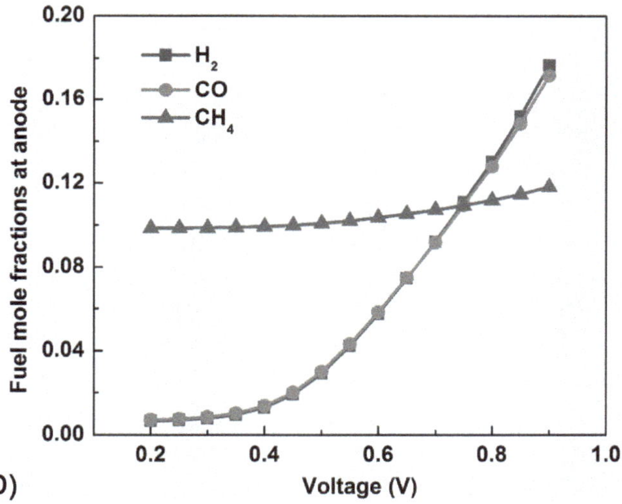

(D)

Fig. 4.11—cont'd

CH_4 is introduced to the anode together with H_2O to prevent methane coking. The produced syngas is then introduced to the F-T reactor for hydrocarbons generation.

4.5.2 Results and discussion

For the SOEC sub-model, experimental data conducted by Luo et al. [16] were adopted, and the same cell structure and material properties are applied

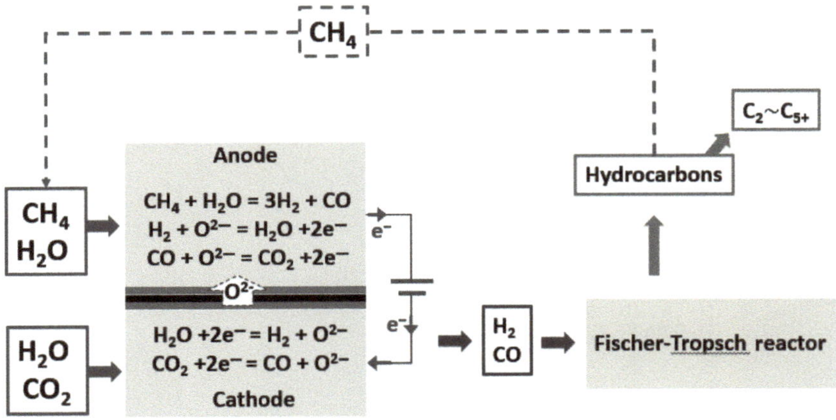

Fig. 4.12 Schematic of combined CH_4-assisted SOEC and F-T reactor system [71].

in the numerical model. The kinetic parameters of the submodel are validated by comparing the *I*–*V* characteristics of the experimental data and simulation results. A good agreement is achieved as shown in Fig. 4.13A. In the Fischer-Tropsch submodel, the kinetic model proposed by Rahimpour and Elekaei [42] was adopted. The comparison between experimental data and numerical results is given in Fig. 4.13B for model validation.

As shown in Figs. 4.14 and 4.15, the proportion of hydrocarbons generated from F-T reactor is very sensitive to the mole fraction of CO_2 at the inlet of the SOEC cathode; Because the CO_2 concentration affects the CO proportion in the syngas, which is generated from the SOEC. The CO proportion further affects the hydrocarbon proportion as determined by the kinetics of reactions in Eq. (4.26) to Eq. (4.33). Generally, the CO/H_2 ratio from the SOEC is improved with the increase in the mole fraction of CO_2 at the inlet. Moreover, a high mole fraction of CO_2 increases its utilization rate but results in less proportion of hydrocarbons in the F-T products because the value of CO/H_2 has great effects on the selectivity of methane and probability of chain growth [72].

The distribution of C atoms in carbon fuels (CO plus C_1 to C_{5+}) is given in Fig. 4.16. The proportions of carbon fuels are calculated as shown in Eq. (4.40).

$$P_c = \frac{C_i}{C_{total}} \tag{4.40}$$

Here, C_i is the number of C atoms in the fuels and C_{total} is the total number of C atoms of all fuels.

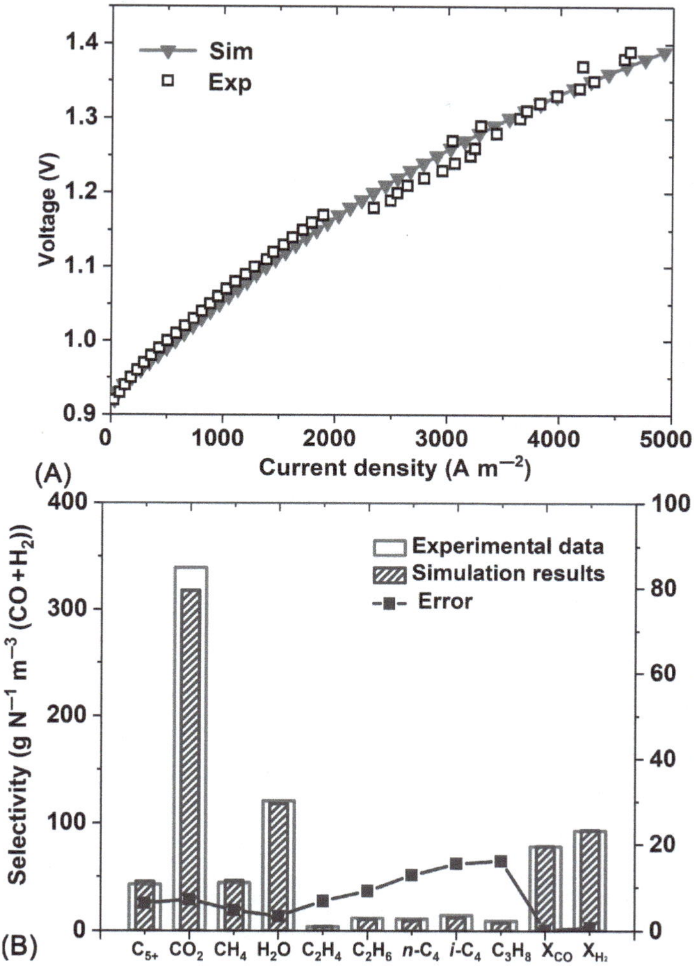

Fig. 4.13 Model validation of the kinetics used in the SOFC model (A) and the photoreactor model (B) [71].

Generally, these proportions are significantly affected by the mole fraction of CO_2 at the inlet of SOEC. Mostly, the C atoms (30%) are in the form of CH_4 at the mole fraction of CO_2 being 0.5. At high inlet CO_2 case (0.65), the C atoms contained in CH_4 decrease dramatically to only 4%. On the other hand, the proportion of CO increases from 9% to 53% with the mole fraction of CO_2 increasing from 0.5 to 0.65 at the inlet. As to the proportion C_{5+}, it keeps relatively stable at different inlet CO_2 mole fractions.

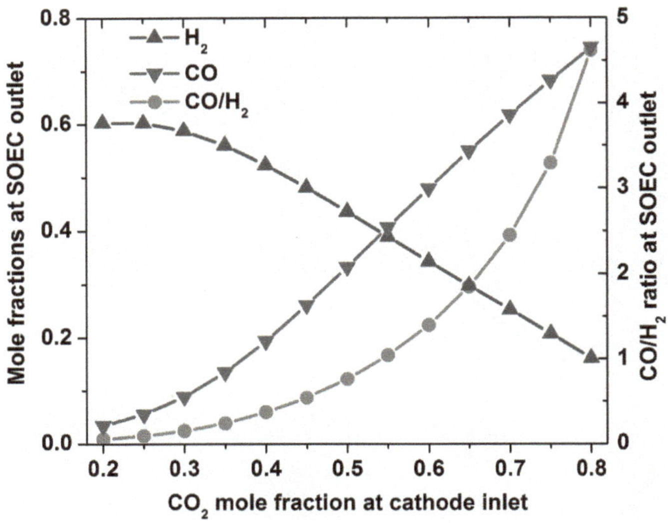

Fig. 4.14 Effects of the mole fraction of CO_2 at the cathode inlet on the mole fractions of H_2 and CO at the outlet of the SOEC [71].

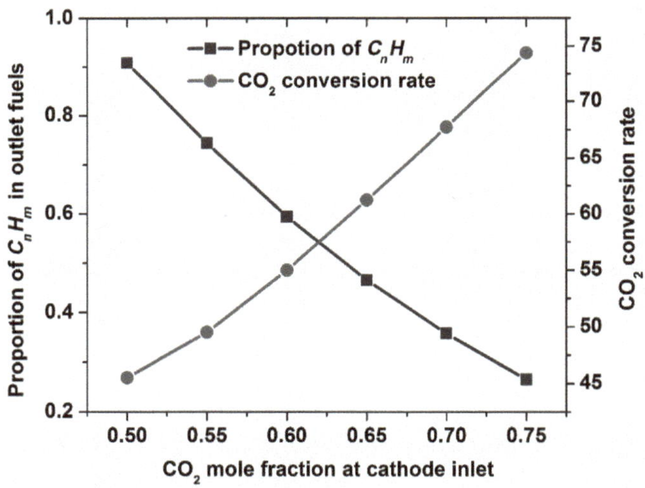

Fig. 4.15 Effects of the mole fraction of CO_2 at the cathode inlet on the proportion of hydrocarbons in the fuels and CO_2 conversion rate at the outlet [71].

4.6 Conclusions

In this chapter, numerical models of hybrid systems combining the solid oxide fuel cells (SOFCs) and other devices are built. Various physical/chemical processes are considered in the numerical simulation, including the chemical/electrochemical reaction, mass/momentum transport, and heat transfer. These models can be used to evaluate the hybrid system in different applications, including the heat management of DCFC, on-site solar-fuel-

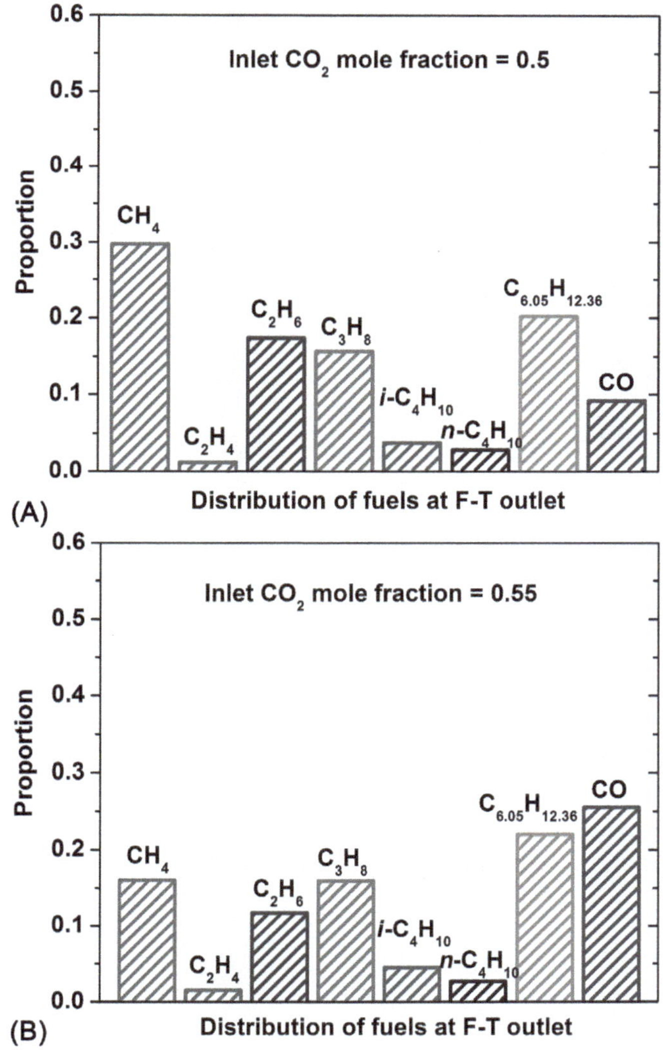

Fig. 4.16 Effects of the mole fraction of CO_2: (A) 0.5, (B) 0.55, *(Continued)*

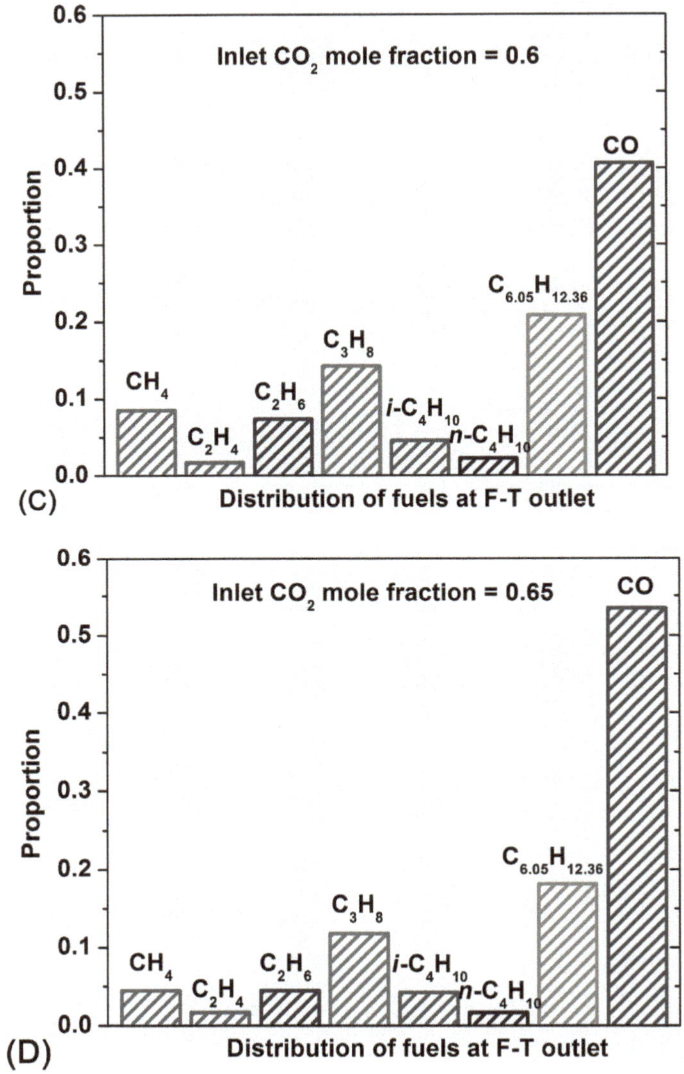

Fig, 4.16, Cont'd (C) 0.6, (D) 0.65, at the inlet of the cathode on the distribution of fuels at the outlet of F-T reactor [71].

to-power generation, and low carbon fuel generation. The effects of different operating parameters on the performance of various systems are discussed. For the SOFC, the operating temperature, applied voltage, gas flow rate, and composition of the feedstock are the key factors. A proper adjust of these parameters can help achieve a high system efficiency. These hybrid systems combine the advantages of different devices and provide alternative solutions to solve our energy and environmental problems, including the energy crisis, global warming, and air pollution. These hybrid systems are promising technologies in achieving a sustainable human civilization.

References

[1] Minh NQ. Solid oxide fuel cell technology—features and applications. Solid State Ion 2004;174:271–7. https://doi.org/10.1016/j.ssi.2004.07.042.

[2] Ormerod RM. Solid oxide fuel cells. Chem Soc Rev 2003;32:17–28. https://doi.org/10.1039/B105764M.

[3] Singhal SC. Advances in solid oxide fuel cell technology. Solid State Ion 2000;135:305–13. https://doi.org/10.1016/S0167-2738(00)00452-5.

[4] Yano M, Tomita A, Sano M, Hibino T. Recent advances in single-chamber solid oxide fuel cells: a review. Solid State Ion 2007;177:3351–9. https://doi.org/10.1016/j.ssi.2006.10.014.

[5] Giddey S, Badwal SPS, Kulkarni A, Munnings C. A comprehensive review of direct carbon fuel cell technology. Prog Energy Combust Sci 2012;38:360–99. https://doi.org/10.1016/j.pecs.2012.01.003.

[6] Kang S, Ahn K-Y. Dynamic modeling of solid oxide fuel cell and engine hybrid system for distributed power generation. Appl Energy 2017;195:1086–99. https://doi.org/10.1016/j.apenergy.2017.03.077.

[7] Wang Y, Yoshiba F, Kawase M, Watanabe T. Performance and effective kinetic models of methane steam reforming over Ni/YSZ anode of planar SOFC. Int J Hydrogen Energy 2009;34:3885–93. https://doi.org/10.1016/j.ijhydene.2009.02.073.

[8] Achenbach E, Riensche E. Methane/steam reforming kinetics for solid oxide fuel cells. J Power Sources 1994;52:283–8. https://doi.org/10.1016/0378-7753(94)02146-5.

[9] Xiao J, Xie Y, Liu J, Liu M. Deactivation of nickel-based anode in solid oxide fuel cells operated on carbon-containing fuels. J Power Sources 2014;268:508–16. https://doi.org/10.1016/j.jpowsour.2014.06.082.

[10] Ni M, Leung DYC, Leung MKH. An improved electrochemical model for the NH_3 fed proton conducting solid oxide fuel cells at intermediate temperatures. J Power Sources 2008;185:233–40. https://doi.org/10.1016/j.jpowsour.2008.07.023.

[11] Pillai M, Lin Y, Zhu H, Kee RJ, Barnett SA. Stability and coking of direct-methane solid oxide fuel cells: effect of CO_2 and air additions. J Power Sources 2010;195:271–9. https://doi.org/10.1016/j.jpowsour.2009.05.032.

[12] Chen Y, deGlee B, Tang Y, Wang Z, Zhao B, Wei Y, et al. A robust fuel cell operated on nearly dry methane at 500°C enabled by synergistic thermal catalysis and electrocatalysis. Nat Energy 2018;3:1042–50. https://doi.org/10.1038/s41560-018-0262-5.

[13] Ebbesen SD, Mogensen M. Electrolysis of carbon dioxide in solid oxide electrolysis cells. J Power Sources 2009;193:349–58. https://doi.org/10.1016/j.jpowsour.2009.02.093.

[14] Chen L, Chen F, Xia C. Direct synthesis of methane from CO_2-H_2O co-electrolysis in tubular solid oxide electrolysis cells. Energ Environ Sci 2014;7:4018–22. https://doi.org/10.1039/C4EE02786H.
[15] Laguna-Bercero MA. Recent advances in high temperature electrolysis using solid oxide fuel cells: a review. J Power Sources 2012;203:4–16. https://doi.org/10.1016/j.jpowsour.2011.12.019.
[16] Luo Y, Shi Y, Li W, Cai N. Comprehensive modeling of tubular solid oxide electrolysis cell for co-electrolysis of steam and carbon dioxide. Energy 2014;70:420–34. https://doi.org/10.1016/j.energy.2014.04.019.
[17] Stempien JP, Ni M, Sun Q, Chan SH. Production of sustainable methane from renewable energy and captured carbon dioxide with the use of solid oxide electrolyzer: a thermodynamic assessment. Energy 2015;82:714–21. https://doi.org/10.1016/j.energy.2015.01.081.
[18] Ni M. 2D thermal modeling of a solid oxide electrolyzer cell (SOEC) for syngas production by H_2O/CO_2 co-electrolysis. Int J Hydrogen Energy 2012;37:6389–99. https://doi.org/10.1016/j.ijhydene.2012.01.072.
[19] Buonomano A, Calise F, d'Accadia MD, Palombo A, Vicidomini M. Hybrid solid oxide fuel cells–gas turbine systems for combined heat and power: a review. Appl Energy 2015;156:32–85. https://doi.org/10.1016/j.apenergy.2015.06.027.
[20] Zhang H, Xu H, Chen B, Dong F, Ni M. Two-stage thermoelectric generators for waste heat recovery from solid oxide fuel cells. Energy 2017;132:280–8. https://doi.org/10.1016/j.energy.2017.05.005.
[21] Bao C, Shi Y, Croiset E, Li C, Cai N. A multi-level simulation platform of natural gas internal reforming solid oxide fuel cell–gas turbine hybrid generation system: part I. Solid oxide fuel cell model library. J Power Sources 2010;195:4871–92. https://doi.org/10.1016/j.jpowsour.2010.01.078.
[22] Kakaç S, Pramuanjaroenkij A, Zhou XY. A review of numerical modeling of solid oxide fuel cells. Int J Hydrogen Energy 2007;32:761–86. https://doi.org/10.1016/j.ijhydene.2006.11.028.
[23] Xu H, Chen B, Liu J, Ni M. Modeling of direct carbon solid oxide fuel cell for CO and electricity cogeneration. Appl Energy 2016;178:353–62. https://doi.org/10.1016/j.apenergy.2016.06.064.
[24] Xu H, Chen B, Tan P, Cai W, He W, Farrusseng D, et al. Modeling of all porous solid oxide fuel cells. Appl Energy 2018;219:105–13. https://doi.org/10.1016/j.apenergy.2018.03.037.
[25] Xu H, Chen B, Zhang H, Sun Q, Yang G, Ni M. Modeling of direct carbon solid oxide fuel cells with H_2O and CO_2 as gasification agents. Int J Hydrogen Energy 2017;42:15641–51. https://doi.org/10.1016/j.ijhydene.2017.05.075.
[26] Stempien JP, Liu Q, Ni M, Sun Q, Chan SH. Physical principles for the calculation of equilibrium potential for co-electrolysis of steam and carbon dioxide in a solid oxide electrolyzer cell (SOEC). Electrochim Acta 2014;147:490–7. https://doi.org/10.1016/j.electacta.2014.09.144.
[27] Xu H, Chen B, Tan P, Zhang Y, He Q, Wu Z, et al. The thermal effects of all porous solid oxide fuel cells. J Power Sources 2019;440:227102. https://doi.org/10.1016/j.jpowsour.2019.227102.
[28] Guo Y, Bessaa M, Aguado S, Steil MC, Rembelski D, Rieu M, et al. An all porous solid oxide fuel cell (SOFC): a bridging technology between dual and single chamber SOFCs. Energ Environ Sci 2013;6:2119–23. https://doi.org/10.1039/C3EE40131F.
[29] Eguchi K, Setoguchi T, Inoue T, Arai H. Electrical-properties of ceria-based oxides and their application to solid oxide fuel-cells. Solid State Ion 1992;52:165–72.
[30] Fan B, Yan J, Yan X. The ionic conductivity, thermal expansion behavior, and chemical compatibility of $La_{0.54}Sr_{0.44}Co_{0.2}Fe_{0.8}O_{3-\delta}$ as SOFC cathode material. Solid State Sci 2011;13:1835–9. https://doi.org/10.1016/j.solidstatesciences.2011.07.007.

[31] Eguchi K. Ceramic materials containing rare earth oxides for solid oxide fuel cell. J Alloys Compd 1997;250:486–91.

[32] Tai LW, Nasrallah MM, Anderson HU, Sparlin DM, Sehlin SR. Structure and electrical properties of $La_{1-x}Sr_xCo_{1-y}Fe_yO_3$. Part 2. The system $La_{1-x}Sr_xCo_{0.2}Fe_{0.8}O_3$. Solid State Ion 1995;76:273–83. https://doi.org/10.1016/0167-2738(94)00245-N.

[33] Chen B, Xu H, Zhang Y, Dong F, Tan P, Zhao T, et al. Combined methane reforming by carbon dioxide and steam in proton conducting solid oxide fuel cells for syngas/power co-generation. Int J Hydrogen Energy 2019;44:15313–21. https://doi.org/10.1016/j.ijhydene.2019.02.244.

[34] Xu H, Chen B, Tan P, Xuan J, Maroto-Valer MM, Farrusseng D, et al. Modeling of all-porous solid oxide fuel cells with a focus on the electrolyte porosity design. Appl Energy 2019;235:602–11. https://doi.org/10.1016/j.apenergy.2018.10.069.

[35] Xu H, Chen B, Zhang H, Tan P, Yang G, Irvine JTS, et al. Experimental and modeling study of high performance direct carbon solid oxide fuel cell with in situ catalytic steam-carbon gasification reaction. J Power Sources 2018;382:135–43. https://doi.org/10.1016/j.jpowsour.2018.02.033.

[36] Smith RJB, Loganathan M, Shantha MS. A review of the water gas shift reaction kinetics. Int J Chem React Eng 2010;8:.

[37] Jin W, Gu X, Li S, Huang P, Xu N, Shi J. Experimental and simulation study on a catalyst packed tubular dense membrane reactor for partial oxidation of methane to syngas. Chem Eng Sci 2000;55:2617–25. https://doi.org/10.1016/S0009-2509(99)00542-4.

[38] Ni M, Leung DYC, Leung MKH. Modeling of methane fed solid oxide fuel cells: comparison between proton conducting electrolyte and oxygen ion conducting electrolyte. J Power Sources 2008;183:133–42. https://doi.org/10.1016/j.jpowsour.2008.04.073.

[39] Huang Z, Zhang J, Zhao Y, Zhang H, Yue G, Suda T, et al. Kinetic studies of char gasification by steam and CO_2 in the presence of H_2 and CO. Fuel Process Technol 2010;91:843–7. https://doi.org/10.1016/j.fuproc.2009.12.020.

[40] Tahir M, Amin NS. Photocatalytic CO_2 reduction with H_2O vapors using montmorillonite/TiO_2 supported microchannel monolith photoreactor. Chem Eng J 2013;230:314–27. https://doi.org/10.1016/j.cej.2013.06.055.

[41] Moazami N, Mahmoudi H, Rahbar K, Panahifar P, Tsolakis A, Wyszynski ML. Catalytic performance of cobalt–silica catalyst for Fischer–Tropsch synthesis: effects of reaction rates on efficiency of liquid synthesis. Chem Eng Sci 2015;134:374–84. https://doi.org/10.1016/j.ces.2015.05.025.

[42] Rahimpour MR, Elekaei H, et al. Fuel Process Technol 2009;90:747–61. https://doi.org/10.1016/j.fuproc.2009.02.011.

[43] Suwanwarangkul R, Croiset E, Fowler MW, Douglas PL, Entchev E, Douglas MA. Performance comparison of Fick's, dusty-gas and Stefan–Maxwell models to predict the concentration overpotential of a SOFC anode. J Power Sources 2003;122:9–18. https://doi.org/10.1016/S0378-7753(02)00724-3.

[44] Rady AC, Giddey S, Badwal SPS, Ladewig BP, Bhattacharya S. Review of fuels for direct carbon fuel cells. Energy Fuel 2012;26:1471–88. https://doi.org/10.1021/ef201694y.

[45] Cao D, Sun Y, Wang G. Direct carbon fuel cell: fundamentals and recent developments. J Power Sources 2007;167:250–7. https://doi.org/10.1016/j.jpowsour.2007.02.034.

[46] Jiang C, Ma J, Corre G, Jain SL, Irvine JTS. Challenges in developing direct carbon fuel cells. Chem Soc Rev 2017; https://doi.org/10.1039/C6CS00784H.

[47] Cai W, Liu J, Xie Y, Xiao J, Liu M. An investigation on the kinetics of direct carbon solid oxide fuel cells. J Solid State Electrochem 2016;20:2207–16. https://doi.org/10.1007/s10008-016-3216-5.

[48] Xie Y, Tang Y, Liu J. A verification of the reaction mechanism of direct carbon solid oxide fuel cells. J Solid State Electrochem 2012;17:121–7. https://doi.org/10.1007/s10008-012-1866-5.

[49] Xu H, Chen B, Zhang H, Kong W, Liang B, Ni M. The thermal effect in direct carbon solid oxide fuel cells. Appl Therm Eng 2017;118:652–62. https://doi.org/10.1016/j.applthermaleng.2017.03.027.
[50] Xu H, Chen B, Tan P, Cai W, Wu Y, Zhang H, et al. A feasible way to handle the heat management of direct carbon solid oxide fuel cells. Appl Energy 2018;226:881–90. https://doi.org/10.1016/j.apenergy.2018.06.039.
[51] Yang Z, Xu H, Chen B, Tan P, Zhang H, Ni M. Numerical modeling of a cogeneration system based on a direct carbon solid oxide fuel cell and a thermophotovoltaic cell. Energ Conver Manage 2018;171:279–86. https://doi.org/10.1016/j.enconman.2018.05.100.
[52] Xu H, Chen B, Tan P, Zhang H, Yuan J, Irvine JTS, et al. Performance improvement of a direct carbon solid oxide fuel cell through integrating an Otto heat engine. Energ Conver Manage 2018;165:761–70. https://doi.org/10.1016/j.enconman.2018.04.008.
[53] Ahmadi MH, Sayyaadi H, Dehghani S, Hosseinzade H. Designing a solar powered Stirling heat engine based on multiple criteria: maximized thermal efficiency and power. Energ Conver Manage 2013;75:282–91. https://doi.org/10.1016/j.enconman.2013.06.025.
[54] Yaqi L, Yaling H, Weiwei W. Optimization of solar-powered Stirling heat engine with finite-time thermodynamics. Renew Energy 2011;36:421–7. https://doi.org/10.1016/j.renene.2010.06.037.
[55] Todd B, Young JB. Thermodynamic and transport properties of gases for use in solid oxide fuel cell modelling. J Power Sources 2002;110:186–200. https://doi.org/10.1016/S0378-7753(02)00277-X.
[56] Ho FH. Graphite design handbook. San Diego, CA: General Atomics; 1988.
[57] Xu H, Chen B, Tan P, Zhang H, Yuan J, Liu J, et al. Performance improvement of a direct carbon solid oxide fuel cell system by combining with a Stirling cycle. Energy 2017;140:979–87. https://doi.org/10.1016/j.energy.2017.09.036.
[58] Lewis NS, Nocera DG. Powering the planet: chemical challenges in solar energy utilization. Proc Natl Acad Sci U S A 2006;103:15729–35. https://doi.org/10.1073/pnas.0603395103.
[59] Lewis NS. Research opportunities to advance solar energy utilization. Science (80-) 2016;351. https://doi.org/10.1126/science.aad1920.
[60] Köktürk G, Tokuç A. Vision for wind energy with a smart grid in Izmir. Renew Sustain Energy Rev 2017;73:332–45. https://doi.org/10.1016/j.rser.2017.01.147.
[61] Sorcar S, Thompson J, Hwang Y, Park YH, Majima T, Grimes CA, et al. High-rate solar-light photoconversion of CO_2 to fuel: controllable transformation from C1 to C2 products. Energ Environ Sci 2018; https://doi.org/10.1039/C8EE00983J.
[62] Roy SC, Varghese OK, Paulose M, Grimes CA. Toward solar fuels: photocatalytic conversion of carbon dioxide to hydrocarbons. ACS Nano 2010;4:1259–78. https://doi.org/10.1021/nn9015423.
[63] Styring S. Artificial photosynthesis for solar fuels. Faraday Discuss 2012;155:357–76. https://doi.org/10.1039/C1FD00113B.
[64] Dhakshinamoorthy A, Navalon S, Corma A, Garcia H. Photocatalytic CO_2 reduction by TiO_2 and related titanium containing solids. Energ Environ Sci 2012;5:9217–33. https://doi.org/10.1039/C2EE21948D.
[65] Fukuzumi S. Production of liquid solar fuels and their use in fuel cells. Joule 2017;1:689–738. https://doi.org/10.1016/j.joule.2017.07.007.
[66] Xu H, Chen B, Tan P, Sun Q, Maroto-Valer MM, Ni M. Modelling of a hybrid system for on-site power generation from solar fuels. Appl Energy 2019; 240:709–18. https://doi.org/10.1016/j.apenergy.2019.02.091.

[67] Krekel D, Samsun RC, Peters R, Stolten D. The separation of CO_2 from ambient air—a techno-economic assessment. Appl Energy 2018;218:361–81. https://doi.org/10.1016/J.APENERGY.2018.02.144.

[68] Chen B, Xu H, Chen L, Li Y, Xia C, Ni M. Modelling of one-step methanation process combining SOECs and Fischer-Tropsch-like reactor. J Electrochem Soc 2016;163: F3001–8. https://doi.org/10.1149/2.0011611jes.

[69] Cao P, Adegbite S, Zhao H, Lester E, Wu T. Tuning dry reforming of methane for F-T syntheses: a thermodynamic approach. Appl Energy 2018;227:190–7. https://doi.org/10.1016/j.apenergy.2017.08.007.

[70] Xu H, Chen B, Irvine J, Ni M. Modeling of CH_4-assisted SOEC for H_2O/CO_2 co-electrolysis. Int J Hydrogen Energy 2016;41:21839–49. https://doi.org/10.1016/j.ijhydene.2016.10.026.

[71] Xu H, Maroto-Valer MM, Ni M, Cao J, Xuan J. Low carbon fuel production from combined solid oxide CO_2 co-electrolysis and Fischer-Tropsch synthesis system: a modelling study. Appl Energy 2019;242:911–8. https://doi.org/10.1016/j.apenergy.2019.03.145.

[72] Malek Abbaslou RM, Soltan Mohammadzadeh JS, Dalai AK. Review on Fischer–Tropsch synthesis in supercritical media. Fuel Process Technol 2009;90:849–56. https://doi.org/10.1016/j.fuproc.2009.03.018.

CHAPTER 5

Toward a holistic approach for energy efficient buildings

Sofia Boulmrharj[a,b], Soukayna Berrabah[a], Mohamed Bakhouya[a], Zineb Bouhssine[a], Radouane Ouladsine[a], and Mohammed Khaidar[b]
[a]College of Engineering, LERMA Lab, International University of Rabat, Sala El Jadida, Morocco
[b]CUR"EnR&SIE, Faculty of Sciences, University Research Center on Renewable Energies & Intelligent Systems for Energy - Chouaïb Doukkali University, El Jadida, Morocco

5.1 Introduction

Building and construction sectors are responsible for almost 40% of energy consumption and more than the third of carbon dioxide emissions (see Fig. 5.1) [1, 2]. These latter result from the manufacturing processes applied to building products and materials, such as steel, glass, cement, and concrete. It is believed that this growth will get even more significant in the upcoming years. This will be induced mainly by the population expansion requiring high standards of living and comfort, which will result in excessive usage of household appliances, and thus an increase in the energy consumption. As a result of the actual situation, electricity consumption increased by 19% in the last decade, which raised CO_2 emissions due to the excessive usage of fuel sources. This electricity covers the space cooling that has increased by 33% since 2010, hot water supply, which has increased by 11%, and appliances and household equipment supply that has raised by 18% [1, 2]. Also, the use of gas for cooking has increased by 8%.

On the other hand, space heating consumption decreased by 20%, thanks to strict actions that were taken in terms of improving building's envelope and construction/insulation materials. In addition, lighting consumption decreased by 17% as a result of using light emitting diode (LED) lamps. It is also worth mentioning that the use of coal has dropped by 10% during the same period [3]. This is attributed to the fact that the use of RES (renewable energy sources) systems has increased by 21%. However, the expenditure related to energy efficiency in building's sector has flattened rather than showing the expected evolution, even if the deployment of RES systems is set to double by 2050 [4]. Therefore, urgent plans are now crucial to decarbonize the buildings' sector so that it can support its full extenuation capacity

Hybrid Energy System Models
https://doi.org/10.1016/B978-0-12-821403-9.00002-0

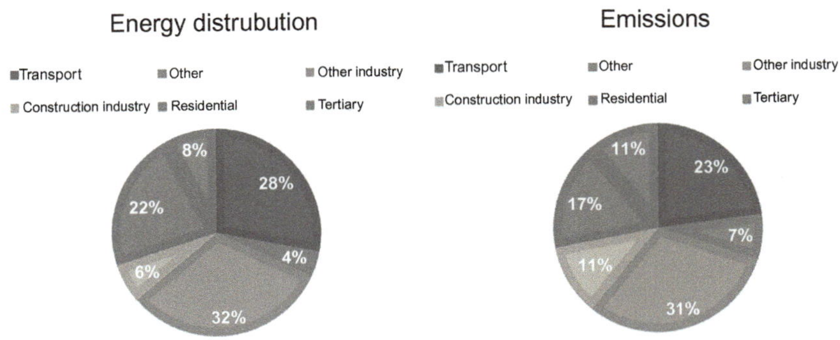

Fig. 5.1 Final energy distribution and CO_2 emissions [1, 2].

and attend the set sustainable development objectives. These actions can be enumerated in the use of RES systems as the main energy producers, reducing the use of fossil products, improving buildings' designs and envelope to reduce cooling and heating needs, and normalizing the use of electric equipment [4].

In this chapter, a holistic approach has been introduced for minimizing the buildings' energy consumption and reducing their greenhouse gas emissions, while maintaining a suitable occupants' comfort. This approach intends to establish a trade-off between the energy consumption's reduction, the occupant's comfort, and the systems' costs, while taking into consideration the three major aspects of the building, mainly *passive building systems*, *active building systems*, and the *integration of the RES systems and storage devices*. Furthermore, a case study was established to show the effectiveness of the proposed holistic approach.

The remainder of this chapter is structured as follows. Section 5.2 introduces the building construction regulations. An overview of the buildings' design and practices is presented in Section 5.3. Section 5.4 presents the proposed holistic approach. The chosen case study together with its simulation results is reported and discussed in Section 5.5. Finally, conclusions and perspectives are given in Section 5.6.

5.2 Building construction regulations

In numerous countries, building codes and regulations are either being introduced for the first time or reinforced. In fact, 73 countries have their own codes. Among these countries, 41 have mandatory residential building codes, 51 with mandatory nonresidential codes, 4 have voluntary residential codes, 12 with voluntary nonresidential codes, and 8 are in the process of developing their codes [4]. These regulations are applied in both developed (e.g., USA and France) and developing (e.g., Morocco, India) countries. Regarding developed countries, in USA, for instance, they are embracing rigorous residential and tertiary building energy codes as a part of achieving effective sustainability. Among the states that have updated their building codes, we can find California, Florida, Idaho, Utah, Wisconsin, Chicago, Illinois, and Vermont [4]. Building regulations in the UK seek to ensure that the codes exhibited in its related constitutions are executed. The building rules applicable in England and Wales are cited in the building act 1984. They mainly concern the design, the construction process, and everything related to the building's envelope and constraints [5]. These regulations are periodically updated and consolidated, and the last version concerns the building regulations 2016. On the other hand, the building regulations of Scotland and Northern Ireland are similar to those of England and Wales with some small differences, which are the legislation related to Building Act 2003 and Building standard regulations 2005 [6]. France has a long-established history of energy regulations to improve energy efficiency in buildings. The first law appeared in 1974 under the name of heating regulation and mainly focused on improving the insulation of all new constructed residential buildings. The last legal act, named RT2012, appeared in 2012 and aimed to a transition to zero-energy consumption in buildings by 2020. Its main objectives were developing positive energy buildings, reducing 38% of energy consumption in existing buildings, and renovating all public administration buildings to reduce energy usage by 40% [7].

Alike industrial countries, emerging countries have developed codes and regulations toward energy efficient buildings. For instance, in India, the national building code first appeared in 1970 and then was revised in 1983. The second version was performed in 2005, to which two amendments were issued in 2015. The code provides guidelines for regulating the building construction activities across the country. The last version of the code was published in 2016 to get along the change in occupancy nature in buildings, types of materials, and so on [8]. Concerning Morocco, the regulation of thermal construction (TRMC) has been signed in 2014. It constitutes a basic document that can be considered as an aid, at the design stage, to the thermal and energy optimization of the building's envelope. This regulation can also be used as a diagnostic tool for existing buildings by providing a benchmark for acceptable thermal insulation levels. In addition, the text of TRMC is proposed to the ministries concerned as a technical document, which can be integrated into a legal text or mandatory standard [9, 10].

It is worth noting that despite these regulations, more sophisticated actions are required to reduce CO_2 emissions and strengthen sobriety in energy consumption in buildings' sector. These actions can be differentiated according to whether the building is new or old and needs to be renovated. For instance, several strategies have to be performed to diminish the energy consumption of existing buildings, such as increasing the renovation rates and increasing the depth of renovation by 30%–50% by making its financing available. Regarding new buildings, developing and establishing buildings' codes is highly required to reduce both CO_2 emissions and buildings' consumption. Finally, multiple operations have to be performed to improve energy management of buildings, for instance, by installing energy management systems (EMS), by using smart controls and sensors (e.g., temperature, humidity, CO_2 rates, occupancy, and movement), by deploying LEDs for lightings, ventilation systems to ensure good air quality, and making information accessible for the EMS to make timely decisions according to occupants' daily actions and living situation [3, 4].

5.3 Overview of buildings' design and practices

In this section, existing approaches and practices used for energy-efficient buildings are reviewed according to three major aspects, architectural design and building's envelope, passive/active systems, and RES and storage systems integration (see Fig. 5.2).

5.3.1 Architectural design and envelope

The building's envelope constitutes the pillar of every successful and thermally efficient structure. Thus, thermal, visual, and acoustic comfort conditions should be evaluated before building's construction. In other words, designers should take into consideration construction regulations and laws while designing the building's envelope. For instance, Oral et al. [11] have developed an approach that takes into consideration construction regulations and building's envelope properties, such as materials, orientation, and dimensions, as inputs to allow the construction of a building with optimal visual, thermal, and audible performances. Therefore, considerable attention has been paid to architectural design, choice of construction and insulation materials, and other constraints (e.g., windows and doors), which are dependent on the construction site, landscape, and weather conditions [12]. In fact, it is advised to use compact shape for buildings, especially the residential ones, to reduce exchange surface between internal and

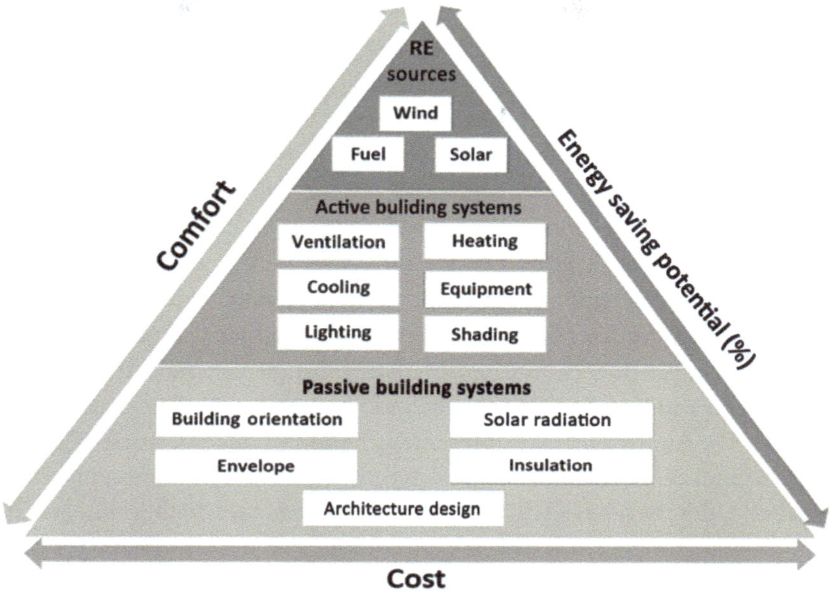

Fig. 5.2 The building as a complex system.

external ambiences. Also, Sozer [13] has shown the importance of using passive solar design techniques to design buildings' envelopes and enhance their efficiency. As a case study, they have taken into consideration a hotel building since its main objective is to ensure occupants' comfort by ensuring good internal ambience. Therefore, they have shown the importance of using appropriate insulation and adequate type of glazing and shading devices as major solutions to reduce heat losses during winter and gains during summer.

Actually, researchers have widely studied these last decades the buildings' envelope and its components to show their influence on buildings' energy efficiency. For instance, the concept of Solar Collection Envelope (SCE) was presented in [14, 15]. It was used to generate a self-shading envelope, which oversees the sun ray's situation within imaginary boundaries derived from the sun's relative motion. In addition, a computer model has been developed to reinforce the cipher of the SCE [16]. This model could allow generating a building's shape having self-shaded facades during a certain period of time, defined by the architect or the designer. This method has shown its efficiency in improving the energy performances of the building. However, it is preferable to use supplement internal/external shading devices, which have to be low emissivity devices, so that the construction of inclined walls is avoided, especially in east and west orientations.

Furthermore, one of the most successful strategies to attend energy efficiency and reduce CO_2 emissions is the use of efficient construction materials. However, it can be easily noticed that most buildings currently use structure material in an inefficient way with a rate of prodigality, which can reach 50% [17]. Therefore, it is widely advised to use local construction and insulation materials, after defining their characteristics. They are known for their low cost and good thermal properties, to generate different building envelope aspects and reduce the gas emissions produced during their production and logistics. In addition, the Phase Change Materials (PCMs) have been studied and developed by many authors worldwide. For example, Da Cunha & De Aguiar [18] have presented the importance of using PCMs to enhance energy efficiency in buildings. They have also presented and described different types of PMCs, their characteristics, and their integration into building components, such as walls, floors, ceiling, and glazed surfaces. Moreover, Wang et al. [19] have shown the importance of phase change building materials and phase change energy storage building envelope in ensuring energy efficiency and reducing buildings' consumption. Bouhssine et al. [20] have studied the effectiveness of PCMs for solar thermal energy storage in buildings.

On another note, the walls' thickness and construction materials have to be carefully selected (with good thermal characteristics). Also, windows, doors, and other openings should be placed to get advantages of natural ventilation without increasing the rate of infiltrations, while enhancing natural lighting and thus reducing electricity consumption. Moreover, the choice of the windows' dimensions and glazing should allow taking advantages from the winter sun while completely blocking summer sun, which contributes extensively to overheating. In addition, it is very important to ensure good acoustic conditions inside the building. For instance, Van Renterghem et al. [21] have shown the importance of building's envelope greening to achieve quietness. Multiple greening solutions have been investigated to show their influence in reaching audible comfort. Also, vertical greening systems, which are integrated into buildings' facades, allow having more sustainable urban designs. This solution also contributes to compensating the lack of green spaces in dense cities [22].

These abovementioned aspects could help in reducing energy consumption and maintaining good occupants' comfort (i.e., indoor air quality, thermal, visual, and acoustic). To further enhance occupants' comfort, the integration of passive and active systems is required. However, this could increase energy consumption with additional technological investment as well.

5.3.2 Passive and active systems

Concerning passive systems, they are systems that unintentionally transfer heat or store it with a little or no consumption at all. The main passive systems are either integrated into the building during the construction phase (e.g., wind tower and air—soil exchange system) or movable devices that can be freely placed (evaporative systems) [23–27]. Due to their importance, researchers are currently granting them attention to enhance their performances by using eco-friendly local materials. For instance, Hosseini et al. [23] have investigated the airflow, thermal comfort, and performances of six different designs of wind towers using computational fluid dynamics CBE tool (Center for the Built Environment). This latter has been developed to predict and evaluate thermal comfort in accordance with Ashrae standard 55 [28]. It is widely known that the low-pressure zone exists at the top of wind catchers as a result of boundary layer separation. This zone is directly placed on top of the tower's inlet opening, which, on the contrary, has a higher pressure. This distribution forces the air away from the tower's inlet. Therefore, Verela-Boydo et al. [24] have presented the importance of integrating inlet extensions to force the air toward the inlet and separate the two different zones from each other. Concerning the direct evaporative pad cooling systems, they have been usually used, for instance, to cool poultry buildings before being integrated as passive systems into residential buildings. In fact, Laknizi et al. [25] have simulated a cellulosic pad cooling system under different climate conditions in Morocco. Results showed that the system can lower outdoor temperature during very hot periods with a performance coefficient, which exceeds 0.8, and an average consumption rate of water higher than 3.3 kg/h. Despite the efficiency of cellulosic pad cooling, they have studied the performance of a new pad cooling, based on pottery material, and experimental results showed to be promising [26].

As for the active building systems, mainly HVAC (Heating, Ventilation, and Air Conditioning), they represent all methods and devices that are required for maintaining the desired occupants' comfort (e.g., thermal and air quality) [29]. However, they are the most consuming equipment in buildings. Therefore, they have to be controlled to reduce their energy consumption while maintaining suitable occupants' comfort. Building's control and automation systems have been extensively developed in the past few decades and have demonstrated the importance of optimal control strategies for reducing the energy consumption of its equipment (e.g., HVAC and lighting) and then improving the overall building energy efficiency while satisfying the occupants' comfort [30–32]. For instance, Lymperopoulos et al. [33] have presented a control approach for HVAC in multizone area, using MATLAB and BCVTB tools. This approach is likely to regulate zone temperature by applying online learning and assuming the exchange of information between neighboring zones. Also, Ren & Cao [34] have created a step-by-step HVAC control strategy, which takes into consideration the indoor air quality and thermal comfort, to create a healthy and energy-efficient environment. Furthermore, recent studies showed that buildings' energy consumption could be reduced by up to 70% with the support of recent ICT-based (Information and Communication Technologies) solutions (e.g., IoT and Big data technologies and real-time machine learning) [32, 35–38]. For instance, Rawte [37] has shown the important role that innovative ICT plays in creating smart energy-efficient and sustainable buildings. Also, David et al. [38] have proposed a three-layered model to integrate solutions for energy control using ICT. These layers are mainly communication technologies, data modeling, and finally tools allowing decision-making.

Despite these advances in building's control and automation techniques, active/passive systems are still studied in a stand-alone manner. However, a holistic framework includes decentralized control mechanisms to allow these systems (e.g., HVAC, ventilation, lighting, and shading), with different and eventually conflicting goals, to operate at different temporal and spatial scales according to the environment changes and buildings occupants' preference and activities. Furthermore, incorporating these systems to further enhance occupants' comfort may lead to energy consumption increases. Using renewable energy (RES and storage) could, therefore, help in supporting the electricity demands from these systems.

5.3.3 RES and storage systems

The RES systems, such as solar energy, wind energy, hydraulic, and biomass, have been widely deployed as promising green energy producers to support the continuous growth of buildings' electricity demands, reduce the greenhouse gas emissions, and valorize the buildings' waste. However, their intermittent nature and the uncontrollable variability of their produced electricity represent the main challenges of their efficient integration into buildings. In fact, the high unpredictability of the energy production and electricity consumption, which is mainly affected by the variability of the actual context (e.g., weather conditions and building's occupancy), make the integration of these systems into buildings a very complex task. The energy storage systems are considered among the most promising technologies that could balance RES energy production with energy consumption. Several energy storage systems exist, such as batteries, superconductors, flywheel energy storage systems, Pumped Hydroelectric Energy Storage (PHES), molten salt, hydrogen, and biogas [39, 40]. For instance, batteries (e.g., lead-acid batteries and Lithium-ion batteries) are the most prominent storage solutions for buildings due to their multiple advantages (e.g., modularity, good energy and efficiency). However, they have to be integrated efficiently because they are the most expensive and sensitive components.

Furthermore, computing the appropriate size of RES systems and storage devices is highly required for continuous electricity supply, especially for rural regions. For that, various parameters, namely, the ambient temperature, the solar irradiation, and the building's consumption, have to be determined [41, 42]. In addition, their modeling has been widely studied in the last few decades to assess and enhance their behavior and performance under various conditions (e.g., ambient temperature and solar irradiation) and make their deployment as easy as possible [42–45]. On the other hand, their deployment is extremely required to validate their corresponding developed models. For instance, many deployed hybrid systems all over the world have been reviewed by authors in [46, 47] to show how important is the deployment of these systems for validation and performance assessment purposes.

In summary, ensuring occupants' comfort while reducing electricity consumption requires an integrated approach together with its supporting framework and tools. In fact, the different aspects of the building have to be explored and controlled in a holistic way, mainly, the shape of the envelope, the thermal and mechanical characteristics of the materials, the orientation of the structure, lighting modes, shading devices, active systems, and the integration of RES and storage systems.

5.4 Holistic approach

The proposed holistic approach intends to establish a trade-off between the energy consumption's reduction, the occupant's comfort, and the systems' costs while taking into consideration the three major aspects of the building, mainly *building's envelope and architecture*, passive/*active building systems*, and the *integration of RES systems and storage devices*.

5.4.1 Building envelope and architecture

Adapting a bioclimatic approach means taking into consideration interactions between the climate and the environment ecosystem. In fact, before launching any construction, its corresponding site must be evaluated to make the right choice of materials, select the best orientation, optimize openings, and predict the future behavior of the construction according to its occupants' behavior and activities. This is really imperative so that the building can be protected from weather changes and getting its benefits as well. In other words, thermal comfort can be increased while reducing energy consumption thanks to an architectural design adapted to the climate and taking into consideration the orientation and thermal insulation of the building. When talking about bioclimatic architecture, we are talking about the most environmentally respectful design, allowing the building to benefit at most from the solar radiation and natural circulation of air. This kind of concepts works on reducing energy needs, maintaining pleasant temperature ambiences, decreasing humidity levels, and prioritizing natural lighting [48]. In other words, bioclimatic architecture takes maximum advantage of the heat of the sun due to the judicious nature of its design, the quality of its components, and optimal natural ventilation to control the supply of air and the degree of humidity.

It should be noted that the architectural choices easily influence the thermal behavior of any structure. It is, therefore, necessary to apply principles allowing the reduction of heat losses and external fenestrations, from the preproject phase, while avoiding overheating during the summer. The strong recommendation concerns the orientation of the building and the presence of openings to optimize the solar gains. The architectural principles and basis, which are simple and easy to implement, are as follows [49]: compact shape, building's layout and orientation, thermal inertia and materials, thermal insulation, opening and bay windows, air circulation, and thermal bridges. For instance, the compact shape allows the reduction of heat losses through the envelope by minimizing the contact with the outside. Regarding the layout and orientation feature, for example, a south orientation allows very good sunshine and, therefore, allows energy consumption reduction of heating and lighting systems. To capture the heat of the sun while avoiding overheating, a solution of opening large windows to the south is more recommended. In this case, thermally insulated glazing must be used to limit heat loss. Also, to be protected from the sun during summer periods, outdoor protection is necessary. The dimensioning of solar protections should take into consideration the latitude of the sun during the summer solstice and winter solstice to take advantage of the sun rays during winter while being fully protected during summer.

The thermal inertia can be simply explained by the accumulation and phase shift phenomena. During hot days, the heavy envelope accumulates heat. During the night, when the outside temperature decreases, all the heat accumulated during the day is then transmitted inside the building avoiding or at least reducing the need for heating. These aspects are represented by thermal conductivity, diffusivity, and effusivity expressing the ability of a material to transmit a variation in temperature and to restore an instantaneous flow of heat. It should be noted that the thermal inertia of the wall is proportional to its thickness and inversely proportional to the coefficient of thermal conductivity of the material. Therefore, a trade-off between the material's conductivity and thickness should be found to avoid oversizing the structure.

The thermal quality of a building from the surface is evaluated by the surface transmission coefficient. For instance, an opaque wall is considered to be efficient as long as its transmission coefficient is small. The overall rate of bay windows (TGBV) is defined by the ratio between the total area of the bay windows on the exterior walls of the heated/cooled spaces and the total gross area of these walls. The same as opaque walls, windows are considered much efficient as long as their transmission coefficient is low. In addition, attention should be paid to its optical properties including the solar factor, which represents the amount of solar energy stored behind windows, which are shaded or directly exposed to solar radiation. Thermal bridges mainly persist on areas with imperfect contact between the components or with coexistence of different materials, resulting in a change in terms of conductivity and then air exhaust. In fact, they represent an important source of heat losses, which can sometimes generate uncomfortable indoor atmospheres due to condensation. It is, therefore, important to reduce them during construction or during rehabilitation.

The remainder of this section will focus on the most important abovementioned aspects, which have to be considered for reducing the building's consumption while maintaining the occupants' comfort.

5.4.1.1 Hygrothermal behavior and choice of materials

The building's envelope is mainly constituted of roof, walls, windows, doors, and foundation. These components provide structural support against internal/external loads and forces, control exchanges between interiors/ exteriors ambiences, and ensure esthetic purposes as well. The foundation is the structural component that transmits the loads from the building to the fundamental bedrock to increase its resilience against external forces [50]. Notwithstanding, the overall structure should be designed to control the transfer of moisture and thermal energy into the interior so that occupants' comfort will not be disturbed. Therefore, it is necessary to pay a lot of attention to the choice of construction material, insulation, and moisture balance and the installation of air barriers and the thermal bridges.

Air barriers are systems of materials that are used to control airflow and can be placed anywhere in building enclosures. They are known for their ability to resist airflow and pressure. Rigid materials, such as gypsum board, exterior sheeting materials, and supported flexible barriers, are typically effective as air barriers, especially if joints and seams are sealed. To be efficient, they should be impermeable to airflow, continuous over time in building enclosure, resilient, and durable. The air barriers and the pressure boundaries (i.e., the location where 50% or more of the air pressure drop across an assembly occurs) are the fire and smoke barriers when placed in interunit separations. They are considered as gas barriers, when placed between a garage and the remainder of the building because they provide the gas-tight separation [51].

Two major categories exist to provide air barriers in residential buildings, namely, Interior-air barriers using drywall and framing or using polyethylene and Exterior air barriers using building wraps or using polyurethane foams. Regarding these latter, it is worth noting that polyurethane foams 'interior moisture-laden' air into assembly cavities during heating periods. Exterior air barriers are easy to install and can control the wind-washing provided by an exterior air seal. However, they cannot control the entry of moist carried by air into cavities. Therefore, the optimal solution is to combine exterior and interior air barriers to address their weaknesses [51].

5.4.1.2 Moisture balance and control for buildings

When designing a building, it is extremely important to take into consideration its interaction with its surrounding environmental conditions, especially, temperature, humidity, rain, as well as its interior climate. These surroundings are considered, by moisture engineering, as environmental loads with principal limit states, such as rot, deterioration, mold, and corrosion. For this, they use iterative and performance metrics to meet moisture-related objectives. In fact, moisture accumulates in the building envelope when the rate of its entry into an assembly exceeds the rate of its removal. In other words, the material can no longer store moist, and thus, problems occur. The storage capacity differs from a material to another, depending on its thermal capacities, temperature, and thickness. If we take steel studs as an example, they absolutely do not have any water storage capacity. In contrast, Gypsum sheathing and wood frame can store approximately 1% and 10% of water, respectively, before the colonization occurs, in average conditions. The wood can also store beyond 10% to reach 16% by weight, but the wood surface is likely to develop mold [52].

Furthermore, convenient moisture control design is very important to ensure energy efficiency in buildings in the construction phase and avoid damage. It is also prerequisite for the restoration of traditional architectures. The key feature to attain desirable results is to make the right choice of control procedures. The latter depend on several parameters, such as local climate conditions, architectural designs, type of activities, and the available budget. Among these measures, we differentiate between (i) control moisture entry, (ii) control moisture accumulation, (iii) removal of moisture, and (iv) a combination of these three measures. These are effective at preventing moisture from entering an assembly; yet, if the building is wet, they will prevent moisture from leaving, and thus can be detrimental. Therefore, the balance between entry and removal is crucial. The major wetting mechanisms in any kind of buildings are liquid flow and capillary suction, which result from groundwater and rain [52].

In addition, building assemblies need to be protected from wetting through air transport and from vapor diffusion by using air barriers, air pressure control, ventilation, and dehumidification. Moist usually goes from warm to cold or from more to less driven by the concentration gradient. Therefore, to choose the most suitable strategy, it is important to define the type of climate as a first step. In cold climates, moisture goes from the interior to the exterior. Therefore, the building needs to be protected from getting wet from the interior. Therefore, air barriers and vapor barriers have to be installed toward the interior warm surfaces and then ventilation systems (dilution) and source control can be added. However, in hot climates, moisture flows from the exterior toward the cooled interior. Therefore, assemblies need to be protected from getting wet from the exterior and must be allowed to dry toward the interior. According to this, air and vapor barriers have to be installed on the exterior surfaces [52].

As for mixed climates, the situation becomes much more complex since the structure should not get wet on both exterior and interior. Therefore, to face the problem, two solutions or strategies are to be adopted. First of all, permeable materials should be used for the construction so that the water does not cumulate. Second, air pressure and moisture control should be adopted as well. The air barrier can be either installed on the inside or outside surfaces. Also, vapor barriers can be installed in the middle of the assembly by installing semi-impermeable sheeting on the exterior of a frame cavity wall. Similar to the first approach, this one as well needs air and moisture control. If the second approach is adopted, it is not recommended to install vapor barriers on the inside since it will prevent the assembly from drying toward the interior during cooling periods [52].

5.4.1.3 Choice of construction materials

Looking for a more sustainable building is not a personal choice anymore. It is now regulated to improve buildings' environmental behavior. Many people are now opting for what is called green building materials, also known as environmentally friendly materials. These are the ones that do not have a big environmental impact during their manufacturing, placement, or maintenance. They are preferred to be local materials so that we can reduce the CO_2 emissions generated during their transportation. They have to be reusable, recyclable, natural, and durable (e.g., soil, stone, mud, clay, rammed earth, wood, adobe, and straw) so that they will not get ruined by cold, heat, moisture, or other environmental hazards [53]. For instance, mud, clay, stone, and fibrous plants (Thatch) are the most used materials all over the world to suit different local weather conditions [54]. To start with the stone, it is broadly used as the basic structural component. As for mud, it is used to fill the gaps, and when mixed with fibrous plants, it can play the role of insulation. The amount of each used material may influence the esthetic of the buildings as well as its thermal efficiency according to the soil's nature. Nowadays, people are being more aware that going back to ancient construction materials and methodologies is the most appropriate way to ensure thermal comfort. Actually, clay is a very good thermal mass. It is mainly good for keeping temperatures at constant levels, i.e., naturally cool in the summer and warm in the winter.

In addition, building's envelope that is made with rammed earth is characterized by its mass inertia, allowing slow conduction of temperature through the wall. As for wood, it is a generic building material that is used in most climates. It is very flexible under load and incredibly resilient. However, woods are not identical, even if they belong to the same species. Therefore, some may be used in certain applications, while others may not. In the late 20th century, concrete blocks appeared and are mainly made from a combination of aggregate and cement in most of the cases. After being mixed, the cement hydrates and hardens into stone-like materials. This type of blocks is usually strengthened by the use of steel bars (reinforced concrete). Finally, grass, more precisely, Thatch, is considered to be a good insulator. It can easily be harvested and is usually placed on roofs with special ridge tiles on top. Each material from the previously mentioned ones has different properties from weight, resilience, cost, and durability. Therefore, they may be suitable for certain type of applications and better be avoided in others. All in all, the choice of materials for construction is based on their durability, cost, and effectiveness to resisting to loads.

5.4.2 Passive/active systems

Passive systems are all devices that could be integrated into a building for the purpose of transferring and storing heat without or with a little energy consumption (e.g., windows, wind towers, and evaporative coolers). Therefore, the design and construction of buildings have to respect the bioclimatic approach by integrating main features, e.g., well insulated and well shaded against solar exposure during summertime. On the other hand, active systems represent all methods and devices (e.g., HVAC) that are required for maintaining the desired occupants' comfort (e.g., thermal and air quality) [29].

5.4.2.1 Passive systems

In this section, we review some passive systems that could be integrated as natural ventilation systems. For instance, a *wind tower* or *wind catcher* is an architectural structure designed to provide natural ventilation and cooling of the building. It first appeared in Iran for many centuries under several designs, mainly unidirectional, bidirectional, and multidirectional [55]. Wind towers are used in hot dry climates and windy sites where they catch cooler air that reigns at higher levels above the ground and conduct it inside the building. This is all due to a change in temperature levels and therefore a change in density of the air around and inside the tower [56]. This variation of density allows the air to either move upward or downward through the tower. However, the tower acts as a chimney in the absence of wind, and in this case, it is the radiation through the walls and roof that refreshes the interior atmosphere. Also, it is worth noting that towers are considered to be more effective than evaporative coolers.

The *evaporative cooler*, also known as desert cooler, is widely used since it is relatively cheap and requires less energy than other cooling systems. It is a device that cools the air through the process of water evaporation. More precisely, water uses the heat of the hot air that passes through the system to evaporate, and thus generates fresh air at the system's outlet. This system is generally used in extremely dry climates since it adds moisture to the air while ensuring thermal comfort. In fact, water takes energy in the form of sensible heat to evaporate and convert it into latent heat without changing the enthalpy of the system [57]. This process leads to a drop in temperature and a noticeable increase in humidity. Actually, water is known to be one of the most substances with high latent heat of evaporation. Accordingly, most of the evaporative cooler systems use it. However, in areas where the air is not very dry, humidity tends to increase a lot, which disturbs the occupant's thermal comfort. Therefore, an indirect evaporative system has been developed and it is widely used in these areas instead of direct evaporative systems; it is also known by the single-stage desert cooler.

Direct evaporative system uses a fan to drag air through a wet pad, providing a larger surface for water evaporation. The pad is always saturated since water is continuously sprayed at its top [58]. The excess of water dripping from the bottom of the pad is collected in a small tank and then recirculated at the top forming a closed cycle. Therefore, water consumption is minimized. However, the system should be often cleaned to ensure optimal performances and avoid the damage of the used pad. In addition, a passive direct evaporative cooling system can be obtained by coupling this classical system to the previously mentioned wind tower and avoiding the use of a fan. In this case, the air coming down from the tower, which is already fresh, is cooled again by the pad cooling and the performances are, therefore, boosted. Unlike direct evaporative systems, indirect evaporative ones are closed circuits used to limit the increase in air humidity [59]. It uses, in addition to the main equipment, a heat exchanger so that the cooled moist air is never in direct contact with the conditioned supply air. The humid stream is released outside to cool other external devices, such as photovoltaic panels to enhance their efficiency [60]. This system is rarely used since it needs special architectural designs and special devices that act as heat exchangers. Also, it consumes a lot of water and can introduce too much risk in terms of water intrusion into the building's structure.

5.4.2.2 Active systems

HVAC (Heating, Ventilation, and Air Conditioning) is a device used for controlling the air temperature, humidity, distribution, and cleanliness.

Therefore, it is considered as the most complicated service utility in the building sector. This process is very simple and operates as follows. The heat is added to the air or extracted from it to produce hot or fresh air, respectively [61]. In fact, the moist air contains three main elements at any given temperature, namely, sensible heat for dry air or water, latent heat of evaporation, and sensible superheat. Thus, both heating and cooling are a kind of energy transfer from the air or space due to a difference in temperature between the source and space or conditioned air. The difference between both of them resides in the fact that heating can take place through direct radiation and free convection, while cooling denotes sensible heat transfer with a decrease in the air temperature. Fig. 5.3 presents an inclusive classification of HVAC systems [62].

Unlike HVAC systems, ventilation systems are used for injecting fresh air into the building to renew its air. In other words, it is mainly used to control the indoor air quality and thus improving the occupants' comfort. It can be classified into two main categories, namely, natural and mechanical ventilation [63]. Regarding the natural ventilation, it occurs when the air movement is not controlled, when there are cracks and infiltrations or when the user opens doors and windows. This type of ventilation depends on outside climate conditions and building's airtightness. Therefore, this either leads to a lack of ventilation or high energy cost if there is high air infiltration. To

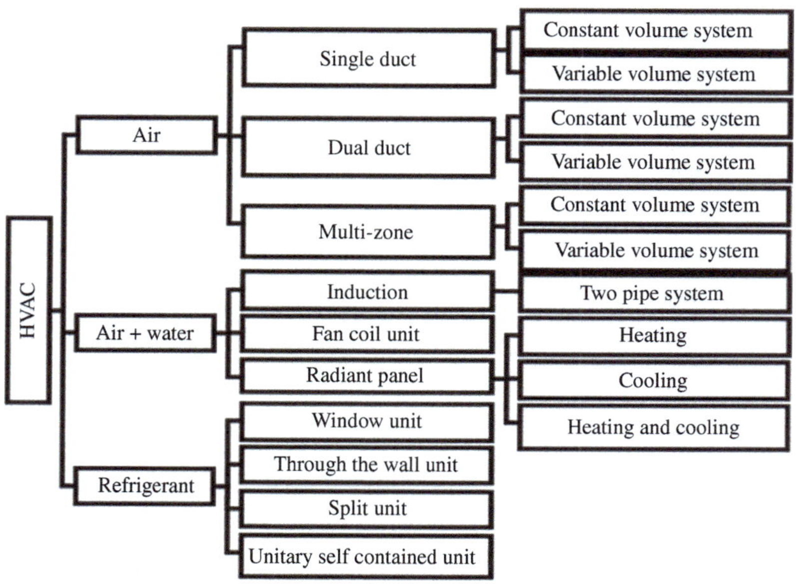

Fig. 5.3 Classification of HVAC systems.

overcome this problem, spot ventilation might be used. It is a device that removes moisture and all kind of air pollutants from their source. This can be done by installing localized exhaust fans, such as the ones used in bathrooms and kitchens. On the other hand, mechanical ventilation is the intentional fan-driven flow of outdoor air into a building [64]. This can be achieved by using one of the four systems, namely, exhaust ventilation, supply ventilation, balanced ventilation, and energy recovered, according to climate conditions [65].

For instance, the exhaust ventilation is composed of a single fan connected to a single exhaust point located in the center of the building. This kind of systems is preferred in cold climates because they reduce the inside air pressure below the outdoor air pressure without drawing moist air into the building. However, they may bring pollutant along with fresh air [66]. Regarding the supply ventilation systems, the mechanical air exhaust system creates a pressure difference over the ventilation openings, and so air is drawn in the building. Therefore, the system is less dependent on climate conditions. Actually, it is advised to place air suppliers as high as possible with the possibility of regulating the inlet grid, so that the draught problems generated by high wind pressure or temperature difference can be avoided.

As for the balance ventilation systems, this kind of systems allows introducing and exhausting equal amounts of fresh outside air and polluted inside air, respectively [67]. A balanced ventilation system is usually composed of two fans and two dust systems. It also uses filters to remove pollutants from air before entering the building. However, as the previously mentioned systems, it cannot remove moist. In fact, this technology is used in both hot and cold climates; however, it remains a little expensive to be installed. Finally, the energy recovery systems are one of the most important systems of mechanical ventilation systems since they have the possibility of heat recovery [68]. The main two equipment used to recover heat is heat wheels and heat exchangers. A heat wheel or regenerator allows, by rotating, absorbing heat and throwing it out. Then, a plate heat exchanger, which is usually made of a set of parallel aluminum plates, separates the airflow. The air (hot and cold) flows through the ducts without being mixed. However, the heat is exchanged with no possible exchange of moist.

5.4.3 RES and storage systems

Actually, the RESs (e.g., solar energy, wind energy, hydraulic, and biomass) have been widely deployed all around the world as green energy producers to support the continuous growth of buildings' electricity demands, reduce the greenhouse gas emissions, and valorize the buildings' waste [69, 70, 71]. Due to their multiple benefits (e.g., abundant and clean), Morocco, for example, puts more emphasis on RES to diminish its reliance on fossil fuels, namely, oil, coal, and natural gas, which is almost 90% [72]. It intends to increase its renewable energy share, particularly solar energy (with an average daily global radiation of around 5 kWh/m^2 [73]), from 42% in 2020 to 52% in 2030 by building several solar and wind power stations, such as NOOR in Ouarzazate [74,75]. However, the intermittent nature and uncontrollable variability of the RES systems' produced electricity represent the main challenges of their efficient integration into buildings. In fact, the high unpredictability of the energy production and electricity consumption, which is mainly affected by the actual context (e.g., weather conditions and building's occupancy), makes the integration of these systems into buildings very complex. Energy storage systems are considered among the most promising technologies that could balance RES energy production with electrical energy consumption. In fact, numerous energy storage systems exist, such as batteries, superconductors, flywheel energy storage systems, Pumped Hydroelectric Energy Storage (PHES), molten salt, hydrogen, and biogas [39,40,76]. Among these storage devices, batteries (e.g., lead-acid batteries and Lithium-ion batteries) are the most deployed in buildings due to their multiple advantages (e.g., modularity and good energy efficiency).

On another note, computing the appropriate size of RES systems and storage devices is mandatory for uninterrupted electricity supply, especially for rural regions. For that, various parameters, namely, the ambient temperature, the solar irradiation, and the building's consumption, have to be determined [41, 42]. In fact, two methods have been developed by the research community for sizing these systems: (i) the first method uses the yearly averaged monthly values of the abovementioned parameters and (ii) the second one uses the worst month's parameters [77]. Moreover, many approaches, such as deterministic and stochastic approaches, software tools (e.g., HOMER and TRNSYS), and algorithmic-based approaches (e.g., artificial intelligence and genetic algorithm), have been reviewed by the authors in [77–87] to optimize the sizing of RES and storage systems for the purpose of assuring the availability of the energy during the whole day, while reducing the cost investment. Furthermore, the modeling of these systems has been widely studied in the last few decades to assess and enhance their behavior and performance under various conditions (e.g., ambient temperature and solar irradiation) and make their deployment as easy as possible [42–45]. Furthermore, numerous simulation software tools, such as HOMER, PVSYST, System Advisor Model (SAM), TRNSYS, and so on, have been developed in the last few decades to simulate the RES and storage systems and also estimate and predict their behaviors and performances [82,88,89]. On the other hand, the deployment of these systems is extremely required to validate their developed models. For instance, many deployed hybrid systems all over the world have been reviewed by the authors in [46, 47] to show how important is the deployment of these systems for the validation of their developed models and the assessment of their performances.

For instance, the hybrid system, which we will study in this chapter, integrates PV panels to generate electricity from the solar irradiation, converters (e.g., DC/DC converters and DC/AC converters), batteries to store the overproduction of the RES systems for supplying the building with electricity when there is no production, and finally thermal panels to supply the building with the needed hot water for shower and so on. The sizing and the modeling of each component of this hybrid system are given with more details in the next sections.

5.4.3.1 Building's load

The sizing of hybrid systems depends on several variables, especially the building's electric energy consumption. Actually, the daily building's energy consumption ($E_{consumption}$) can be approximated using Eq. (5.1) [90]:

$$E_{consumption} = \sum_{i=1}^{n} N_i P_i \Delta t_i. \tag{5.1}$$

where n is the number of equipment types, N_i stands for the number of each equipment type, P_i represents the power of each equipment type (the equipment power can be obtained from the manufacturer datasheet), and finally Δt_i is the number of hours of operation of each equipment type.

However, this method gives only an approximation of the average building energy consumption since the latter is highly affected by the occupants' behavior and the environmental conditions. Therefore, accurate measurement and estimation of it are highly required. Regarding its measurement, several instrumentation and sensors have to be deployed to measure the consumption of the building in real time [91–95]. For example, the authors in [92, 93] have used a set of current and voltage sensors to monitor the daily building consumption. Also, Al-Rabghi et al. [91] have used a PC board from Highland Technology (model AT-180) to determine the building consumption. As for the estimation of the building energy consumption, several software tools, such as DOE-2, EnergyPlus, TRNSYS, and algorithm approaches (e.g., Neural Networks, Support vector machines, and Genetic algorithms), have been developed for accurately estimating the electric energy consumption of a building [96–101]. In addition, many authors have used the hybrid methods, which are a combination of software tools and algorithmic-based approaches, to overcome the limitations of each method and thus come up with precise estimated energy consumption [98]. Furthermore, the authors in [91] have compared the measured electric consumption with the estimated one to validate the developed models. Finally, the prediction and forecasting of the building electric consumption has attracted much interest in the last few decades. In fact, many authors have used several algorithm approaches, such as Neural Networks, Support vector machines, Genetic algorithms, ARIMA, and SARIMA or a combination of them (hybrid models) to accurately predict and forecast the energy consumption of a building [93,102].

5.4.3.2 PV modeling and sizing

PV panels, which are among the most promising RESs that are capable of satisfying the buildings' needs, have been broadly deployed in the last few decades to convert the solar irradiation into electricity. When it comes to the PV panels, various technologies have been developed, namely, crystalline silicon (c-Si) solar cells, thin film cells, and multijunction cells. Currently, the most used and manufactured PV panels worldwide, with over 90% of the available PV panels, are the crystalline silicon ones [103]. The latter are classified into three types, namely, monocrystalline silicon (mc-Si), polycrystalline silicon (pc-Si), and amorphous silicon (a-Si) [103]. For instance, the pc-Si PV panels are still the most used compared to the monocrystalline and amorphous ones. Regarding the PV panels' sizing, their peak power P_{PV} (W) could be calculated using Eq. (5.2) [41,69,92]:

$$P_{PV} = \frac{E_{consumption} * I_r}{E_r * PR}. \tag{5.2}$$

where $E_{consumption}$ represents the maximum electrical daily consumption (kWh/day), I_r is the solar irradiance at STC (standard conditions), which are 25°C and 1000 W/m^2, E_r stands for the lowest monthly average daily solar irradiation throughout the year (kWh/m^2.day), and *PR* represents the performance ratio, which is a correction coefficient for both stand-alone and grid-connected PV systems. This *PR*, which varies between 0.55 and 0.75 for stand-alone systems and between 0.7 and 0.85 for grid-connected PV systems, represents the fraction between the real and the ideally possible generated energy, and it depends on several characteristics, mainly the battery's efficiency, the converter's efficiency, and the voltage drop in the cables [41, 42, 92]. Therefore, to meet the electricity demand throughout the whole year, the number of required PV modules (N_{PV}) is calculated using Eq. (5.3) [41,42,92]:

$$N_{PV} = \frac{P_{PV}}{P_{module}}. \tag{5.3}$$

where P_{module} represents the peak power of one PV module (W). In addition, the number of branches (N_p) and the number of series modules in each branch (N_s) are computed using Eqs. (5.4) and (5.5) [41,42,92]:

$$N_p = \frac{I_{max,\,converter}}{I_{sc}}, \tag{5.4}$$

$$N_s = \frac{V_{max,\,converter}}{V_{oc}}. \tag{5.5}$$

where $I_{max,\ converter}$ represents the maximum current at the input of the converter (A), I_{sc} is the short circuit current of the PV module (A), $V_{max,\ converter}$ stands for the maximum voltage at the input of the converter (V), and finally V_{oc} represents the open circuit voltage of the PV module (V). As for the PV panels' modeling, a mathematical model of their electrical equivalent circuit has been proposed and developed by the research community to simulate and predict the behavior of the PV panels and evaluate their performance under several weather conditions, such as the ambient temperature and the solar irradiation [42,70,104]. In fact, the simplest model, which is named the ideal solar cells' model, is composed of a current source in parallel to a diode. This model has been improved by the addition of a series resistance [105], which represents the cells' internal and contact resistances and a shunt resistance [106], which may result from imperfections on the surface of the PV cell as well as from leakage currents through its edges, to build the single-diode model [42,104,107]. The latter has been enhanced by the addition of a second diode to simulate the exact behavior of the PV modules [108]. However, the single-diode model (see Fig. 5.4) is still the most used in simulations due to its simplicity and because it gives a good approximation of the real-world scenarios [70].

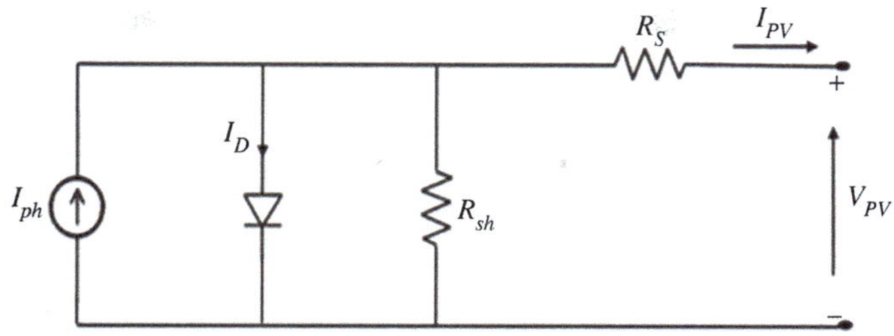

Fig. 5.4 The single-diode equivalent circuit of the PV cell.

The equation of the $I_{PV}-V_{PV}$ characteristic of the PV cell, which is derived from its electrical equivalent circuit, is given in Eq. (5.6) [42, 70]:

$$I_{PV} = I_{ph} - I_D - I_{sh}. \tag{5.6}$$

where I_{ph} represents the photocurrent, I_D stands for the diode current, and I_{sh} is the shunt current. In addition, the Shockley diode equation, the photocurrent, and the shunt current are expressed in Eqs. (5.7), (5.8), and (5.9) respectively [42]:

$$I_{ph} = \frac{I_r}{I_{r0}} * \left[I_{sc} + K_I\left(T - T_{ref}\right)\right], \tag{5.7}$$

$$I_D = I_s\left[\exp\left(\frac{q\left(V_{PV} + I_{PV}R_s\right)}{nKT}\right) - 1\right], \tag{5.8}$$

$$I_{sh} = \frac{V_{PV} + I_{PV}\,R_s}{R_{sh}}. \tag{5.9}$$

where I_r represents the solar irradiance (W/m^2), I_{r0} is the irradiance at STC (1000 W/m^2), I_{sc} stands for the cell's short circuit current at STC (A), K_I is the temperature coefficient of the short-circuit current (A/K),- T represents the cell's temperature (K), T_{ref} is the cell's reference temperature (K), q stands for the electron charge (1.6×10^{-19}C), n is the diode's ideality factor (it varies between 1.2 and 1.6 for crystalline silicon), and K represents the Boltzmann constant (1.38×10^{-23} J/K). Regarding the diode saturation current (I_s), it could be calculated using Eqs. (5.10) and (5.11) [42,70,104]:

$$I_s = I_{rs}\left(\frac{T}{T_{ref}}\right)^3 \exp\left[\frac{q\,E_g}{n\,K}\left(\frac{1}{T_{ref}} - \frac{1}{T}\right)\right], \tag{5.10}$$

$$I_{rs} = \frac{I_{sc}}{\exp\left[\dfrac{q\,V_{oc}}{N_s\,K\,n\,T_{ref}}\right] - 1}. \tag{5.11}$$

where I_{rs} represents the diode reverse saturation current (A), E_g stands for the cell's semiconductor bandgap energy (eV), V_{oc} is the cell's open-circuit voltage at STC (V), and N_s represents the series cells' number in a module. Concerning the series and the shunt resistances (R_s and R_{sh}, respectively), they could be estimated using the expressions given by the authors in [104,109]. Finally, Eq. (5.6) could be expressed for a PV module by Eq. (5.12) [42]:

$$I_{PV} = N_p I_{ph} - N_p I_s \left[\exp\left(\frac{q\left(\dfrac{V_{PV}}{N_s} + \dfrac{I_{PV} R_s}{N_p}\right)}{nKT}\right) - 1\right] - \frac{\dfrac{N_p V_{PV}}{N_s} + I_{PV} R_s}{R_{sh}}. \tag{5.12}$$

where N_p represents the number of parallel cells in a module. On another note, the performance of the PV system has to be analyzed since it is highly influenced by many factors, especially the weather conditions and the PV modules' technology [110]. Therefore, numerous studies have been carried out for the purpose of evaluating the performance of several PV systems under different circumstances (e.g., weather conditions, stand-alone or connected to the grid, and PV modules' technology) [111].

5.4.3.3 Converter modeling and sizing

In hybrid systems, converters have been widely used to connect the different RES systems to the storage devices as well as the building's appliances. They are used to convert and adapt the RES systems' variable voltages to the appliances' voltages as well as control the energy flow of the RES systems on the basis of the appliances' consumption [112]. There are many types of converters, mainly DC/DC converters, DC/AC converters (inverters), AC/DC converters, and AC/AC converters [113,114]. For instance, the DC/DC converters are used to adapt the RES system's voltages to the appliances' voltages as well as regulate the batteries' charge and discharge for avoiding their damage and aging [42,92]. The DC/AC converters, also named inverters, are used to convert the DC current produced by the RES systems into an AC current, which could be used to supply the building's appliances.

Concerning the AC/DC converters, they are used to convert the AC produced, for example, by the wind tower into a DC to supply the DC loads. Finally, the AC/AC converters are used to convert the AC waves to another AC wave with different frequencies and magnitude for adapting them to the appliances' voltages.

Concerning these converters, four categories exist, namely, the buck converter, the boost converter, the buck-boost converter, and the boost-buck converter [45,115]. For instance, the buck and the boost converters provide an output voltage lower or higher than the input voltage, respectively [42]. However, the buck-boost and boost-buck converters, which are a combination of the buck and boost converters, are used for the applications that need an output voltage that might be lower or higher than the input voltage [42]. Besides, most of these converters are controlled with an MPPT (Maximum Power Point Tracking) to maximize the power production of the RES systems [116]. Actually, several MPPT control methods have been reported in the literature, such as the Fractional Open Circuit voltage (FOC), the Fractional Short-Circuit Current (FSC), the constant voltage control method, the Incremental Conductance MPPT algorithm (IC), and the Perturb and Observe MPPT algorithm (P&O) [117]. The P&O is the most used, especially with PV systems, due to its simplicity and ease of implementation [117].

Concerning their sizing, several methods, optimization approaches, and software tools have been proposed, developed, and reported in the literature to determine the size of the power converters based on the power of the RES systems, the voltage and the capacity of the batteries, and the appliance power [42,92,118]. As for their modeling, many models have been developed to simulate the behavior of the power converters as well as assess their performances under various conditions [42,119]. In fact, they are usually represented by their equivalent circuits, which are mainly composed of controlled current and voltage sources (i.e., the position and the type of controlled source depend on the type of power converter), diodes, capacitors, and thyristors [115,119]. In addition, filters (i.e., passive, active, and hybrid) are extremely needed for enhancing their power's quality. However, to overcome the reasonable losses and expensive cost of these filters [115], a new variety of power converters have been developed using new solid-state self-commutating devices, mainly MOSFETs and IGBTs. The use of these new solid-state self-commutating devices, instead of thyristors and filters, has increased the converters' efficiency and power quality while reducing their costs [115].

5.4.3.4 Battery modeling and sizing

Storage devices, such as batteries, superconductor, hydrogen, flywheels, and Pumped Hydroelectric Energy Storage (PHES), have attracted plenty of interest in the last few decades due to the strong appearance of hybrid systems and the intermittent nature of RES [40,120,121]. Their main objective is to store the RES systems' overproduction for using it when there is no production. Actually, batteries are the most used storage device in hybrid systems since they have several benefits, such as fast response, good energy efficiency, and modularity [121]. As a matter of fact, several varieties of batteries are available, such as Lead-acid batteries, Lithium-ion batteries, nickel–cadmium batteries, vanadium flow batteries, and sodium–sulfur batteries [40,120]. Actually, the sizing, modeling, and performance evaluation of batteries are mandatory when it comes to their integration into hybrid systems to have an idea about the optimal size to install and evaluate their behavior under different circumstances in simulations before installing them in real-sitting scenarios. Regarding the sizing of batteries, their capacity (C_{bat}) can be computed using Eq. (5.13) [41,69,92]:

$$C_{bat} = \frac{E_{consumption} \times N_{Aut}}{V_{bat} \times \eta_{bat} \times DoD}. \tag{5.13}$$

where $E_{consumption}$ represents the maximum electrical daily consumption (kWh/day), N_{Aut} is the number of days of autonomy (day), V_{bat} represents the battery's voltage (V), η_{bat} stands for the battery's efficiency (≈80%), and finally DoD is the Depth of Discharge (≈60% for the Lead-acid batteries). The main aim of the sizing of batteries is to ensure the building appliances' operation when there is no production. Therefore, the computation of the batteries' capacity requires the determination of the number of days of autonomy, which varies according to the geographical location, and thus the weather conditions [41,92]. On the other hand, the necessary capacity to install (C_{10} in Ah) is calculated using Eq. (5.14):

$$C_{10} = \frac{C_{bat}}{F_{C_{bat}}}. \tag{5.14}$$

where F_{Cbat} represents a correction factor, which is used to compute the capacity C_{10}. This factor is equal to 1.25 if the number of days of autonomy (N_{Aut}) varies between 1 and 4 days [41]. In addition, the number of series batteries (N_s) depends on the used battery's voltage (V_{bat} in V) and the converter's voltage ($V_{converter}$ in V), as expressed in Eq. (5.15), while the number of branches (N_p) has to verify Eq. (5.16), where C_{10bat} is the battery's capacity in C_{10} (Ah) [41]:

$$N_s = \frac{V_{converter}}{V_{bat}}, \tag{5.15}$$

$$N_p \geq \frac{C_{10}}{C_{10_{bat}}}. \tag{5.16}$$

As for the battery modeling, their complicated behavior makes their modeling, their performance evaluation, and their State of Charge (SoC) estimation and prediction very difficult [43]. This complicated behavior is attributed to the variation of several battery parameters, such as current, voltage, capacity, and temperature, during the charge/discharge cycles. In fact, the research community has developed numerous battery models, mainly the electrochemical models and the electrical-circuit models, for estimating and evaluating its behavior under different circumstances, namely, the charging/discharging current and the temperature [43,122]. For instance, the electrochemical models are still the most accurate models. Alas, the presentation of the batteries' dynamic behavior, the identification of their parameters, and the estimation of their SoC using these models are too much difficult because of their strong dependence on temperature as well as their complex electrochemistry [123]. As an alternative, the electrical-circuit models, mainly the ideal battery model and the Thevenin battery models, are extensively used, particularly when the batteries are combined with RES systems, for estimating, in analogy with electricity, the batteries' dynamic behavior due to the fact that they provide an appropriate accuracy while maintaining an acceptable computation complexity [122]. They are constituted of a voltage source, resistors, and capacitors. These models are more accurate when the order of the model (i.e., the number of RC networks) increases [124]. On the other hand, the battery parameters have to be accurately identified for obtaining a precise battery model, which could simulate its real dynamic behavior. The number of battery parameters depends on the used battery model. In fact, the battery parameters are affected by several conditions, such as the charge/discharge cycles, the temperature, the Depth of Discharge (DoD), and the overcharge. Therefore, they have to be identified in real time instead of computing fixed parameters that could lead to an important model error. Actually, the research community has proposed and developed numerous methodologies for precisely identifying the battery parameters, such as Recursive Least Squares (RLS) method, impedance spectroscopy method, OCV tests, Kalman Filter (KF), Artificial Neural Network (ANN) algorithms, and Hybrid Pulse Power Characterization (HPPC) test [125–127]. For instance, the impedance spectroscopy method and the OCV tests are used to extract fixed battery parameters [124,125], while the other methods (i.e., RLS, KF, ANN, and HPPC) are used for identifying the battery parameters in real time [126,127]. It is worth noting that RLS is the most used for identifying the battery parameters in real time due to its accuracy and simplicity [127]. This method aims at minimizing the error, which is the difference between the experimental and simulation results, to enhance the battery model.

On another note, to protect the battery against the overcharge and the deep discharge, its SoC is the main indicator. It is used to determine the remaining capacity in the battery; however, there are no sensors that could measure it, which leads to the inevitability of its estimation. Actually, it can be accurately estimated by numerous mathematical methods, which have been proposed, developed, and reported in the literature, namely, the direct measurement methods, the model-based methods, the artificial intelligence methods, and the hybrid methods [128]. For instance, the direct measurement methods, mainly the coulomb counting method, the Open Circuit Voltage (OCV) method, and the Electrochemical Impedance Spectroscopy (EIS), estimate the battery's SoC based on the battery's dynamic measurement (e.g., voltage, current, and impedance) [129]. Alas, their accuracy is not too much important because of the measurement noises and the inexact initial SoC. In addition, the model-based methods, mainly the Luenberger observer and the Kalman Filter, as well as the artificial intelligence methods, such as the Neural Network and the Fuzzy Logic, are widely used to accurately estimate and predict the battery's SoC [130]. Finally, hybrid methods aim at using the benefits of each method to improve the accuracy of the battery's SoC estimation [131]. In fact, they combine different methods, such as the Coulomb Counting with Kalman Filter methods and the Kalman Filter with Neural Network methods.

5.4.3.5 Thermal panel modeling and sizing

Thermal panels have been used for centuries for the purposes of heating buildings, water for swimming pools, showering, and so on. In this section, we are mainly focusing on heating water for showering and hygiene purposes. Thermal panels are used to convert solar energy into heat. In fact, atoms composing the receiver's material are excited when exposed to the sun. Then, by harvesting a part of this energy, they change their energy state to create thermal agitation. The excess of energy is then released by atoms in forms of thermal energy, manifesting in the form of heat [132]. The latter will be acquired by water or another coolant (e.g., oil and gas). Once heated, it will be able to gradually distribute its heat or store it for ulterior usage. More explicitly, the solar irradiation propagates through glazing (transparent cover) arriving at the absorber. Then, the solar energy attacks the surface of the absorber, which is a darkened receptor of solar radiation characterized by its high absorptivity. Thus, a large part of energy is absorbed and transferred to a heat transfer fluid flowing inside the conductive thermal conduits. The latter are generally made of copper or aluminum, but particularly of metal. Then, the heat transfer fluid is transported for storage or for direct usage. To minimize heat losses, it is necessary to isolate the bottom of the absorbent plate and the two sides of the collector.

Thermal panels, also named solar water heaters, are classified into two categories, namely, solar heating and the Individual Solar Water Heaters (ISWH). Regarding solar heating, it is one of the most used thermal panels. In fact, the panel is connected to several circuits to provide domestic hot water, water radiator, and parquet floor [133]. Theoretically speaking, this kind of technology is very attractive. However, the complexity of its installation as well as its disappointing results puts the brake of its development. As for the ISWH, on which we will focus in this section, it is generally the most suitable system for individual buildings. The ISWH is used to capture sun rays for the main aim of heating water, which will then be used in the building for either showering or heating. It is constituted of two main components, namely, the thermal panel and the water storage tank as well as some other important accessories [134]. Regarding the thermal panel, it is used to capture the sun rays and it is composed of several components, mainly transparent cover, absorbent plate, fluid tubes, coolant fluid, and collector housing and insulation. The transparent layers, which are usually made of glass, are used to transmit the solar radiation to the absorbent plate; however, they reduce radiative and convective heat loss of this surface. This is attributed to the fact that the glass is transparent to short wave solar radiation and almost opaque to long-wave thermal radiation. This phenomenon is known as the greenhouse effect. The black absorbent plate, to which the fluid tubes are fixed, is used to convert solar radiation into heat. It is mainly characterized by its high absorption coefficient. These fluid tubes are used to transport the coolant fluid from the inlet of the collector to its outlet, and so it can be heated. The coolant fluid is a fluid that can capture the generated heat and transport it between two points. A collector housing is also needed to fix and protect the panel's components against the external conditions, such as dust and humidity. Finally, insulation is extremely required to minimize heat losses. As for the water storage tank, it is mainly used to store the produced hot water for eventual usage. Finally, some other accessories, such as pumps, circulators, regulators, and heat exchangers, are highly needed to link between the thermal panel and the water storage tank.

In fact, three ISWH technologies exist, namely, water heaters with electric pumps, water heaters with thermosyphon, and the Monoblock water heaters [134]. Regarding the first technology, which uses an electric pump, its cost as well as the cost of its maintenance is very high; however, it is the most used in some European countries, such as in France. It is also worth noting that its storage tank can be placed whenever (e.g., roof and garage). As for the water heaters with thermosyphon, it does not use a pump, and thus, it is crucial to place the storage tank above the thermal panel. This system is considered to be one of the best systems in terms of quality/price [135]. Finally, the Monoblock water heaters are similar to the water heaters with thermosyphon; however, the overall system is positioned in the same place. Therefore, the deployment of the system becomes easier and the price is more affordable. It is also worth mentioning that in this kind of technologies, the water temperature has to be maintained at 65 °C because of the following reasons: (i) limiting the evolution of many microbes and bacteria (e.g., legionellosis), which evolve at a temperature between 40°C and 60°C, (ii) limiting the water consumption, i.e., if the water is too hot, it must be eventually mixed with a large amount of cold water, which increases the water consumption, (iii) limiting the risk of burns that too hot water could cause when leaving the tap, and (iv) limiting the deformation of the balloon and pipes, which ends up damaging the installation.

The solar water heaters can normally cover 100% of hot water needs in summer, depending on the regions. However, during winter and when the sun is barely present, the panels cannot produce enough. Therefore, it is very important to connect other components to the water tank, such as a backup water heater, to meet the hot water needs during these critical periods. Furthermore, before installing a water heater, it would be better to make a feasibility study to see if it is profitable or not. In other words, the roof has to be studied to analyze the panels' inclination and orientation. The panels' inclination can be computed using its average value during the least sunny and most sunny period. In fact, the panels' inclination should be between 32° and 35° for installation-oriented south. It is also very crucial to check the level of the sun in the area where the panel is to be installed in such a way that panels are exposed to the sun for as long as possible. It is worth noting that the system should not be undersized so that it can cover the hot water needs, nor oversized so that space would not be lost unnecessarily. Moreover, focusing on the size of the water tank, it is calculated according to the hot water needs, which is on average 40–70 L/person. For example, a family of 4–5 persons will at least need a balloon of 300 L. But if the family opts for a larger model, the tank will only take more space. In addition, stagnant water will create bacteria and electricity will be unnecessarily consumed. As for the area of the solar panels, it has to be computed to easily heat the stored water in the tank. This is dependent on the region and the panels' orientation. As a matter of fact, if the panels are oriented south, the area is limited to 3–4 m^2 for a 300 L tank, but not less than 5–6 m^2 for the panels that are oriented north.

The equations managing the operation of the thermal panels are explained below [136]. Eqs. (5.17)–(5.22) explicit the thermal balance of the absorbent wall, the absorbed flow, the flux stored in the panel, the useful flow (when the coolant fluid does not undergo any change), the heat losses, and the yield of the collector (η), respectively [137–139]:

$$\rho_{sa} = \rho_{p\rightarrow} + \rho_u + \rho_{st}, \tag{5.17}$$

$$\rho_{s\rightarrow p} = \tau_{cs}\alpha_{ps} G^*(i, \gamma) S, \tag{5.18}$$

$$\rho_{st} = M_w C_w \frac{\partial T}{\partial t}, \tag{5.19}$$

$$\rho_u = q_{cf}\left(T_{f,out} - T_{f,in}\right), \tag{5.20}$$

$$\rho_{p\rightarrow} = h_p\left(T_{pm} - T_a\right) S, \tag{5.21}$$

$$\eta = \frac{\rho_u}{G^*{}_{(i,\gamma)} S}. \tag{5.22}$$

where ρ_{sa} is the absorbed solar flow, $\rho_{p\rightarrow}$ represents the lost flow through the absorbent layer, ρ_u is the useful flow transmitted to the heat transfer fluid, ρ_{st} represents the flux stored in the panel, $\rho_{s\rightarrow p}$ is the absorbed solar flow from the exposed surface, τ_{cs} stands for the transmission coefficient of the transparent cover, α_{ps} is the absorption coefficient of the absorbent layer, $G^*(i, \gamma)$ is the flux density incident on the collector, S represents the area of the absorbent layer, M_w stands for the mass of water in the panel, which is defined as $\sum m_i c_i = M_w C_w$, with i being the different elements composing the panel and C_w being the concentration of water in the panel, T represents the mean temperature of the panel, t is the time, q_{cf} stands for the heat flow rate of the coolant, $T_{f,\ out}$ is the coolant temperature at the outlet of the absorbent, $T_{f,\ in}$ is the coolant temperature at the inlet of the absorbent, h_p represents the overall coefficient of panel losses, T_{pm} being the average absorbent layer temperature, and finally T_a represents the outdoor air temperature.

5.5 Case study

In this section, a family two-story residential building has been designed and modeled. Its simulation has been conducted and based on its results, the building has been optimized. The simulation results of the optimized version of the building have been afterward reported and showed that the materials' change, the addition of the insulation, and the enhancement of the glazing have led to a decrease in the demand of heating and cooling by 76.66% and 72.433%, respectively. Then, its consumption has been computed to use it for figuring out the size of the hybrid system's components. Once the hybrid system is sized, it has been modeled and simulated. The simulation results have been reported to show the effectiveness of the sizing methodology and the behavior and performance of the hybrid system under different conditions (e.g., solar irradiation, ambient temperature, and building's consumption).

5.5.1 Design, modeling, simulation results, and optimization

In this part, we have modeled a two-story residential building housing a family of 4–5 people. The building is basically constructed of aerated bricks, usually constructed from clay, sand, and other additives. We chose this material for an initial study since it becomes preponderant in new constructions due to its durability, low thermal conductivity, and lower density, despite its resistance to external effects. The simulation was launched using the Design-Builder software to assess the building's performances. In the second part, we have incorporated some modifications concerning the construction materials, mainly replacement of aerated blocks with thermoblocks, to reduce heating and cooling needs of the building and ensure the occupants' comfort. These blocks are sandwich blocks with insulation inside; in other words, they are self-insulated. These materials are considered to be innovative since they can keep the heat during the winter and cool in summer. It is worth mentioning that HVAC systems are installed in the living room, master bedroom, and the two other bedrooms. Mechanical ventilation is assumed to be installed into the WCs to get rid of odors. The rest of the building is mainly aerated by natural ventilation. In addition, the glazing at first was simple with no shadowing means. However, in the optimization part, in addition to the change of construction materials, the glazing has changed from simple to double clear 6-13 mm air with internal curtains with high reflectivity. Furthermore, the building is located in Rabat, Morocco, and it is oriented south to allow a good performance of energy equipment, mainly photovoltaic and thermal panels, which will be installed on its roof.

5.5.1.1 Hypothesis and parameters

To design and simulate our building, we have taken into consideration the following hypothesis. The geographical and meteorological data of Rabat-Sale, Morocco, where the building will be built, are as follows: (i) latitude: 34°02′60"N; (ii) longitude: 6°46′12"W; (iii) ASHRAE climate zone: 3A, elevation above sea level: 79 m; (iv) normal exposure to the wind; (v) time zone: GMT; (vii) winter period: October–March; and (viii) summer period: April–September. Moreover, lighting levels in this building should be adapted to the type of activities, which will take place in each zone. Recommended values according to the type oflocal and activities are disclosed by several norms, such as EN 124641 for working spaces, others for residential spaces, and so on. Optimal intensity values are presented in Table 5.1. Furthermore, the internal gains are taken as follows: (i) Metabolism: metabolic contribution according to the activity level on each space. This may depend on the weight and height of a person as well (men = 1, women = 0.85, and Kids = 0.75). (ii) The computer gains are equal to 7 W/m^2. In addition, to maintain the occupants' thermal comfort, the following hypothesis has been considered: (i) Heating setpoint: 18°C. (ii) Minimum limit value for heating: 12°C. (iii) Cooling setpoint: 25°C. (iv) Minimum limit value for cooling: 28°C. (v) No shading devices are used

Table 5.1 Optimal intensity values.

Hall, corridor, and toilettes	50–100 Lux
Sanitary	200–300 Lux
Kitchen	200–500 Lux
Office	500 Lux
Living space	100–300 Lux
Dining room	100–200 Lux
Bedrooms	100–200 Lux
Store room	50–100 Lux

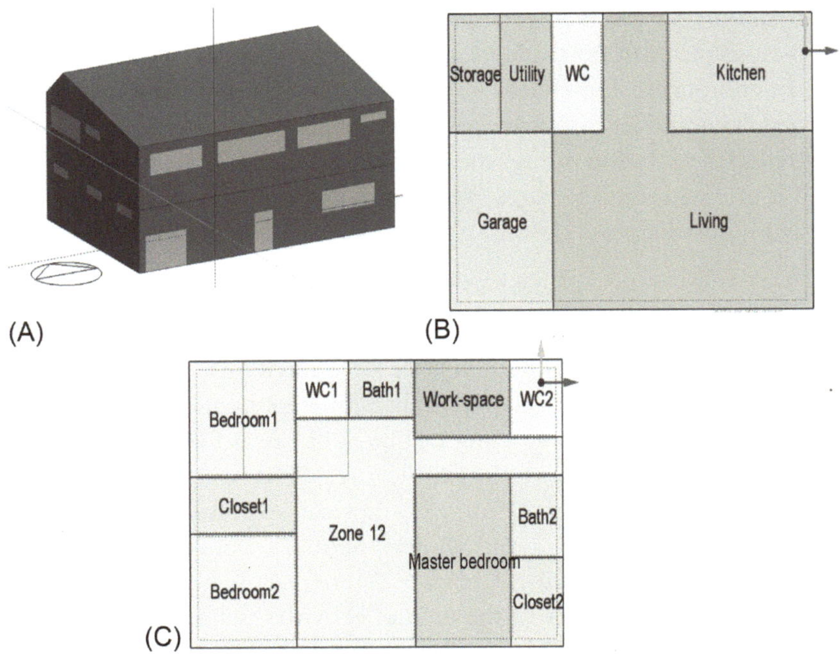

Fig. 5.5 (A) Overall view of the building East-South, (B) Ground floor plan and zone distribution, and (C) First floor plan and zone distribution.

for the first case of study. We have also taken a minimum fresh air of about 3 L/s-person (the value should be set between 2 and 4).

The building that we have chosen to study is constituted of two floors including the ground with a total surface of 150 m^2 and a roof tilted by 30° and oriented south. Fig. 5.5 shows more details about the design and zone distribution.

As for the construction materials, Table 5.2 shows detailed information about the different layers, thermal properties, thicknesses, and calculated transmission coefficients.

5.5.1.2 Simulation results

Based on real weather data of the city of Rabat, and thanks to Design Builder, we were able to follow the course of the sun around the building during the sunniest and least sunny periods. Fig. 5.6 presents this path on 15th August as well as 15th November.

On another note, Design Builder radiance simulation grants detailed multizone physics of illuminance level calculus on the working plan of

Table 5.2 Construction materials and thermal properties, with Th: thickness (m), Cd: conductivity (W/mK), D: density (kg/m^3), SH: specific heat (J/kg.K), R: resistance (m^2.K/W), and TC: transmissivity coefficient (W/m^2.K).

Materials		Th	Cd	D	SH	R	TC
External walls	Painting	0.0001	0.314	1.53	100	0.0031	1.8
	Mortar	0.02	1.8	2500	1000	0.011	
	Aerated brick	0.07	0.3	1000	840	0.15	
	Air gap	0.1	0.3	1000	1000	0.18	
	Aerated brick	0.07	0.3	1000	840	0.15	
	Mortar	0.02	1.8	2500	1000	0.011	
	Painting	0.0001	0.314	1.53	100	0.0031	
Roof	Mortar blocks	0.03	1.8	2500	1000	0.011	2.19
	Pure Bitumen	0.01	0.17	1050	1000	0.15	
	Reinforced concrete	0.05	2	2450	1000	0.025	
	Slab	0.2	1176	1456	1000	0.17	
	Gypsum plastering	0.02	0.4	1000	1000	0.15	
Internal partitions	Painting	0.0001	0.314	1.53	100	0.0031	1.8
	Mortar	0.02	1.8	2500	1000	0.011	
	Aerated brick	0.07	0.3	1000	840	0.15	
	Mortar	0.02	1.8	2500	1000	0.011	
	Painting	0.0001	0.314	1.53	100	0.0031	
Intermediate floor	Mosaic	0.01	0.3	1000	1000	0.15	1.4
	Mortar	0.05	1.8	2500	1000	0.011	
	Slab	0.2	1176	1456	1000	0.17	
	Air gap	0.1	0.3	1000	1000	0.18	
	Gypsum plastering	0.02	0.4	1000	1000	0.15	
Ground	Stone-granite	0.3	3.49	2880	840	0.15	2.73
	Reinforced concrete	0.05	2	2450	1000	0.025	
	Mortar	0.02	1.8	2500	1000	0.011	
	Mosaic	0.01	0.3	1000	1000	0.15	

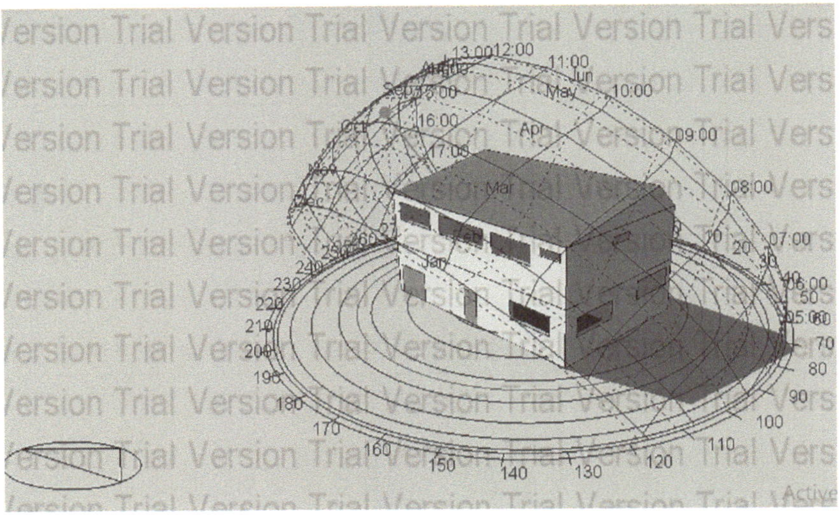

(A)

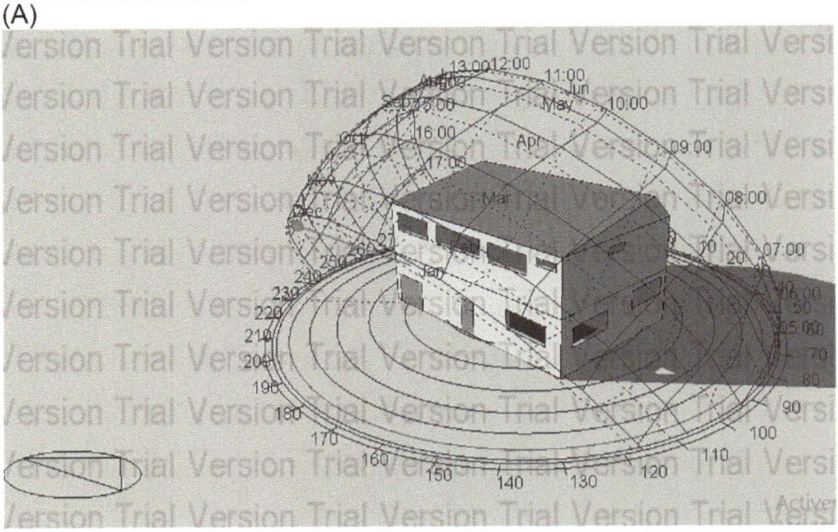

(B)

Fig. 5.6 Course of the sun during: (A) 15th August at 3 pm and (B) 15th November at 3 pm.

the concerned building. These calculations allow light transmission through windows (external/internal), the shading effects, and reflection of local sun shading components [140]. In our case, only external windows are taken into consideration. Fig. 5.7 shows the intensity distribution on the ground and first floors, respectively. In both cases, we notice that higher illuminance levels are near the glazings, especially the big windows. Illuminance values range from 200 to 1000 Lux and the daylight factors from 2 to 10.

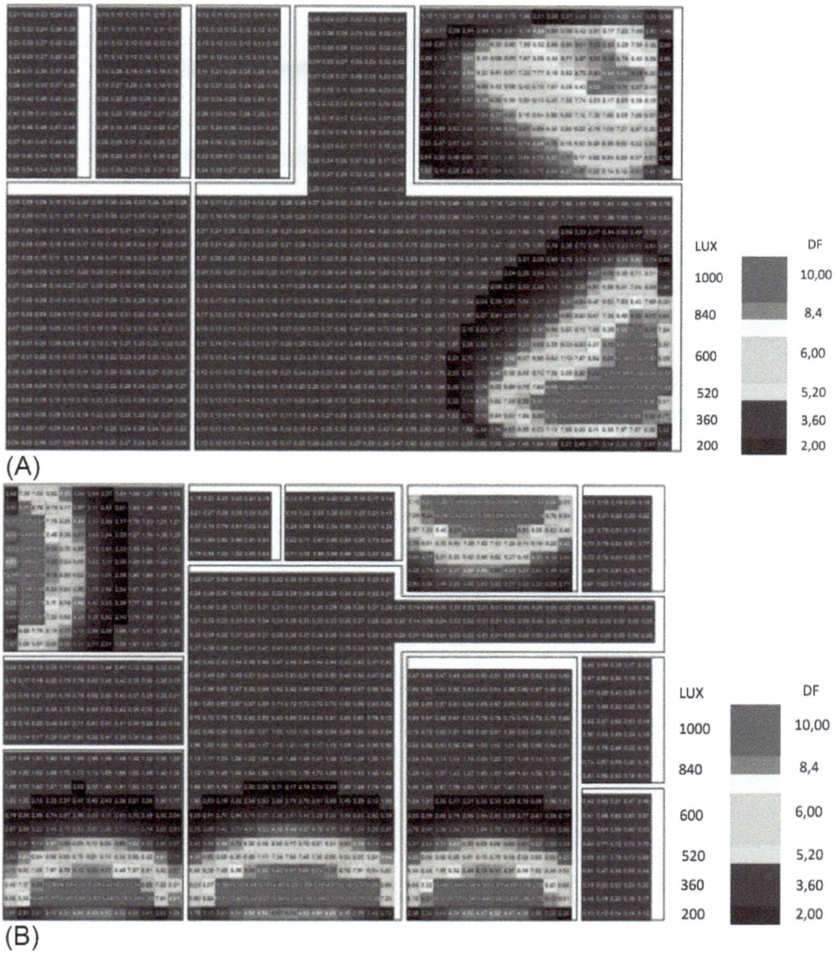

Fig. 5.7 Distribution of daylight intensity and factors: (A) ground floor's level and (B) first floor's level.

Heating design calculations are made to fix the heating equipment's capacity (generally an HVAC) according to weather conditions and winter conception (the coldest winter day) related to the location [141]. These calculations are normally done following steady-state models as the ones proposed by ASHRAE [28]. Figs. 5.8–5.10 show the comfort conditions in the building, fabric and ventilation heat losses and gains, and internal and solar gains, respectively. We notice that, from Fig. 5.8, internal air temperature and relative humidity values are 12.87°C and 61.23%, respectively. When analyzing Fig. 5.9, we notice that most of the heat losses come from glazes (1.75 kW), external walls (3.39 kW), and the roof (1.06 kW). Therefore,

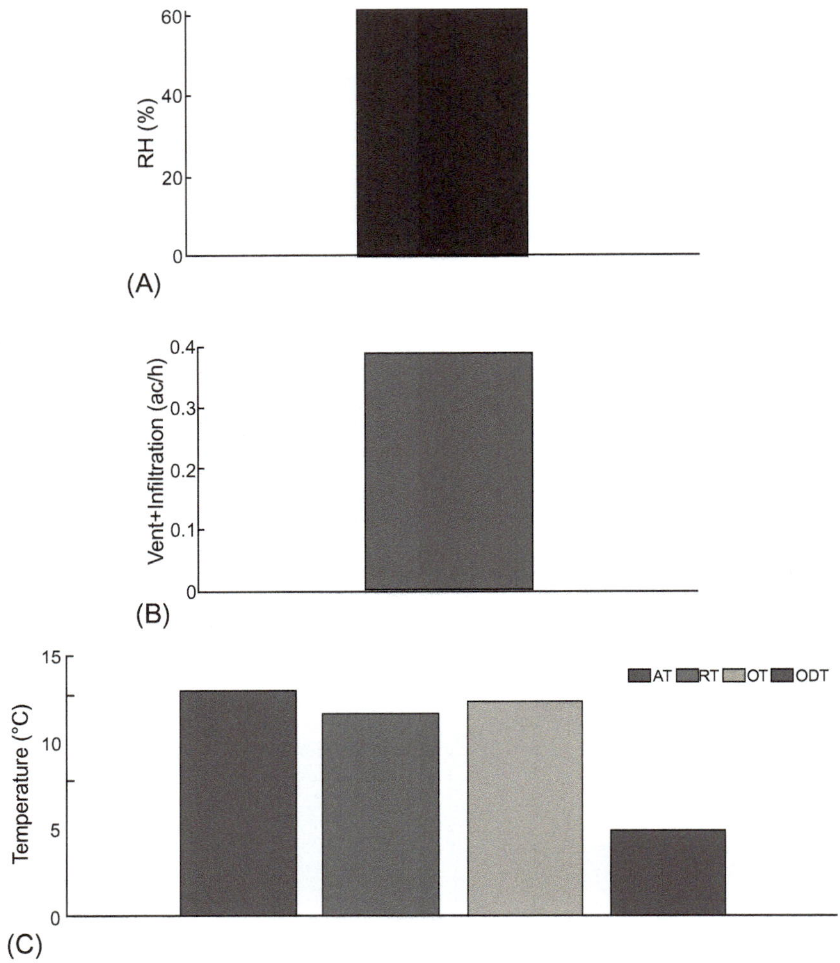

Fig. 5.8 (A) Relative humidity, (B) ventilation and infiltration rates, and (C) temperature distribution issued from comfort calculation used in heating design, with AT: air temperature, RT: radiant temperature, OT: operative temperature, and ODT: outside dry bulb temperature.

Fig. 5.9 Fabric and ventilation heat losses and gains issued from heating design calculations, with G: glazing, W: walls, C: ceiling, F: floor, GF: ground floor, P: partitions, R: roof, and EV: external ventilation.

Fig. 5.10 Building's internal and solar gains issued from heating design calculations.

actions should be taken to enhance these structures' performances and reduce heat perditions. According to Fig. 5.10, internal and solar gains inside the building are fixed at 4.55 kW. When talking about internal gains, we are talking about the contributions of occupants, the lighting, the heat diffused by electronic equipment, heat and air conditioning systems, and so on.

Cooling design calculations allow determining the capacity of the mechanical cooling device according to meteorological data and summer conception days (the hottest day during summer) of the building's site location. If some areas in the building are not cooled mechanically, temperature values are then calculated according to natural or mechanical ventilation [142]. Calculations are adapted to periodical data in steady state as the ones mentioned on the ASHRAE code [28]. Figs. 5.11–5.13 depict comfort conditions according to cooling design, heat losses and gains through fabrics and ventilation, and internal and solar gains, respectively.

Fig. 5.11 depicts periodic calculated exterior temperatures and interior comfort temperature according to meteorological data of summer conception days. Accordingly, we notice that the relative humidity inside the building ranges from 50.57% to 66.62%. As for air and radiant temperatures, they vary between a minimum value of 26°C and a maximum value of 35°C, controlled by the variation of operative temperature. Fig. 5.12 shows the distribution of gains and losses through fabrics and ventilations during the summer period. We notice that losses persist on wall partitions, intermediate floor, and ground floor during the day. However, at the level of external infiltration, glazing, walls, and roofs, we notice positive gains in terms of heating. Among these gains, the most relevant one concerns the roofs; therefore, it should be insulated to avoid overheating during summer and reduce cooling needs during the day. Fig. 5.13 presents the heating balance in the building according to internal loads, such as occupants, lighting, equipment, and so on. It also presents two types of cooling loads, which are sensible and

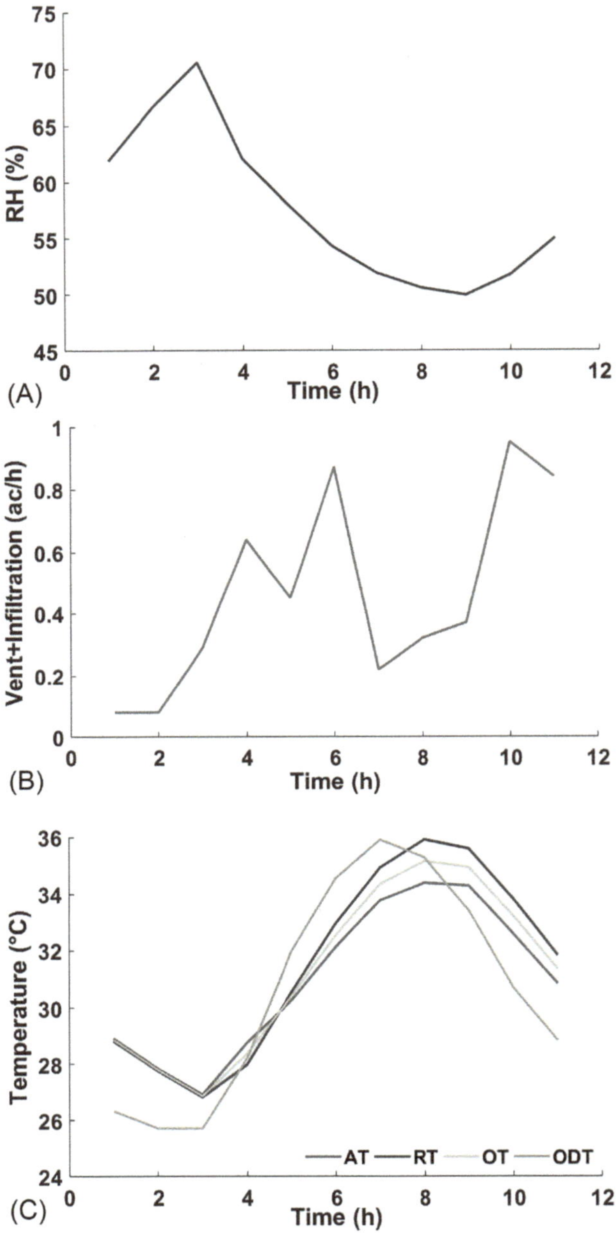

Fig. 5.11 (A) Relative humidity, (B) ventilation and infiltration rates, and (C) temperature distribution issued from comfort calculation used in cooling design, with AT: air temperature, RT: radiant temperature, OT: operative temperature, and ODT: outside dry bulb temperature.

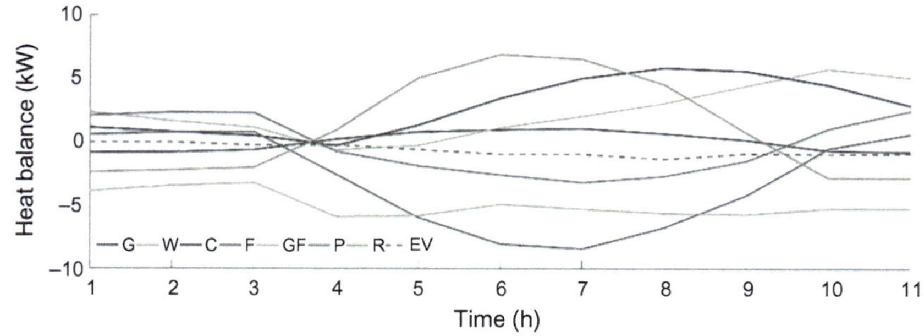

Fig. 5.12 Fabric and ventilation heat losses and gains according to cooling design for the building, with G: glazing, W: walls, C: ceiling, F: floor, GF: ground floor, P: partitions, R: roof, and EV: external ventilation.

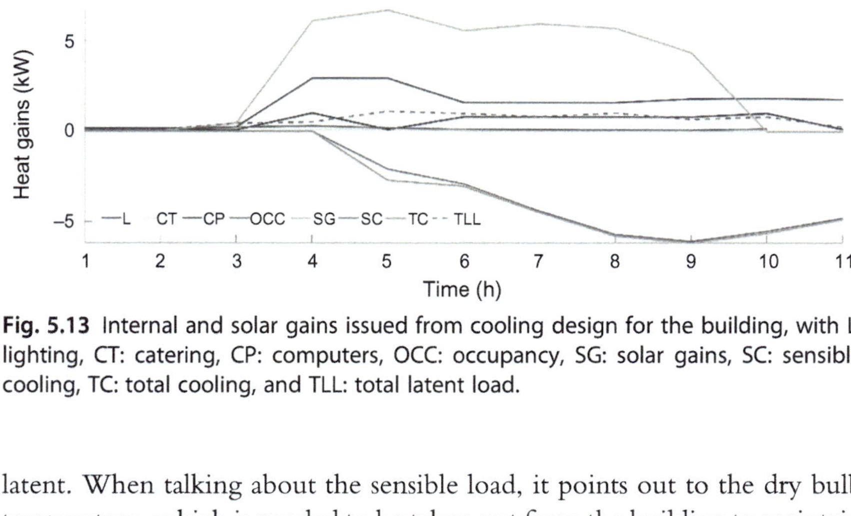

Fig. 5.13 Internal and solar gains issued from cooling design for the building, with L: lighting, CT: catering, CP: computers, OCC: occupancy, SG: solar gains, SC: sensible cooling, TC: total cooling, and TLL: total latent load.

latent. When talking about the sensible load, it points out to the dry bulb temperature, which is needed to be taken out from the building to maintain comfort without affecting the humidity. As for the latent load, it refers to wet bulb temperature that can alter humidity of air [143].

5.5.1.3 Optimization and results

To reduce the energy demand of the building, it was important to apply some changes to its envelope, such as adding insulation materials. When talking about insulation, it is either placed on the external side, internal or intermediate. In this case, we chose to avoid placing it on the external side since it is not very common in Morocco. Table 5.3 depicts the changes made at the level of each layer from external walls, roof, and intermediate

Table 5.3 The added materials to the building with their thermal properties, with Th: thickness (m), Cd: conductivity (W/mK), D: density (kg/m^3), SH: specific heat (J/kg.K), R: resistance (m^2.K/W), and TC: transmissivity coefficient (W/m^2.K).

Materials		Th	Cd	D	SH	R	TC
External walls	Painting	0.0001	0.314	1.53	100	0.0031	0.621
	Mortar	0.02	1.8	2500	1000	0.011	
	1 in. thermal block	0.0254	1	1000	1000	0.6164	
	Air gap	0.1	0.3	1000	1000	0.18	
	1 in. thermal block	0.0254	1	1000	1000	0.6164	
	Mortar	0.02	1.8	2500	1000	0.011	
	Painting	0.0001	0.314	1.53	100	0.0031	
Roof	Mortar blocks	0.03	1.8	2500	1000	0.011	0.560
	Pure Bitumen	0.01	0.17	1050	1000	0.15	
	Mineral fiber/wool-wool	0.05	0.03	140	840	0.15	
	Reinforced concrete	0.05	2	2450	1000	0.025	
	Slab	0.2	1176	1456	1000	0.17	
	Gypsum plastering	0.02	0.4	1000	1000	0.15	
Intermediate floor	Mosaic	0.01	0.3	1000	1000	0.15	0.493
	Mortar	0.05	1.8	2500	1000	0.011	
	Mineral fiber/Wool-Wool	0.05	0.03	140	840	0.15	
	Slab	0.2	1176	1456	1000	0.17	
	Air gap	0.1	0.3	1000	1000	0.18	
	Gypsum plastering	0.02	0.4	1000	1000	0.15	

floor to reduce energy demand. According to these changes, we can observe that the transmission coefficient reduced ensuring good insulation. As for the glazing, we chose to work with double clear 6 mm/13 mm air glazing. Shading was chosen as blinds with high reflectivity slats positioned inside. These blinds are placed in spaces with big window surfaces (e.g., living and kitchen).

According to Fig. 5.14, we can notice that external air and radiant temperature values are 13.5°C and 12.61°C, respectively, leading to an operative temperature of 13.05°C. As for humidity, it increased to attend 58.48%. This value exceeds the one fixed by the regulations (should not exceed 50%); however, this can be easily overcome by using air barriers to control moisture.

As for Fig. 5.15, which depicts heat losses and gains through fabrics, we notice that heat losses through the glazing, walls, and intermediate floor decreased to −1.05, −2.13, and − 1.87, respectively. Therefore, the choice of new construction materials was successful. Concerning internal gains, presented in Fig. 5.16, we notice that they dropped to reach 3.28 kW.

Figs. 5.17–5.19 depict comfort conditions according to cooling design, heat losses and gains through fabrics and ventilation, and internal and solar gains, respectively. Fig. 5.17 depicts periodically calculated exterior temperatures and interior comfort temperature according to meteorological data of summer conception days. Accordingly, we notice that the relative humidity inside the building ranges from 51.77% to 75.05%. As for air and radiant temperatures, they vary between a minimum value of 25°C and a maximum value of 35°C, controlling by this the variation of operative temperature. Last but not least, the mean comfort temperature according to the cooling design is 30°C.

Figs. 5.18 and 5.19 show the distribution of gains and losses through fabrics and ventilations during the summer period as well as the heating balance in the building according to internal loads, such as occupants, lighting, equipment, and so on. Fig. 5.19 also presents two types of cooling loads, which are sensible and latent. When talking about the sensible load, it points out to the dry bulb temperature, which is needed to be taken out from the building to maintain comfort without affecting the humidity. As for the latent load, it refers to wet bulb temperature that can alter humidity of air [143].

From the results we have obtained, it seems that the energy efficiency of a building can only be achieved by ensuring good performance of the thermal envelope. This can be achieved by making a good choice of the construction

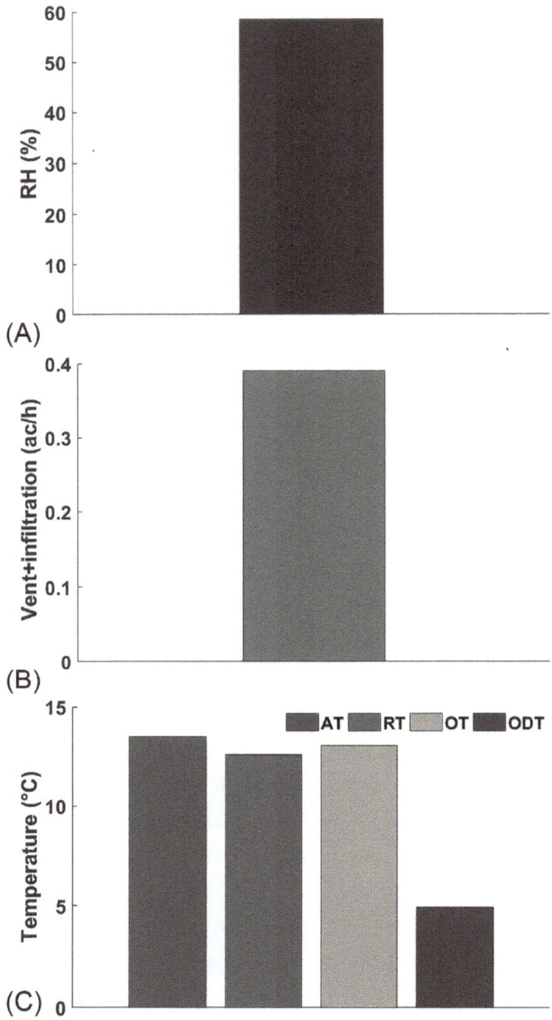

Fig. 5.14 (A) Relative humidity, (B) ventilation and infiltration rates, and (c) temperature distribution issued from comfort calculation used in heating design, with AT: air temperature, RT: radiant temperature, OT: operative temperature, and ODT: outside dry bulb temperature.

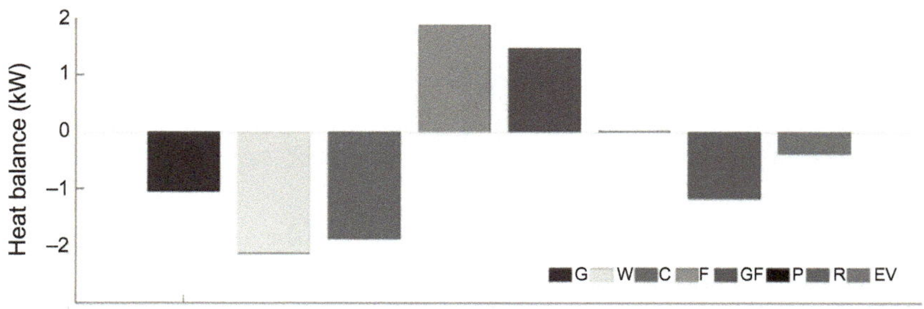

Fig. 5.15 Fabric and ventilation heat losses and gains issued from heating design calculations, with G: glazing, W: walls, C: ceiling, F: floor, GF: ground floor, P: partitions, R: roof, and EV: external ventilation.

Fig. 5.16 Building's internal and solar gains issued from heating design calculations.

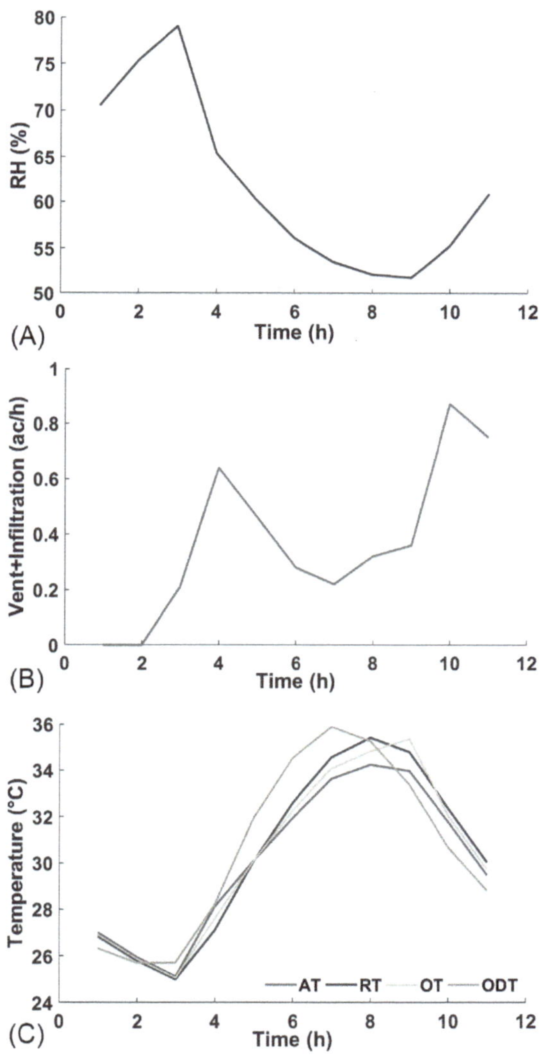

Fig. 5.17 (A) Relative humidity, (B) ventilation and infiltration rates, and (C) temperature distribution issued from comfort calculation used in cooling design, with AT: air temperature, RT: radiant temperature, OT: operative temperature, and ODT: outside dry bulb temperature.

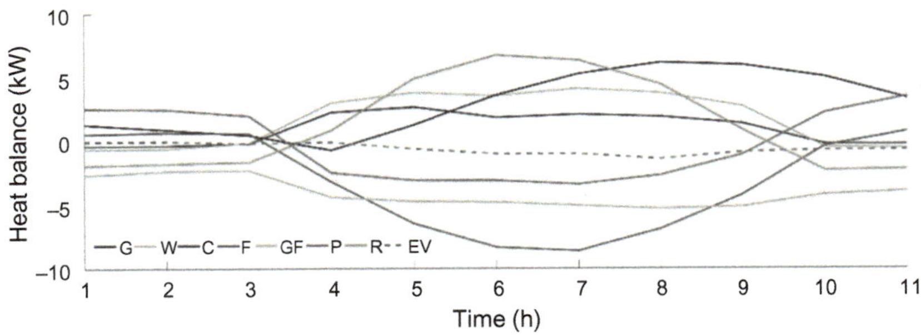

Fig. 5.18 Fabric and ventilation heat losses and gains according to cooling design for the building, with G: glazing, W: walls, C: ceiling, F: floor, GF: ground floor, P: partitions, R: roof, and EV: external ventilation.

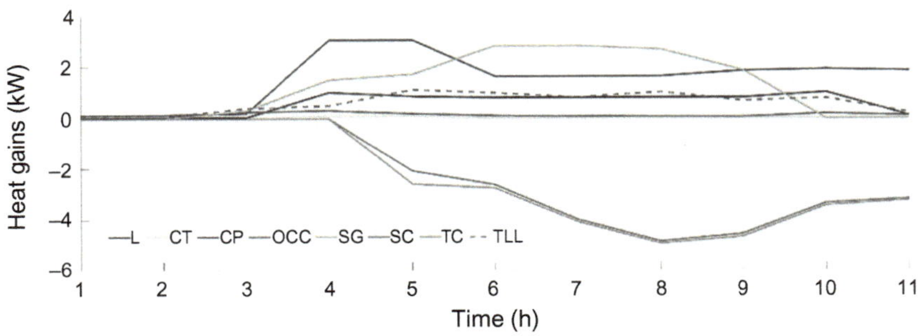

Fig. 5.19 Internal and solar gains issued from cooling design for the building, with L: lighting, CT: catering, CP: computers, OCC: occupancy, SG: solar gains, SC: sensible cooling, TC: total cooling, and TLL: total latent load.

Table 5.4 Summary of heating and cooling needs in the building.

Building's state	Heating needs (kW)	Cooling needs (kW)
Initial	5.68	9.12
Optimized	4.11	8.38

site, materials, insulation, and type of glazing as well as prioritizing passive solutions. According to Table 5.4, we notice that a good choice of construction materials as well as the glazing lowered energy demands in terms of heating and cooling. It is worth noting that a better design of the envelope could decrease effectively the energy consumption of a building. Also, incorporating RES and storage devices could help in filling the gap toward net-zero-energy buildings. Thus, the next section focuses on RES and storage systems integration to fulfill this purpose.

5.5.2 Hybrid system's description, sizing, modeling, and simulation results

In this section, the sizing and modeling of a hybrid system have been performed. Then, its simulation has been conducted under MATLAB. The simulation results have been afterward reported to show the effectiveness of the sizing methodology and the behavior and performance of the hybrid system under different conditions (e.g., solar irradiation, ambient temperature, and consumption). The hybrid system, which is depicted in Fig. 5.20, is composed of PV panels for electricity generation from solar irradiation, batteries to store the PV panels' overproduction for using it during the nights or on cloudy days, hybrid inverter to convert the PV produced DC into an AC, which could be used by the building's appliances and to regulate the batteries' charge/discharge, thermal panels to produce hot water, hot water tank for storing the produced hot water, and finally the designed and optimized building.

This hybrid system has been sized to have a stand-alone system (i.e., zero consumption from the grid). Its sizing is highly dependent on several variables, mainly the solar irradiation and the building's consumption. Concerning the solar irradiation, we have used the geographical coordinates of Rabat (Latitude 34°02′60" N, Longitude 6°46′12"W, Altitude 79 m) to determine the solar irradiation data on the roof surface of our building, which is tilted by 33°, using the PVGIS website. We found that the lowest daily solar irradiation throughout the year is approximately equal to 4.064 kWh/m^2.day.

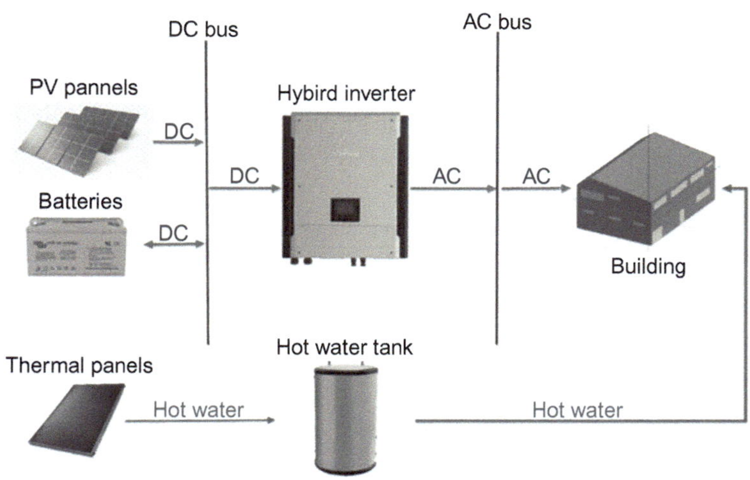

Fig. 5.20 Hybrid system's architecture.

As for the heating and cooling systems' size, it has been provided by the Design Builder software (see Table 5.4). However, the consumption of the other building's appliances (e.g., refrigerator, cooking plate, oven, cloth washer, dishwasher, computers, phone chargers, TV, lamps, and fans) has been estimated using Eq. (5.1). The total daily electricity consumption of the building has been found to be approximately equal to 43.639 kWh/day. Then, the peak power of the PV panels has been computed using Eq. (5.2) with a performance ratio (PR) equal to 0.70. The required PV panels' peak power to be installed to supply the building with the needed electricity was found to be equal to 15.34 kW. In our case, we have chosen to install polycrystalline PV modules, which are manufactured by Amerisolar AS-6P30, with a peak power of 265 W each (see more PV characteristics in Table 5.5).

Thus, the required number of PV modules has been calculated using Eq. (5.3) and has been found to be equal to 57.88 modules. As a result, 58 PV modules are needed to operate the building's appliances throughout the year, with a total peak power of 15.37 kW and a total surface of about 94.36 m^2. These PV panels will be installed on the roof of the building, which is tilted by 33° and oriented south, and has a total area of around 125 m^2. Regarding the hybrid inverters, which will be installed in the building's utility room, their power values have to be similar or around the power of the PV panels. Therefore, we have chosen to install four hybrid inverters manufactured by Must with a rated power of 3.6 kW each, a maximal PV input voltage of about 580 V_{DC}, a maximal PV input current of about 11 A_{DC}, an MPP voltage range of around 120–550 V_{DC}, and a nominal battery DC voltage of about 48 V_{DC}. Thus, the number of PV branches has been computed using Eq. (5.4) and found to be equal to 1 branch in each inverter. Also, the number of series modules in each branch, which has been calculated using Eq. (5.5), was found to be equal to 15 modules. As a result, 15 PV modules (1 branch with

Table 5.5 The PV module's characteristics.

Maximum power At STC (P_{mpp})	265 W
Maximum power point voltage (V_{mpp})	30.9 V
Maximum power current (I_{mpp})	8.58 A
Open circuit voltage (V_{oc})	38.3 V
Short circuit current (I_{sc})	8.98 A
Solar cell number	60 cells
Dimensions	1640*992*40 mm^3
Temperature coefficient of I_{sc}	0.05%/°C

15 series modules) will be connected to the first inverter, while another 15 PV modules with the same configuration will be linked to the second inverter. In addition, 14 PV modules (1 branch with 14 series modules) will be connected to the third inverter, while another 14 PV modules with the same configuration will be linked to the fourth inverter.

As for the PV panels' modeling, Eq. (5.12) has been implemented into MATLAB for the purpose of building their model. This model has been already validated in our previous work under different weather conditions (e.g., solar irradiation and ambient temperature) using experimental data [42]. As for the inverter, its modeling has been described with more details in our foregoing work [42]. Therefore, several simulations have been carried out and the results of both the least sunny month of the year, which is November, and the sunniest month of the year, which is August, have been reported to show the behavior of the PV panels as well as their electricity production during two different seasons. Fig. 5.21 depicts the weather conditions as well as the daily average production of the PV panels of the least sunny month of the year. Concerning the weather conditions, which are the daily average solar irradiation and the daily average ambient temperature, they have been provided by the PVGIS website using the geographical coordinates of Rabat and the tilt of the PV panels (33°). These weather conditions have been used as inputs to our PV model to simulate the PV panels' production. As for the daily average PV panels' production, we observe that the PV production starts to increase from 7 AM until noon, when it reaches 9.274 kW, and then decreases until 6 PM. We also notice that the PV production in this day is very affected by the solar irradiation (i.e., its maximum is very far from the PV peak power (15.37 kW)), which does not exceed 700 W/m^2. In addition, the energy produced by the PV panels is approximately equal to 60.061 kWh/day, which is higher than the building's consumption. The PV overproduction will be stored into the batteries to supply the building with the needed electricity at night and on cloudy days.

Furthermore, Fig. 5.22 depicts the weather conditions as well as the daily average production of the PV panels of the sunniest month of the year. Concerning the weather conditions, which are the daily average solar irradiation and the daily average ambient temperature, they have been provided by the PVGIS website using the geographical coordinates of Rabat and the tilt of the PV panels (33°). These weather conditions have been used as inputs to our PV model to simulate the PV panels' production. As for the daily average PV panels' production, we notice that the PV production starts to rise from 6 AM until noon, when it achieves 13.37 kW, and then drops until almost

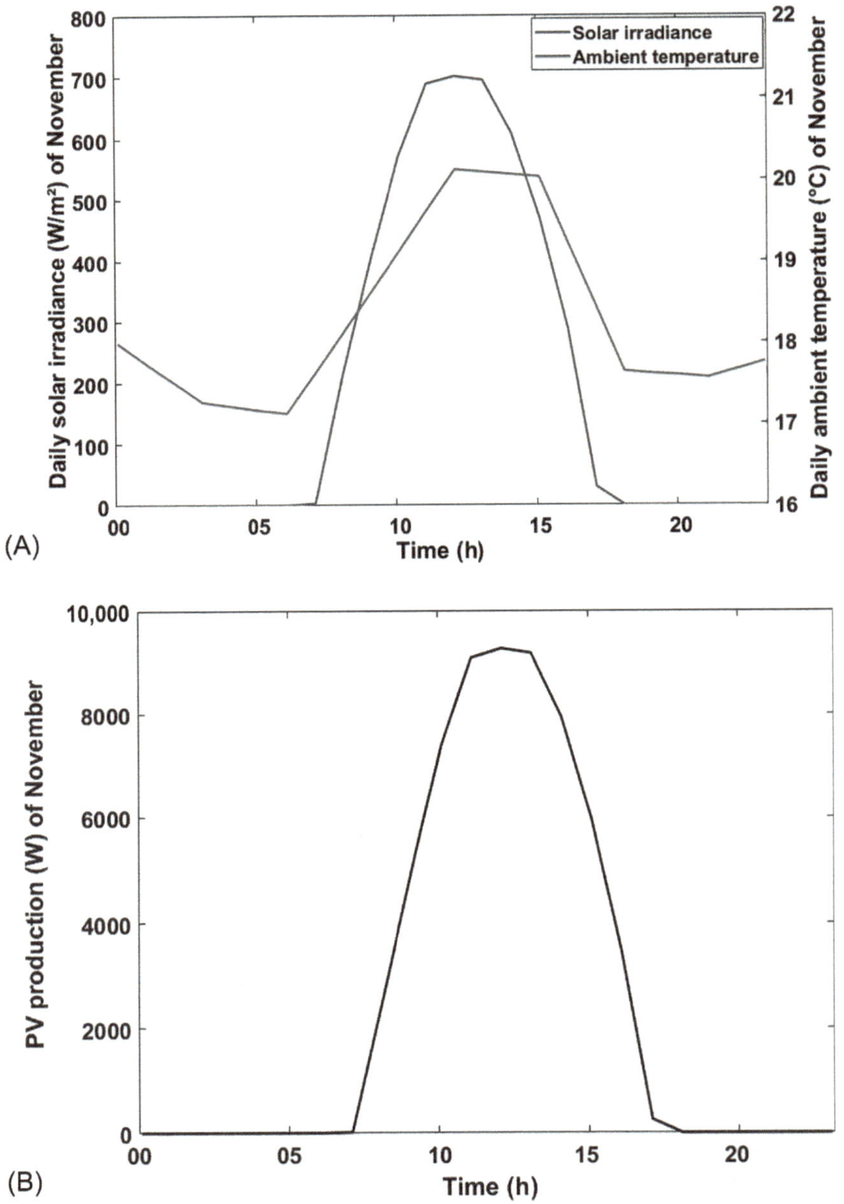

Fig. 5.21 (A) The daily average solar irradiance (W/m^2) and ambient temperature (°C) and (B) the daily average PV production (W) of the least sunny month of the year (November).

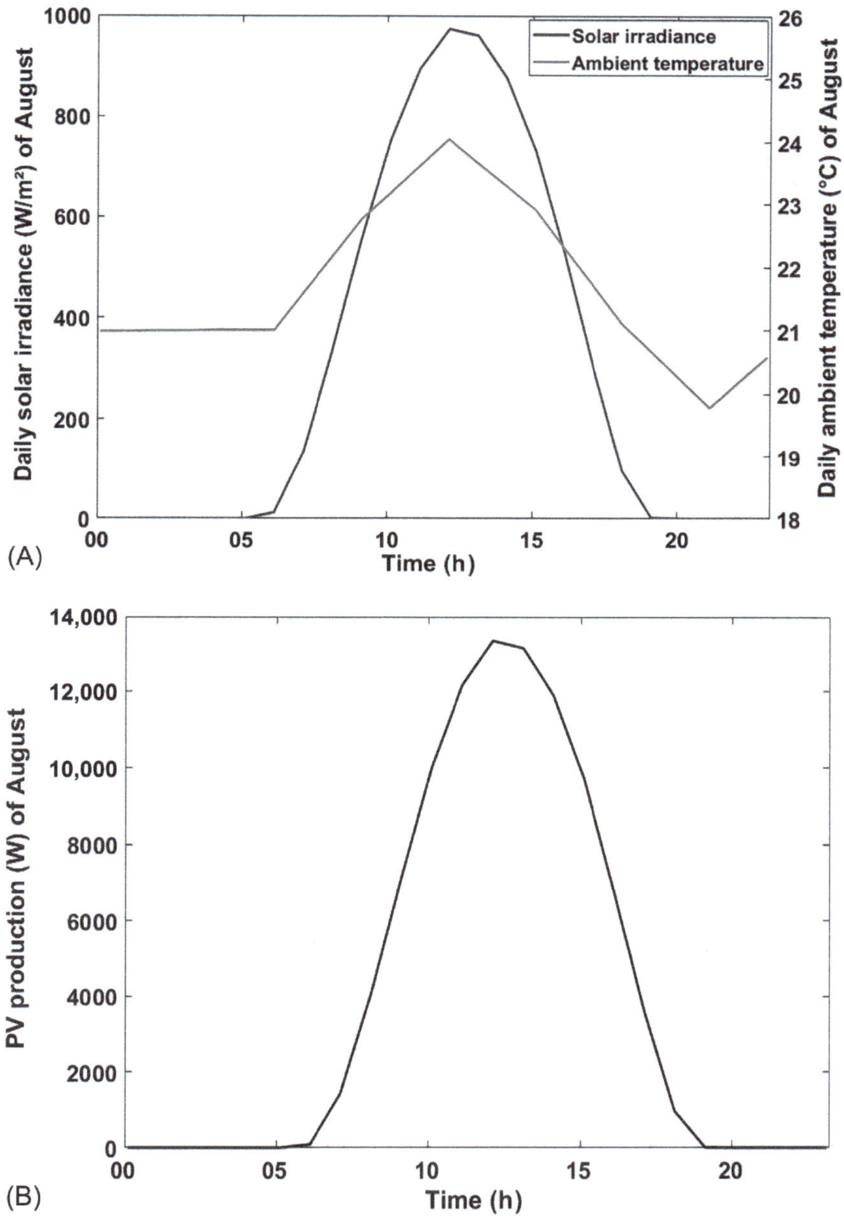

Fig. 5.22 (A) The daily average solar irradiance (W/m^2) and ambient temperature (°C) and (B) the daily average PV production (W) of the sunniest month of the year (August).

8 PM. We also observe that the maximum PV production in this day is close to the PV peak power (15.37 kW) since the solar irradiation is very close to 1000 W/m^2 (its maximum reaches 974 W/m^2). In addition, the energy produced by the PV panels is approximately equal to 94.414 kWh/day, which is very higher than the building's consumption. The PV overproduction will be stored in the batteries to supply the building with the needed electricity at night and on cloudy days. From these results, we conclude that the PV panels can ensure the operation of the building's consumption throughout the year, which show the effectiveness of the sizing method and also validate the developed model.

Concerning the batteries, we have chosen to work with GEL batteries that are manufactured by Victron energy, with a nominal voltage of 12 V and a nominal capacity of 220 Ah. Thus, their total capacity has been calculated using Eqs. (5.13) and (5.14), with 1 day of autonomy (in Rabat, the number of days of autonomy varies between 1 and 3 days). As a result, 6060.97 Ah, which is equivalent to 28 batteries, has to be considered to ensure the building's appliance operation during the night and on cloudy days while maintaining the occupants' comfort. Using Eqs. (5.15) and (5.16), we conclude that 28 batteries (7 branches with 4 series batteries in each branch) have to be set up. As for their SoC, which is an important indicator of the remaining capacity in the battery, it has been estimated using the Coulomb Counting method with an initial SoC of approximately 70%. Thus, using the PV as well as the building's appliance current, the batteries' current has been computed during 24 h, and then the batteries' SoC has been estimated. Fig. 5.23 depicts the batteries' SoC in the least sunny and the sunniest months of the year. We can clearly observe that the building's appliances are totally supplied by the batteries (i.e., their SoC does not exceed 40%) during the nights of both the least sunny and the sunniest days. We can also notice that the batteries are fully charged during both the least sunny and the sunniest days. From these results, we can conclude the accuracy of the sizing method and its efficiency in sizing stand-alone PV systems.

In regard to the thermal panels, we decided to use a monobloc thermal panel with water as a heat transfer fluid to ensure the coverage of hot water needs in the building. As mentioned in the above theoretical study, a panel of 4 m^2 is sufficient to cover the needs of a family of 4–5 persons. Therefore, a simulation model under MATLAB has been used to study and analyze the performance of the system as well as predict the hot water outlet temperature in December (cold period). First of all, we made an assumption of water needs during the day in both summer and winter seasons (see Fig. 5.24). This assumption was made according to Moroccan citizens' testimonies [144].

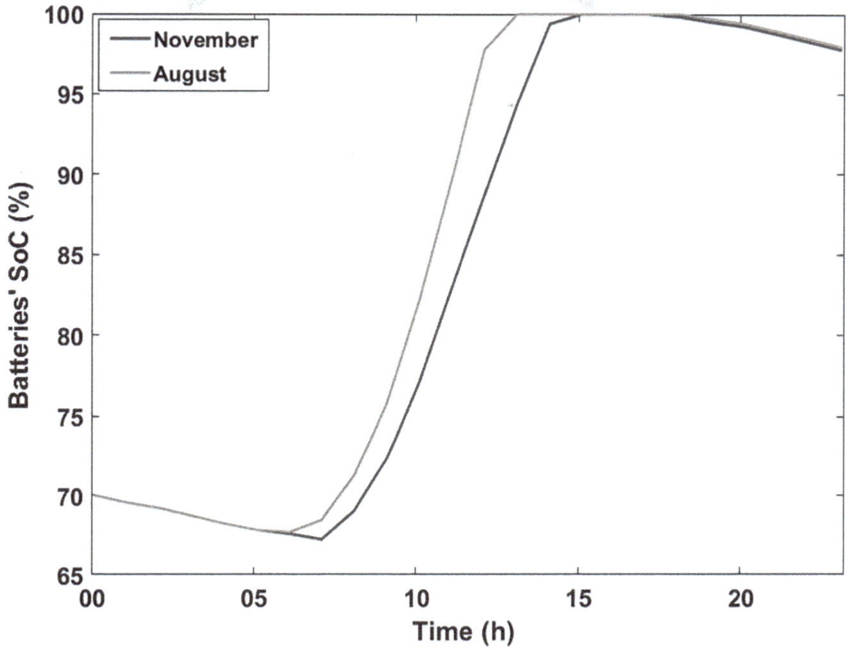

Fig. 5.23 The batteries' SoC (%) in the least sunny and the sunniest months (November and August, respectively) of the year.

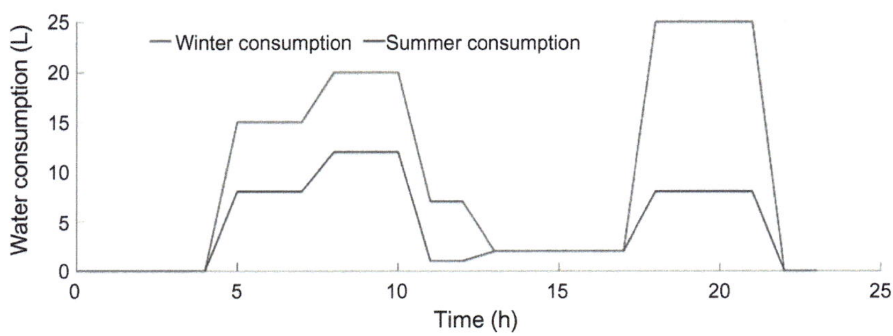

Fig. 5.24 Daily hot water consumption (L).

The choice of inclination is important for our study to optimize the irradiation of direct and diffuse radiation received by the solar collectors. For this, we chose an angle of inclination equal to the average between the least sunny month (November) and the sunniest month (August). In our case, we find it equal to 32.5°, which is close to the latitude of the city (Rabat). Simulation results are presented below. Fig. 5.25 depicts the water outlet temperature and efficiency. We notice that the efficiency of the panel remains

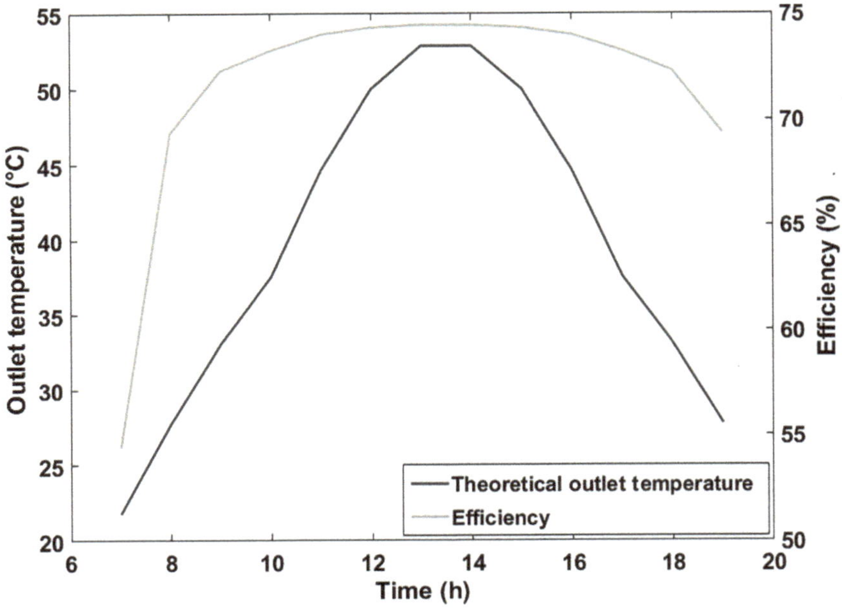

Fig. 5.25 The flat plate solar collector's outlet temperature (°C) and efficiency (%).

constant during the day and represents almost 75%. As for the outlet temperature, it attends its maximum (about 52°C) between 12PM and 3PM, where irradiations are at their maximum. Therefore, according to the theoretical study as well as the simulation, we conclude that a 4m^2 thermal panel with a 300L storage tank is more than enough to cover the needs of the building.

5.5.3 Summary and discussions

In this section, a simple case study has been established and detailed to show the effectiveness of the proposed holistic approach. First, a family two-story residential building has been designed and modeled. Its simulation has been conducted and based on preliminary results, the building has been optimized. The simulation results of the optimized version of the building have been afterward reported and showed that a good choice of construction materials as well as the glazing lowered energy demands in terms of heating and cooling by 76.66% and 72.433%, respectively. Thus, it is worth noting that, as stated above, a better design of the envelope could decrease effectively the energy consumption of a building. Also, incorporating RES and storage systems could help in filling the gap toward net-zero-energy buildings. Therefore, a hybrid system has been sized according to the building's consumption, modeled, and simulated. Then, the simulation results have been reported and showed the effectiveness of the sizing methodology as well as the behavior of the system under different conditions (e.g., solar irradiation and ambient temperature).

In the case study, we have chosen to work with batteries as storage devices since they are widely deployed in buildings, and this is attributed to their multiple benefits, such as modularity and good energy efficiency. As an alternative, the PV overproduction can be stored in the form of hydrogen, which has attracted much interest in the last few decades for the purpose of deploying it into both buildings and vehicles due to its advantages, mainly, cleanliness, efficiency, and abundance [145,146]. Actually, hydrogen is the most abundant gas in nature; however, most of this element only exists as part of other molecules (due to the strong covalent bond), such as water. This fuel could be generated from different raw materials, namely, water, biomass, biofuels, and fossil fuels [147], using several resources, mainly, water electrolysis and natural gas reforming [76,148]. Currently, more than 96% of the total hydrogen production in the world is based on fossil fuels (i.e., oil, natural gas, and coal), while less than 4% is produced by the electrolysis of water [147]. However, for the aim of integration with RES systems and superseding fossil fuels, the water electrolysis is the cleanest method for producing hydrogen while minimizing the greenhouse gases [147]. This is attributed to its various advantages, mainly, water availability, cleanliness, and high purity of the generated hydrogen [147].

In addition, the produced hydrogen can be stored in different states, namely, the gaseous state, the liquid state, and the solid state, in accordance with its application [149]. This hydrogen could be used by fuel cells, which have various advantages (e.g., higher efficiency that can reach 85–90%, cleanliness, modularity, and low noise level), to produce electricity and thermal energy [76,150]. Therefore, the produced electricity can be used to operate the building's appliances, while the produced thermal energy could be harnessed for either heating the building in winter time using an air to air exchanger or cooling it during summer period using a refrigeration process, thus reducing the building's electricity demand. The efficiency of this whole system (i.e., PV-hydrogen electrolyzer-fuel cell) can achieve 60% as described in our recent work [145]. In fact, this system has been extensively deployed all around the world to show its effectiveness and assess its performance when operating as a Combined Heat and Power (CHP) system to provide the building with both electricity and thermal energy [151–156].

However, batteries are still the most used in buildings compared to hydrogen storage since they present the best compromise between performance and cost, and they are a more mature technology [40,156,157]. The presented case study will be then enriched, in our forthcoming work, by integrating hydrogen as an auxiliary storage system and investigating its assessment in terms of energy efficiency, cost, and CO_2 reduction.

5.6 Conclusions and perspectives

In this chapter, we introduced a holistic approach for energy-efficient buildings, which could be seen as complex systems composed of different heterogeneous subsystems and components. We mainly highlighted approaches, which aim to deliver, implement, and optimize the building's concepts to further reduce energy consumption while maintaining good indoor quality. To shed more light on the effectiveness of the holistic approach, we presented a case study by considering a family two-story residential building. It has been designed, modeled, simulated, and optimized taking both the envelope and RES and storage systems integration. Reported simulation results showed that the proposed hybrid system is a stand-alone system that could cover the electricity and hot water needs of the considered building, toward a net-zero-energy building, while maintaining the occupants' comfort.

The proposed approach intends to establish the different trade-offs between the four main objective functions: (i) minimizing the buildings' energy consumption, (ii) reducing greenhouse gas emissions, (iii) maximizing occupants' comfort, and (iv) minimizing investment costs. It could be carried out by taking into consideration the three major aspects of the building, mainly the architectural design and the envelope, the passive building systems, the active building systems, and the integration of RES and storage systems. However, most of the existing tools take into consideration only a part of today's buildings (e.g., envelope, active systems, and RES and storage systems). Thus, an integrated and supporting tool is required to carry out the proposed holistic approach. This tool could be used to investigate the building as a complex system by considering, in a holistic way, its physical properties (e.g., geography, building type, and location), its equipment (e.g., heating, ventilation, and air-conditioning system), outdoor environment (i.e., weather condition), occupants' behavior and activities (e.g., reading, eating, and watching TV), and integration of RES and storage systems for electricity and hot water production (e.g., solar panel, batteries, and hydrogen storage).

Acknowledgments

This work was supported by MIGRID project (grant 5-398, 2017–2019), which was funded by USAID under the PEER program. It was also partially supported by HOLSYS project, which was funded by IRESEN.

References

[1] IEA. World energy statistics and balances (database), www.iea.org/statistics; 2019. [Accessed 10 February 2020].
[2] IEA. Energy technology perspectives, buildings model, www.iea.org/buildings; 2019. [Accessed 10 February 2020].
[3] IEA. Perspectives for clean energy transition, the critical role of buildings, https://www.iea.org/reports/the-critical-role-of-buildings; 2019.
[4] IEA. Global status report for buildings and construction. Towards a zero-emissions, efficient and resilient buildings and construction sector, https://www.iea.org/reports/global-status-report-for-buildings-and-construction-2019; 2019.
[5] Building regulations in the United Kingdom, https://en.wikipedia.org/wiki/Building_regulations_in_the_United_Kingdom. [Accessed 1 February 2020].
[6] Building Act, https://en.wikipedia.org/wiki/Building_Act_1984; 1984. [Accessed 1 February 2020].
[7] Policy measure fact sheet France Heating Regulation 2012.
[8] National building code of India. volume 1, https://bis.gov.in/index.php/standards/technical-department/national-building-code/. [Accessed 2 February 2020].
[9] Aderee. Thermal regulation of Moroccan construction, simplified version, Morocco; 2014.
[10] Agence urbaine de Skhirat-Témara. Decrees of thermal regulation of constructions, http://www.aust.ma/images/aust/reglementation/Decrets/R%C3%A8glement_thermique_de_construction_au_Maroc.pdf; 2019. [Accessed 2 February 2020].
[11] Oral GK, Yener AK, Bayazit NT. Building envelope design with the objective to ensure thermal, visual and acoustic comfort conditions. Build Environ 2004;39(3):281–7. https://doi.org/10.1016/S0360-1323(03)00141-0.
[12] Pacheco R, Ordóñez J, Martínez G. Energy efficient design of building: a review. Renew Sust Energ Rev 2012;16(6):3559–73. https://doi.org/10.1016/j.rser.2012.03.045.
[13] Sozer H. Improving energy efficiency through the design of the building envelope. Build Environ 2010;45(12):2581–93. https://doi.org/10.1016/j.buildenv.2010.05.004.
[14] Capeluto IG. Energy performance of the self-shading building envelope. Energ Buildings 2003;35(3):327–36. https://doi.org/10.1016/S0378-7788(02)00105-6.
[15] Knowles RL. The solar envelope: its meaning for energy and buildings. Energ Buildings 2003;35(1):15–25. https://doi.org/10.1016/S0378-7788(02)00076-2.
[16] Sadineni SB, Madala S, Boehm RF. Passive building energy savings: a review of building envelope components. Renew Sust Energ Rev 2011;15(8):3617–31. https://doi.org/10.1016/j.rser.2011.07.014.
[17] Orr J, Drewniok MP, Walker I, Ibell T, Copping A, Emmitt S. Minimising energy in construction: practitioners' views on material efficiency. Resour Conserv Recycl 2019;140:125–36. https://doi.org/10.1016/j.resconrec.2018.09.015.
[18] da Cunha SRL, de Aguiar JLB. Phase change materials and energy efficiency of buildings: a review of knowledge. J Energ Storage 2020;27:101083. https://doi.org/10.1016/j.est.2019.101083.

[19] Wang X, Zhang Y, Xiao W, Zeng R, Zhang Q, Di H. Review on thermal performance of phase change energy storage building envelope. Chin Sci Bull 2009;54 (6):920–8. https://doi.org/10.1007/s11434-009-0120-8.
[20] Bouhssine Z, Najam M, El Alami M. Phase change material for solar thermal energy storage in buildings: numerical study. J Sol Energy Eng 2016;138(6). https://doi.org/10.1115/1.4034518.
[21] Van Renterghem T, Hornikx M, Forssen J, Botteldooren D. The potential of building envelope greening to achieve quietness. Build Environ 2013;61:34–44. https://doi.org/10.1016/j.buildenv.2012.12.001.
[22] Perini K, Ottelé M, Haas EM, Raiteri R. Greening the building envelope, façade greening and living wall systems. Open J Ecol 2011;1(01):1. https://doi.org/10.4236/oje.2011.11001.
[23] Hosseini SH, Shokry E, Hosseini AA, Ahmadi G, Calautit JK. Evaluation of airflow and thermal comfort in buildings ventilated with wind catchers: simulation of conditions in Yazd City, Iran. Energy Sustain Dev 2016;35:7–24. https://doi.org/10.1016/j.esd.2016.09.005.
[24] Varela-Boydo CA, Moya SL. Inlet extensions for wind towers to improve natural ventilation in buildings. Sustain Cities Soc 2020;53:101933. https://doi.org/10.1016/j.scs.2019.101933.
[25] Laknizi A, Mahdaoui M, Abdellah AB, Anoune K, Bakhouya M, Ezbakhe H. Performance analysis and optimal parameters of a direct evaporative pad cooling system under the climate conditions of Morocco. Case Stud Therm Eng 2019;13:100362. https://doi.org/10.1016/j.csite.2018.11.013.
[26] Laknizi A, Ben Abdellah A, Faqir M, Essadiqi E, Dhimdi S. Performance characterization of a direct evaporative cooling pad based on pottery material. Int J Sustain Eng 2019;1–11. https://doi.org/10.1080/19397038.2019.1677800.
[27] Kharbouch A, El Maakoul A, Bakhouya M, El Ouadghiri D. Modeling and performance evaluation of an air-soil exchange system in energy efficient buildings. In: 2018 6th International renewable and sustainable Energy conference (IRSEC). IEEE; 2018. p. 1–6. https://doi.org/10.1109/IRSEC.2018.8702975.
[28] ASHRAE handbook. Residential cooling and heating load calculations; 2007. Chapter 17.
[29] Bradshaw V. The building environment: Active and passive control systems. John Wiley & Sons; 2010.
[30] Nguyen TA, Aiello M. Energy intelligent buildings based on user activity: a survey. Energ Buildings 2013;56:244–57. https://doi.org/10.1016/j.enbuild.2012.09.005.
[31] Kamienski C, Borelli F, Biondi G, Rosa W, Pinheiro I, Zyrianoff I, Sadok D, Pramudianto F. Context-aware energy efficiency management for smart buildings. In: 2015 IEEE 2nd world forum on internet of things (WF-IoT). IEEE; 2015. p. 699–704. https://doi.org/10.1109/WF-IoT.2015.7389139.
[32] Lachhab F, Essaaidi M, Bakhouya M, Ouladsine R. A state-feedback approach for controlling ventilation systems in energy efficient buildings. In: 2015 3rd International renewable and sustainable energy conference (IRSEC). IEEE; 2015. p. 1–6. https://doi.org/10.1109/IRSEC.2015.7454986.
[33] Lymperopoulos G, Ioannou P. Building temperature regulation in a multi-zone HVAC system using distributed adaptive control. Energ Buildings 2020;109825. https://doi.org/10.1016/j.enbuild.2020.109825.
[34] Ren C, Cao SJ. Development and application of linear ventilation and temperature models for indoor environmental prediction and HVAC systems control. Sustain Cities Soc 2019;51:101673. https://doi.org/10.1016/j.scs.2019.101673.
[35] Energy Efficient Buildings, https://ec.europa.eu/information_society/activities/sustainable_growth/buildings/index_en.htm. [Accessed 19 February 2020].

[36] Berouine A, Ouladsine R, Bakhouya M, Lachhab F, Essaaidi M. A model predictive approach for ventilation system control in energy efficient buildings. In: 2019 4th World conference on complex systems (WCCS). IEEE; 2019. p. 1–6. https://doi.org/10.1109/ICoCS.2019.8930739.
[37] Rawte R. The role of ICT in creating intelligent, energy efficient buildings. Energy Procedia 2017;143:150–3. https://doi.org/10.1016/j.egypro.2017.12.663.
[38] David M, Aubry A, Derigent W. Towards energy efficient buildings: how ICTs can convert advances? IFAC-Papers OnLine 2018;51(11):758–63. https://doi.org/10.1016/j.ifacol.2018.08.410.
[39] Bajracharya Q. Dynamic modeling, monitoring and control of energy storage system. Dissertation. Karlstad University, Faculty of Technology and Science; 2013. http://urn.kb.se/resolve?urn=urn:nbn:se:kau:diva-26521.
[40] Amrouche SO, Rekioua D, Rekioua T, Bacha S. Overview of energy storage in renewable energy systems. Int J Hydrog Energy 2016;41(45):20914–27. https://doi.org/10.1016/j.ijhydene.2016.06.243.
[41] Djafour A, Aida MS, Azoui B. Photovoltaic assisted fuel cell power systems. Energy Procedia 2014;50:306–13. https://doi.org/10.1016/j.egypro.2014.06.037.
[42] Boulmrharj S, Rabeh R, Felix V, Ouladsine R, Bakhouya M, Zine-Dine K, Khaidar M, Sinit M, Abid MR. Modeling and dimensioning of grid-connected photovoltaic systems. In: 2017 International renewable and sustainable Energy conference (IRSEC). IEEE; 2017. p. 1–6. https://doi.org/10.1109/IRSEC.2017.8477392.
[43] Seaman A, Dao TS, McPhee J. A survey of mathematics-based equivalent-circuit and electrochemical battery models for hybrid and electric vehicle simulation. J Power Sources 2014;256:410–23. https://doi.org/10.1016/j.jpowsour.2014.01.057.
[44] Zhou T, Francois B. Modeling and control design of hydrogen production process for an active hydrogen/wind hybrid power system. Int J Hydrog Energy 2009;34 (1):21–30. https://doi.org/10.1016/j.ijhydene.2008.10.030.
[45] Lajnef T, Abid S, Ammous A. Modeling, control, and simulation of a solar hydrogen/fuel cell hybrid energy system for grid-connected applications. Adv Power Electron 2013;2013. https://doi.org/10.1155/2013/352765.
[46] Lidula NWA, Rajapakse AD. Microgrids research: a review of experimental microgrids and test systems. Renew Sust Energ Rev 2011;15(1):186–202. https://doi.org/10.1016/j.rser.2010.09.041.
[47] Nayar C, Tang M, Suponthana W. Wind/PV/diesel micro grid system implemented in remote islands in the Republic of Maldives. In: 2008 IEEE international conference on sustainable energy technologies. IEEE; 2008. p. 1076–80. https://doi.org/10.1109/ICSET.2008.4747166.
[48] Attia S, Lacombe T, Rakotondramiarana HT, Garde F, Roshan G. Analysis tool for bioclimatic design strategies in hot humid climates. Sustain Cities Soc 2019;45:8–24. https://doi.org/10.1016/j.scs.2018.11.025.
[49] Zuhairy AA, Sayigh AAM. The development of the bioclimatic concept in building design. Renew Energy 1993;3(4–5):521–33. https://doi.org/10.1016/0960-1481(93)90118-Z.
[50] Ouyang Z, Spence SM. A performance-based damage estimation framework for the building envelope of wind-excited engineered structures. J Wind Eng Ind Aerodyn 2019;186:139–54. https://doi.org/10.1016/j.jweia.2019.01.001.
[51] Lstiburek J. Air barriers. Building Science Corporation; 2004. https://www.buildingscience.com/documents/reports/rr-0403-air-barriers/view.
[52] Lstiburek J. Moisture control for buildings. ASHRAE J 2002;44(2):36–41. http://sweets.construction.com/swts_content_files/2109/676715.pdf.
[53] Samuel DL, Dharmasastha K, Nagendra SS, Maiya MP. Thermal comfort in traditional buildings composed of local and modern construction materials. Int J Sustain Built Environ 2017;6(2):463–75. https://doi.org/10.1016/j.ijsbe.2017.08.001.

[54] Allen E, Iano J. Fundamentals of building construction: Materials and methods. John Wiley & Sons; 2019.
[55] Bahadori MN, Mazidi M, Dehghani AR. Experimental investigation of new designs of wind towers. Renew Energy 2008;33(10):2273–81. https://doi.org/10.1016/j.renene.2007.12.018.
[56] Kabir IFSA, Kanagalingam S, Safiyullah F. Performance evaluation of air flow and thermal comfort in the room with wind-catcher using different CFD techniques under neutral atmospheric boundary layer. Energy Procedia 2017;143:199–203. https://doi.org/10.1016/j.egypro.2017.12.671.
[57] Yang Y, Cui G, Lan CQ. Developments in evaporative cooling and enhanced evaporative cooling—a review. Renew Sust Energ Rev 2019;113:109230. https://doi.org/10.1016/j.rser.2019.06.037.
[58] Nada SA, Fouda A, Mahmoud MA, Elattar HF. Experimental investigation of energy and exergy performance of a direct evaporative cooler using a new pad type. Energ Buildings 2019;203:109449. https://doi.org/10.1016/j.enbuild.2019.109449.
[59] Jiménez-Arreola M, Wieland C, Romagnoli A. Direct vs indirect evaporation in organic Rankine cycle (ORC) systems: a comparison of the dynamic behavior for waste heat recovery of engine exhaust. Appl Energy 2019;242:439–52. https://doi.org/10.1016/j.apenergy.2019.03.011.
[60] Porumb B, Ungureşan P, Tutunaru LF, Şerban A, Bălan M. A review of indirect evaporative cooling technology. Energy Procedia 2016;85:461–71. https://doi.org/10.1016/j.egypro.2015.12.228.
[61] Resch G, Held A, Faber T, Panzer C, Toro F, Haas R. Potentials and prospects for renewable energies at global scale. Energy Policy 2008;36(11):4048–56. https://doi.org/10.1016/j.enpol.2008.06.029.
[62] Vedavarz A, Kumar S, Nawaz MH. HVAC: handbook of heating, ventilation and air conditioning for design and implementation. Industrial Press Inc; 2007.
[63] Sha H, Qi D. A review of high-rise ventilation for energy efficiency and safety. Sustain Cities Soc 2019;101971. https://doi.org/10.1016/j.scs.2019.101971.
[64] Chen Q. Ventilation performance prediction for buildings: a method overview and recent applications. Build Environ 2009;44(4):848–58. https://doi.org/10.1016/j.buildenv.2008.05.025.
[65] Evola G, Gagliano A, Marletta L, Nocera F. Controlled mechanical ventilation systems in residential buildings: primary energy balances and financial issues. J Build Eng 2017;11:96–107. https://doi.org/10.1016/j.jobe.2017.04.010.
[66] Meng X, Wang Y, Xing X, Xu Y. Experimental study on the performance of hybrid buoyancy-driven natural ventilation with a mechanical exhaust system in an industrial building. Energ Buildings 2020;208:109674. https://doi.org/10.1016/j.enbuild.2019.109674.
[67] Jing G, Cai W, Zhang X, Cui C, Yin X, Xian H. An energy-saving oriented air balancing strategy for multi-zone demand-controlled ventilation system. Energy 2019;172:1053–65. https://doi.org/10.1016/j.energy.2019.02.044.
[68] Wang L, Ma G, Zhou F, Liu Y, Tian T. Multicriteria decision-making approach for selecting ventilation heat recovery devices based on the attributes of buildings and the preferences of decision makers. Sustain Cities Soc 2019;51:101753. https://doi.org/10.1016/j.scs.2019.101753.
[69] Boulmrharj S, NaitMalek Y, ElMouatamid A, Bakhouya M, Ouladsine R, Zine-Dine K, Khaidar M, Abid R. Approach for dimensioning stand-alone photovoltaic systems. Energy Procedia 2018;153:56–61. https://doi.org/10.1016/j.egypro.2018.10.058.
[70] Elibrahimi M, Elmouatamid A, Bakhouya M, Feddi K, Ouladsine R. Performance evaluation of fixed and Sun tracking photovoltaic systems. In: 2018 6th International renewable and sustainable energy conference (IRSEC). IEEE; 2018. p. 1–6. https://doi.org/10.1109/IRSEC.2018.8702932.

[71] El Ibrahimi M, Khay I, El Maakoul A, Moussa MO, Barkaoui A, Bakhouya M. Anaerobic co-digestion in a liquid recirculation pilot-scale reactor: Thermal and hydraulic study. Energy Reports 2020;6:496–502. https://doi.org/10.1016/j.egyr.2019.09.014.
[72] Fossil fuel energy consumption (% of total), https://data.worldbank.org/indicator/eg.use.comm.fo.zs?end=2015&start=1960. [Accessed 30 January 2020].
[73] Othieno H, Awange J. Energy resources in Africa. Basel, Switzerland: Springer International Publishing; 2016. https://doi.org/10.1007/978-3-319-25187-5.
[74] National Energy Strategy, http://giz-energy.ma/energy-context/national-energy-strategy/?lang=en. [Accessed 30 January 2020].
[75] Ouarzazate Solar Power Station, https://en.wikipedia.org/wiki/Ouarzazate_Solar_Power_Station. [Accessed 30 January 2020].
[76] Boulmrharj S, Khaidar M, Siniti M, Bakhouya M, Zine-dine K. Towards performance assessment of fuel cell integration into buildings. Energy Rep 2020;6:288–93. https://doi.org/10.1016/j.egyr.2019.08.058.
[77] Anoune K, Bouya M, Astito A, Abdellah AB. Sizing methods and optimization techniques for PV-wind based hybrid renewable energy system: a review. Renew Sust Energ Rev 2018;93:652–73. https://doi.org/10.1016/j.rser.2018.05.032.
[78] Gupta BL, Bhatnagar M, Mathur J. Optimum sizing of PV panel, battery capacity and insulation thickness for a photovoltaic operated domestic refrigerator. Sustainable Energy Technol Assess 2014;7:55–67. https://doi.org/10.1016/j.seta.2014.03.005.
[79] Hocaoğlu FO, Gerek ÖN, Kurban M. A novel hybrid (wind–photovoltaic) system sizing procedure. Sol Energy 2009;83(11):2019–28. https://doi.org/10.1016/j.solener.2009.07.010.
[80] Xu D, Kang L, Chang L, Cao B. Optimal sizing of standalone hybrid wind/PV power systems using genetic algorithms. In: Canadian conference on electrical and computer engineering, 2005. IEEE; 2005. p. 1722–5. https://doi.org/10.1109/CCECE.2005.1557315.
[81] Khatib T, Ibrahim IA, Mohamed A. A review on sizing methodologies of photovoltaic array and storage battery in a standalone photovoltaic system. Energy Convers Manag 2016;120:430–48. https://doi.org/10.1016/j.enconman.2016.05.011.
[82] Bernal-Agustín JL, Dufo-Lopez R. Simulation and optimization of stand-alone hybrid renewable energy systems. Renew Sust Energ Rev 2009;13(8):2111–8. https://doi.org/10.1016/j.rser.2009.01.010.
[83] Zolfaghari M, Ghaffarzadeh N, Ardakani AJ. Optimal sizing of battery energy storage systems in off-grid micro grids using convex optimization. J Energ Storage 2019;23:44–56. https://doi.org/10.1016/j.est.2019.02.027.
[84] Sarma U, Ganguly S. Determination of the component sizing for the PEM fuel cell-battery hybrid energy system for locomotive application using particle swarm optimization. J Energ Storage 2018;19:247–59. https://doi.org/10.1016/j.est.2018.08.008.
[85] Zhou W, Lou C, Li Z, Lu L, Yang H. Current status of research on optimum sizing of stand-alone hybrid solar–wind power generation systems. Appl Energy 2010;87(2):380–9. https://doi.org/10.1016/j.apenergy.2009.08.012.
[86] Al-Falahi MD, Jayasinghe SDG, Enshaei HJEC. A review on recent size optimization methodologies for standalone solar and wind hybrid renewable energy system. Energy Convers Manag 2017;143:252–74. https://doi.org/10.1016/j.enconman.2017.04.019.
[87] Fossati JP, Galarza A, Martín-Villate A, Fontan L. A method for optimal sizing energy storage systems for microgrids. Renew Energy 2015;77:539–49. https://doi.org/10.1016/j.renene.2014.12.039.
[88] Connolly D, Lund H, Mathiesen BV, Leahy M. A review of computer tools for analysing the integration of renewable energy into various energy systems. Appl Energy 2010;87(4):1059–82. https://doi.org/10.1016/j.apenergy.2009.09.026.

[89] Allegrini J, Orehounig K, Mavromatidis G, Ruesch F, Dorer V, Evins R. A review of modelling approaches and tools for the simulation of district-scale energy systems. Renew Sust Energ Rev 2015;52:1391–404. https://doi.org/10.1016/j.rser.2015.07.123.
[90] Menezes AC, Cripps A, Buswell RA, Wright J, Bouchlaghem D. Estimating the energy consumption and power demand of small power equipment in office buildings. Energ Buildings 2014;75:199–209. https://doi.org/10.1016/j.enbuild.2014.02.011.
[91] Al-Rabghi OM, Al-Beirutty MH, Fathalah KA. Estimation and measurement of electric energy consumption due to air conditioning cooling load. Energy Convers Manag 1999;40(14):1527–42. https://doi.org/10.1016/S0196-8904(99)00045-X.
[92] Boulmrharj S, NaitMalek Y, Elmouatamid A, Bakhouya M, Ouladsine R, Zine-Dine K, Khaidar M, Siniti M. Battery characterization and dimensioning approaches for micro-grid systems. Energies 2019;12(7):1305. https://doi.org/10.3390/en12071305.
[93] Hadri S, Naitmalek Y, Najib M, Bakhouya M, Fakhri Y, Elaroussi M. A comparative study of predictive approaches for load forecasting in smart buildings. Procedia Comp Sci 2019;160:173–80. https://doi.org/10.1016/j.procs.2019.09.458.
[94] Bakhouya M, NaitMalek Y, Elmouatamid A, Lachhab F, Berouine A, Boulmrharj S, Ouladsine R, Felix V, Zinedine K, Khaidar M, Elkamoune N. Towards a context-driven platform using IoT and big data technologies for energy efficient buildings. In: 2017 3rd International conference of cloud computing technologies and applications (CloudTech). Rabat, Morocco: IEEE; 2017. p. 1–5. https://doi.org/10.1109/CloudTech.2017.8284744.
[95] Berouine A, Lachhab F, Malek YN, Bakhouya M, Ouladsine R. A smart metering platform using big data and IoT technologies. In: 2017 3rd International conference of cloud computing technologies and applications (CloudTech). Rabat, Morocco: IEEE; 2017. p. 1–6. https://doi.org/10.1109/CloudTech.2017.8284729.
[96] Swan LG, Ugursal VI. Modeling of end-use energy consumption in the residential sector: a review of modeling techniques. Renew Sust Energ Rev 2009;13(8):1819–35. https://doi.org/10.1016/j.rser.2008.09.033.
[97] Zhao HX, Magoulès F. A review on the prediction of building energy consumption. Renew Sust Energ Rev 2012;16(6):3586–92. https://doi.org/10.1016/j.rser.2012.02.049.
[98] Fumo N. A review on the basics of building energy estimation. Renew Sust Energ Rev 2014;31:53–60. https://doi.org/10.1016/j.rser.2013.11.040.
[99] Swan LG, Ugursal VI, Beausoleil-Morrison I. Occupant related household energy consumption in Canada: estimation using a bottom-up neural-network technique. Energ Buildings 2011;43(2–3):326–37. https://doi.org/10.1016/j.enbuild.2010.09.021.
[100] RSMeans. Green building: project planning and cost estimating. Wiley; 2010.
[101] Gruber JK, Prodanovic M, Alonso R. Estimation and analysis of building energy demand and supply costs. Energy Procedia 2015;83:216–25. https://doi.org/10.1016/j.egypro.2015.12.176.
[102] Biswas MR, Robinson MD, Fumo N. Prediction of residential building energy consumption: a neural network approach. Energy 2016;117:84–92. https://doi.org/10.1016/j.energy.2016.10.066.
[103] Ogbomo OO, Amalu EH, Ekere NN, Olagbegi PO. A review of photovoltaic module technologies for increased performance in tropical climate. Renew Sust Energ Rev 2017;75:1225–38. https://doi.org/10.1016/j.rser.2016.11.109.
[104] Jazayeri M, Uysal S, Jazayeri K. High capacity optical networks and emerging/enabling technologies; 2013. https://doi.org/10.1109/HONET.2013.6729755.
[105] El-Adawi MK, Al-Nuaim IA. A method to determine the solar cell series resistance from a single I–V. Characteristic curve considering its shunt resistance—new approach. Vacuum 2001;64(1):33–6. https://doi.org/10.1016/S0042-207X(01)00370-0.

[106] McMahon TJ, Basso TS, Rummel SR. Cell shunt resistance and photovoltaic module performance. In: Conference record of the twenty fifth IEEE photovoltaic specialists Conference-1996. IEEE; 1996. p. 1291–4. https://doi.org/10.1109/PVSC.1996.564369.
[107] Villalva MG, Gazoli JR, Ruppert Filho E. Comprehensive approach to modeling and simulation of photovoltaic arrays. IEEE Trans Power Electron 2009;24(5):1198–208. https://doi.org/10.1109/TPEL.2009.2013862.
[108] Gow JA, Manning CD. Development of a photovoltaic array model for use in power-electronics simulation studies. IEE Proc-Electr Power Appl 1999;146(2):193–200. https://doi.org/10.1049/ip-epa:19990116.
[109] De Soto W, Klein SA, Beckman WA. Improvement and validation of a model for photovoltaic array performance. Sol Energy 2006;80(1):78–88. https://doi.org/10.1016/j.solener.2005.06.010.
[110] Skoplaki E, Palyvos JA. On the temperature dependence of photovoltaic module electrical performance: a review of efficiency/power correlations. Sol Energy 2009;83 (5):614–24. https://doi.org/10.1016/j.solener.2008.10.008.
[111] Quansah DA, Adaramola MS, Appiah GK, Edwin IA. Performance analysis of different grid-connected solar photovoltaic (PV) system technologies with combined capacity of 20 kW located in humid tropical climate. Int J Hydrog Energy 2017;42 (7):4626–35. https://doi.org/10.1016/j.ijhydene.2016.10.119.
[112] Planas E, Andreu J, Gárate JI, De Alegría IM, Ibarra E. AC and DC technology in microgrids: a review. Renew Sust Energ Rev 2015;43:726–49. https://doi.org/10.1016/j.rser.2014.11.067.
[113] Jamal A, Venkata D, Andrew M. A review of power converter topologies for wind generators. Renew Energy 2007;32(14):2369–85. https://doi.org/10.1016/j.renene.2006.12.002.
[114] Das V, Padmanaban S, Venkitusamy K, Selvamuthukumaran R, Blaabjerg F, Siano P. Recent advances and challenges of fuel cell based power system architectures and control–a review. Renew Sust Energ Rev 2017;73:10–8. https://doi.org/10.1016/j.rser.2017.01.148.
[115] Singh B, Singh BN, Chandra A, Al-Haddad K, Pandey A, Kothari DP. A review of single-phase improved power quality AC-DC converters. IEEE Trans Ind Electron 2003;50(5):962–81. https://doi.org/10.1109/TIE.2003.817609.
[116] Salas V, Olias E, Barrado A, Lazaro A. Review of the maximum power point tracking algorithms for stand-alone photovoltaic systems. Sol Energy Mater Sol Cells 2006;90 (11):1555–78. https://doi.org/10.1016/j.solmat.2005.10.023.
[117] Femia N, Petrone G, Spagnuolo G, Vitelli M. Optimization of perturb and observe maximum power point tracking method. IEEE Trans Power Electron 2005;20 (4):963–73. https://doi.org/10.1109/TPEL.2005.850975.
[118] Takahashi N, Mejbri H, Ammous K, Abid S, Morel H, Ammous A. Bi-objective sizing optimization of power converter using genetic algorithms. COMPEL 2014. https://doi.org/10.1108/COMPEL-03-2012-0029.
[119] Castañeda M, Cano A, Jurado F, Sánchez H, Fernández LM. Sizing optimization, dynamic modeling and energy management strategies of a stand-alone PV/hydrogen/battery-based hybrid system. Int J Hydrog Energy 2013;38(10):3830–45. https://doi.org/10.1016/j.ijhydene.2013.01.080.
[120] Broussely M, Pistoia G, editors. Industrial applications of batteries: from cars to aerospace and energy storage. Elsevier; 2007.
[121] Najafi-Shad S, Barakati SM, Yazdani A. An effective hybrid wind-photovoltaic system including battery energy storage with reducing control loops and omitting PV converter. J Energ Storage 2020;27:101088. https://doi.org/10.1016/j.est.2019.101088.

[122] Boulmrharj S, NaitMalek Y, El Mouatamid A, Ouladsine R, Bakhouya M, Ouldmoussa M, Abid R. Towards a battery characterization methodology for performance evaluation of micro-grid systems. In: 2018 International conference on smart Energy systems and technologies (SEST). Seville, Spain: IEEE; 2018. p. 1–6. https://doi.org/10.1109/SEST.2018.8495829.
[123] Westerhoff U, Kurbach K, Lienesch F, Kurrat M. Analysis of lithium-ion battery models based on electrochemical impedance spectroscopy. Energ Technol 2016;4(12):1620–30. https://doi.org/10.1002/ente.201600154.
[124] Tremblay O, Dessaint LA. Experimental validation of a battery dynamic model for EV applications. World Electr Vehicle J 2009;3(2):289–98. https://doi.org/10.3390/wevj3020289.
[125] Mauracher P, Karden E. Dynamic modelling of lead/acid batteries using impedance spectroscopy for parameter identification. J Power Sources 1997;67(1–2):69–84. https://doi.org/10.1016/S0378-7753(97)02498-1.
[126] Plett GL. Extended Kalman filtering for battery management systems of LiPB-based HEV battery packs: part 3. State and parameter estimation. J Power Sources 2004;134(2):277–92. https://doi.org/10.1016/j.jpowsour.2004.02.033.
[127] Xia B, Lao Z, Zhang R, Tian Y, Chen G, Sun Z, Wang H. Online parameter identification and state of charge estimation of lithium-ion batteries based on forgetting factor recursive least squares and nonlinear Kalman filter. Energies 2017;11(1):1–23. https://doi.org/10.3390/en11010003.
[128] Rivera-Barrera JP, Muñoz-Galeano N, Sarmiento-Maldonado HO. SoC estimation for lithium-ion batteries: review and future challenges. Electronics 2017;6(4):102. https://doi.org/10.3390/electronics6040102.
[129] Zhang C, Jiang J, Zhang L, Liu S, Wang L, Loh PC. A generalized SOC-OCV model for lithium-ion batteries and the SOC estimation for LNMCO battery. Energies 2016;9(11):900. https://doi.org/10.3390/en9110900.
[130] Charkhgard M, Farrokhi M. State-of-charge estimation for lithium-ion batteries using neural networks and EKF. IEEE Trans Ind Electron 2010;57(12):4178–87. https://doi.org/10.1109/TIE.2010.2043035.
[131] Baccouche I, Jemmali S, Manai B, Omar N, Amara NEB. Improved OCV model of a Li-ion NMC battery for online SOC estimation using the extended Kalman filter. Energies 2017;10(6):764. https://doi.org/10.3390/en10060764.
[132] Tiwari GN, Tiwari A. Handbook of solar energy; 2006. https://doi.org/10.1007/978-981-10-0807-8.
[133] Deutsche Gesellschaft für Sonnenenergie. Planning and installing solar thermal systems: a guide for installers, architects, and engineers. Earthscan; 2005.
[134] Han J, Mol AP, Lu Y. Solar water heaters in China: a new day dawning. Energy Policy 2010;38(1):383–91. https://doi.org/10.1016/j.enpol.2009.09.029.
[135] Chien CC, Kung CK, Chang CC, Lee WS, Jwo CS, Chen SL. Theoretical and experimental investigations of a two-phase thermosyphon solar water heater. Energy 2011;36(1):415–23. https://doi.org/10.1016/j.energy.2010.10.023.
[136] Methods of testing and determine the thermal performance of solar collectors, n.d. ANSI/ASHRAE 93-1986 (RA91).
[137] Jannot Y. Thermique solaire; 2003. p. 30–70.
[138] Duffie JA, Beckman WA, Blair N. Solar engineering of thermal processes, photovoltaics and wind. John Wiley & Sons; 2020.
[139] Akpinar EK, Koçyiğit F. Energy and exergy analysis of a new flat-plate solar air heater having different obstacles on absorber plates. Appl Energy 2010;87(11):3438–50. https://doi.org/10.1016/j.apenergy.2010.05.017.
[140] Calculs d'Éclairement naturel, http://www.batisim.net/aide/index.html?daylightingcalculations.htm. [Accessed 15 February 2020].

[141] Conception de chauffage, http://www.batisim.net/aide/index.html?_heating_design_simulation.htm. [Accessed 15 February 2020].
[142] Conception de la climatisation, http://www.batisim.net/aide/index.html?_cooling_design_simulation.htm. [Accessed 15 February 2020].
[143] Ansermet JP, Bréchet SD. Principles of thermodynamics. Cambridge University Press; 2019.
[144] Maroc: 70 litres d'eau par habitant et par jour. USA: 600 litres!, https://www.lavieeco.com/economie/maroc-70-litres-deau-par-habitant-et-par-jour-usa-600-litres-22228/. [Accessed 25 February 2020].
[145] Boulmrharj S, Bakhouya M, Zine-dine K, Siniti M, Khaidar M. Performance analysis of a grid-connected PV panels - electrolyzer - fuel cell system for cogeneration in energy efficient buildings. In: 2019 International renewable and sustainable energy conference (IRSEC). IEEE; 2020. p. 1–6 [In press].
[146] Clarke RE, Giddey S, Ciacchi FT, Badwal SPS, Paul B, Andrews J. Direct coupling of an electrolyser to a solar PV system for generating hydrogen. Int J Hydrog Energy 2009;34(6):2531–42. https://doi.org/10.1016/j.ijhydene.2009.01.053.
[147] Mohammadi A, Mehrpooya M. A comprehensive review on coupling different types of electrolyzer to renewable energy sources. Energy 2018;158:632–55. https://doi.org/10.1016/j.energy.2018.06.073.
[148] Dutta S. A review on production, storage of hydrogen and its utilization as an energy resource. J Ind Eng Chem 2014;20(4):1148–56. https://doi.org/10.1016/j.jiec.2013.07.037.
[149] Abe JO, Ajenifuja E, Popoola OM. Hydrogen energy, economy and storage: review and recommendation. Int J Hydrog Energy 2019. https://doi.org/10.1016/j.ijhydene.2019.04.068.
[150] Peighambardoust SJ, Rowshanzamir S, Amjadi M. Review of the proton exchange membranes for fuel cell applications. Int J Hydrog Energy 2010;35(17):9349–84. https://doi.org/10.1016/j.ijhydene.2010.05.017.
[151] Widera B. Renewable hydrogen implementations for combined energy storage, transportation and stationary applications. Therm Sci Eng Prog 2020;16:100460. https://doi.org/10.1016/j.tsep.2019.100460.
[152] IMPACT ULHyS, http://lue.univ-lorraine.fr/fr/article/impact-ulhys. [Accessed 11 April 2020].
[153] Fuel Cell Energy. Annual report, http://www.annualreports.com/HostedData/AnnualReportArchive/f/NASDAQ_FCEL_2013.pdf; 2013. [Accessed 11 April 2020].
[154] Power SU. FuelCell Energy wins power plant order for London office tower. Fuel Cells Bull 2012. https://doi.org/10.1016/S1464-2859(12)70346-2.
[155] Le projet MHyRABEL. fruit lorrain de la transition énergétique, https://www.servirlepublic.fr/2016/05/le-projet-mhyrabel-fruit-lorrain-de-la-transition-energetique/. [Accessed 11 April 2020].
[156] Belmonte N, Luetto C, Staulo S, Rizzi P, Baricco M. Case studies of energy storage with fuel cells and batteries for stationary and mobile applications. Challenges 2017;8(1):9. https://doi.org/10.3390/challe8010009.
[157] Ibrahim H, Ilinca A, Perron J. Energy storage systems—characteristics and comparisons. Renew Sust Energ Rev 2008;12(5):1221–50. https://doi.org/10.1016/j.rser.2007.01.023.

CHAPTER 6

Performance and energetic modeling of hybrid PV systems coupled with battery energy storage

Arechkik Ameur[a,b], Asmae Berrada[c], Khalid Loudiyi[b], and Raymond Adomatis[d]

[a]Faculty of Science, Ibn Tofail University, Kenitra, Morocco
[b]School of Science and Engineering, Al Akhawayn University, Ifrane, Morocco
[c]College of Engineering, LERMA Lab, International University of Rabat, Sala El Jadida, Morocco
[d]Department of Chemical and Biomolecular Engineering, Institute for Systems Research, University of Maryland, College Park, MD, United States

6.1 Introduction

Energy production is a major challenge facing the electricity sector as countries need more and more energy to carry out their development. Today, much of the world's energy is produced from fossil sources [1]. The consumption of these sources results in greenhouse gas emissions and therefore causes an increase in pollution. Additionally, the increasing consumption of natural resources decreases fossil fuel reserves in a dangerous manner for future generations [2].

Now, renewable energy sources are gradually becoming fully fledged energy sources, competing with fossil fuels in terms of cost and performance [3, 4]. Photovoltaic energy is currently experiencing a strong development in the world, given its ecological character [5]. Photovoltaic electricity can be produced close to the consumption site, in a decentralized way, which makes it accessible to a large part of the world population [6]. However, it will, in particular, be necessary to overcome the intermittent nature of renewable energy supply. Following sudden and frequent changes in climatic conditions (temperature and irradiation), the operating point (intersection of the charge curve with that of the photovoltaic generator) changes significantly, which makes the identification of the maximum power point more than necessary to minimize power losses of the photogenerator. Photovoltaic generators are generally coupled to an electrical

Hybrid Energy System Models
https://doi.org/10.1016/B978-0-12-821403-9.00008-1

network or a storage system to ensure the continuous availability of energy. The type of storage often used by these hybrid systems is batteries [7, 8].

In this chapter, a photovoltaic system with storage batteries is studied. The first part of this chapter is devoted to some generalities on photovoltaic systems and energy storage technologies. In the second part, the different characteristics and mathematical models of the aforementioned system existing in literature are discussed. The last part proposes a simulation model of a residential hybrid PV system with energy storage.

The energy produced from photovoltaic systems is unlimited, nonpolluting, and has the least ecological damage. This energy can be used as an isolated energy source or connected to the grid [9]. One of the main disadvantages of solar energy is its intermittent nature [10, 11]. For permanent use, it is therefore necessary to store part of the energy produced. Therefore, there is a need for electrical storage whenever the electrical demand is shifted in time with respect to the solar gain. There are several storage methods which can be, for example, in the form of water, hydrogen; or stored in a flywheel, in an electrochemical battery (lead-acid, lithium), or in a supercapacitor [12–14].

6.2 Brief history of photovoltaic systems

The photovoltaic effect transforms solar energy into electricity [15]. The word "photovoltaic" comes from the Greek word "photos" (light) and from "volta" name of the Italian physicist who discovered the electric battery.

1839: Alexandre-Edmond Becquerel (1820–1891) discovered the electrical effects produced by the sun's rays in a battery made up of platinum and oxidized copper electrodes and immersed in an acidic electrolytic solution.

1873: The American engineer Willoughby Smith discovers the photosensitive properties of selenium.

1877: The photovoltaic effect of silicon has been explained by W.G. Adams and R.E. Day.

1883: The first cell in silicon and gold has been built by Charles Fritts. It achieved a yield of around 1%.

1905: Albert Einstein publishes on a heuristic point of view concerning the production and transformation of light. He earned the Nobel Prize in Physics in 1922.

1911: Wilhelm Ostwald, Nobel Prize in Chemistry in 1909, "in recognition of his work on catalysis and for his research on the fundamental principles governing chemical equilibrium and reaction rates."

1918: The scientist Polish Jan Czochralski develops a process to create monocrystalline silicon.

1939: Russel Ohl in (1898–1987) discovers the P-N junction.

1954: Gerald Pearson, Darryl Chapin, and Calvin Fuller develop a silicon photovoltaic cell for the Bell laboratories.

1955: American researchers (Chapin, Fuller, Pearson, and Prince) working for Bell Telephone Laboratories (now Alcatel-Lucent Bell Labs) develop photovoltaic cell with high efficiency of 6%.

1958: First use of solar cells in the American Vanguard satellite.

1958: The Sputnik III satellite is equipped with photovoltaic cells.

1971: Elliot Berman creates the company Solarpower to develop terrestrial applications for photovoltaic modules.

1973: The first house powered by photovoltaic cells is built at the University of Delaware.

1983: The first car powered by PV energy travels a distance of 4000 km in Australia.

Space was the test bench for photovoltaic technology. The high manufacturing costs of the cells and their poor yields do not allow them during their beginning to be used on a large scale. It was necessary to wait until the 1970s for the government and industry to invest in photovoltaic systems. Today, these systems are affordable with higher efficiency [16].

6.3 Photovoltaic cell operation

The PV cell is the smallest part of a photovoltaic installation. It is made of semiconductor materials, and it directly transforms light energy into electrical energy. The operating principle of this cell uses the properties of radiation and those of semiconductors [17, 18]. A P-N junction is created by bringing an N-doped semiconductor (electron donor) into contact with a P-doped semiconductor (composed of holes). The interface between these two regions is called a junction [17, 18]. When the free surface of the cell is illuminated, photons of energies greater than the width of the bandgap E_g can excite electron–hole pairs in the neutral regions P and N.

The electrons and holes created in the P and N regions, respectively, diffuse and reach the space charge zone, accelerated by the internal electric field; they pass through the transition zone.

The N region receives electrons and charges negatively; region P accepts holes and charges positively. If we connect the sides of the junction to a load, a current I flows through it and a potential difference appears [17, 18].

6.3.1 Current-voltage characteristic of a photovoltaic cell

The characteristic curve of a PV cell illustrates the variation of the current as a function of the voltage across the PV cell from the short circuit (which represents the zero-voltage corresponding to the maximum produced current) to the open circuit (which represents the zero-current for a maximum voltage across the cell) [19]. The function of photovoltaic cells depends on the irradiance and temperature conditions on the surface of the cell. Thus, each current–voltage curve corresponds to specific operating conditions [19] (Fig. 6.1).

where V_{oc} is the open-circuit voltage, I_{sc} is the short-circuit current, V_{mpp} is the voltage at the maximum power point, I_{mpp} represents the current at the maximum power point, and P_{mpp} is the power at the maximum power point.

6.3.2 Factors influencing the operation of a photovoltaic cell

The short current Isc varies directly with solar radiation (the relationship is linear), while the voltage remains relatively constant. This is interesting, especially for charging a battery [20]. The temperature has a slight effect on the current, but an increase in temperature causes an apparent decrease in the open-circuit voltage [20].

6.4 PV modules

A photovoltaic generator or module comprises a set of photovoltaic cells connected in series and/or in parallel to obtain the required electrical performances such as power, short-circuit current, and open-circuit voltage [19, 21].

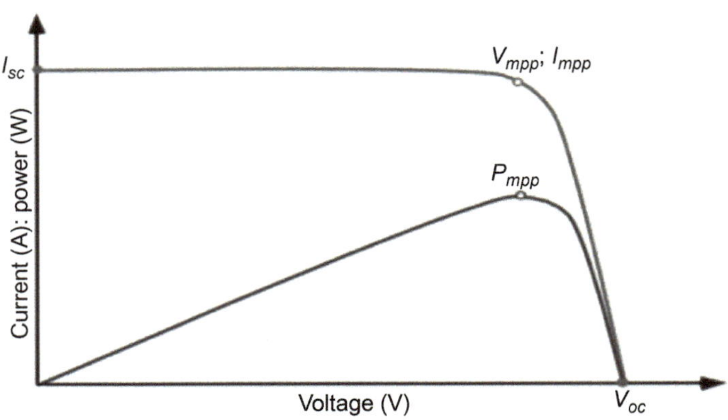

Fig. 6.1 Photovoltaic cell power-voltage and current-voltage characteristics [18].

6.4.1 Classic architecture of a photovoltaic module

6.4.1.1 Series connection

In series grouping, the cells are crossed by the same current and the resulting characteristic of a series association is obtained by adding the voltages at a given current [21]. The characteristic of a grouping of N_s identical PV cells is presented in Fig. 6.2.

Here, $V_{SOC} = N_s.V_{OC}$; and $I_{SSC} = I_{SC}$

With V_{SOC} being the sum of the open-circuit voltages of N_s cells in series and I_{SSC} is the short-circuit current of N_s cells in series.

6.4.1.2 Parallel connection

Using a parallel connection, the photovoltaic cells are subject to the same voltage. The resulting current corresponds to the sum of the currents generated by each cell [21] (Fig. 6.3).

Where the N_p cells in parallel, V_{POC} ($V_{POC} = V_{OC}$), is the same as each cell open-circuit voltage, and the short-circuit current, I_{PSC} ($I_{PSC} = N_p.I_{SC}$), is the sum of N_p cells open-circuit currents

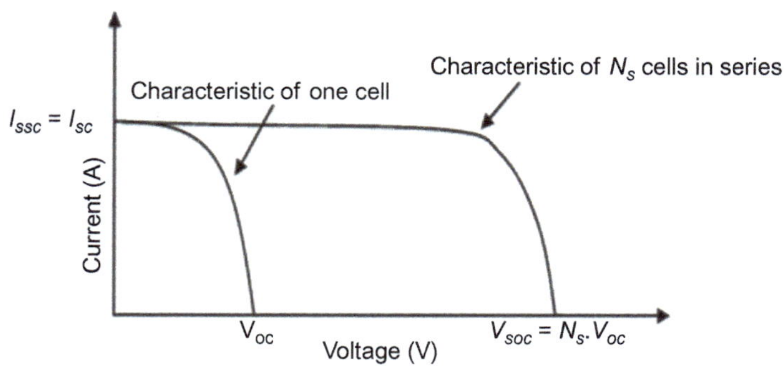

Fig. 6.2 Characteristic of a N_s identical cells in series.

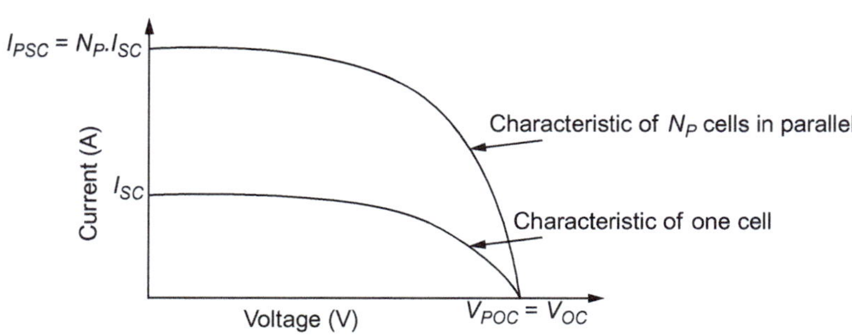

Fig. 6.3 Characteristic of a N_p identical cells in parallel.

6.4.2 Hybrid association (in series and in parallel)

According to the association in series and/or parallel of the PV cells, the total short-circuit current and the total open-circuit voltage are given by Eqs. (6.1) and (6.2), respectively:

$$I_{SC}^{t}=N_p I_{SC}, \tag{6.1}$$

$$V_{OC}^{t}=N_s V_{OC}, \tag{6.2}$$

where N_p is the number of cells in parallel, and N_s is the number of cells in series (Fig. 6.4).

6.5 Maximal power point tracker

For a PV cell, the short-circuit current I_{SC} and the open-circuit voltage V_{OC} are defined. The current I_{SC} corresponds to the value of the current delivered when the cell is short-circuited while V_{OC} is the voltage at its terminals in the absence of a load [19, 21, 22].

When plotting the power curve, a P_{MPP} operating point can be obtained. This represents the point at which the power is maximum. In photovoltaic applications, the delivery of the available power at the level of the PV generator is a very strong desire. The presence of converters (DC/DC/and DC/AC) allows the development of algorithms for identifying the maximum power point [19, 21, 22].

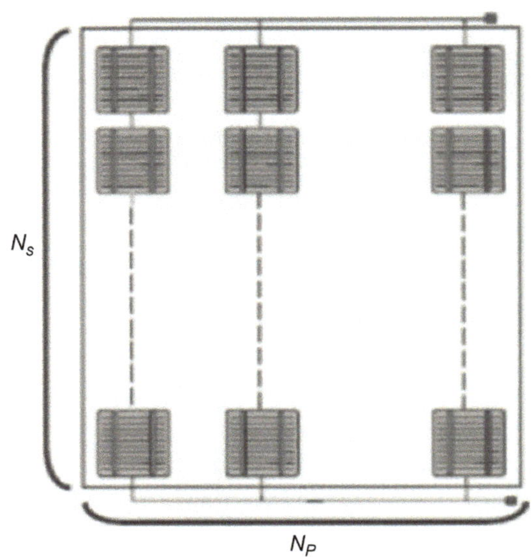

Fig. 6.4 Association of N_s identical cells and N_p identical cells in parallel [21].

6.6 Efficiency and fill factor

The efficiency of a photovoltaic cell is the quotient of the maximum output power and the light power received by the cell. This light power corresponds to the product of the area of the cell and the solar radiation.

The Fill factor is the performance factor of a solar cell. The closer it is to unity; the cell's performance is better. It compares the maximum power delivered (P_{MPP}) to the product of the short-circuit (I_{SC}) and the open-circuit voltage (V_{OC}) [21].

6.7 Concept of peak power and operating temperature

The peak power (STC) of a photovoltaic module is expressed in Watt-peak. It is the power that can be provided by the PV if it operates close to its optimal load with 1000 W/m solar radiation at 25°C temperature [21].

The cell operating temperature (NOCT) is the temperature reached by a cell encapsulated in a module subject to the sunlight of 800 W/m^2, at an ambient temperature of 20°C, at an inclination of 45°, and a 1 m/s wind speed under open-circuit conditions [21].

6.8 Photovoltaic technologies based on silicon

The basic material currently used in photovoltaic systems is silicon. It is found in very large quantities on our planet since it constitutes about 28% of the Earth's crust. It is mainly found in the form of silicon dioxide (SiO_2) which is the main constituent of sand. Silicon is abundant and inexpensive.

There are several types of photovoltaic cells. Each type is characterized by its own characteristics and performance. Nevertheless, whatever the PV type, the efficiency of PV cells remains quite low; it is between 8% and 25%. The most used PV cells include monocrystalline, polycrystalline, and amorphous silicon cells [21, 23].

6.8.1 Monocrystalline silicon

Monocrystalline cells are solar cells made from silicon crystallized into a single crystal. Their efficiency is 15%–24%, but their manufacturing is complex and expensive [23].

6.8.2 Polycrystalline silicon

Polycrystalline PV cells are consisting of a block of crystallized silicon in the form of multiple crystals. Their average efficiency is 13%–20%, and their production cost is slightly lower than that of monocrystalline cells [23].

6.8.3 Amorphous silicon

Amorphous cells are made up of very thin layers of silicon applied to a support made of glass, flexible plastic, or metal. Initially, their performance was low (6%–12%), but the technology is evolving rapidly. They operate with poor or diffuse lighting (even in overcast weather, including under artificial lighting from 2 to 3000 lx) [21, 23].

6.9 Types of photovoltaic systems

There are three main applications of PV systems which include autonomous, hybrid, and grid-connected systems [22, 24].

6.9.1 Autonomous photovoltaic system

It is a photovoltaic system completely independent of any other energy source; It supplies the user with electricity without being connected to the electric grid. In most cases, autonomous systems require batteries to store energy. They are typically used to power homes in remote locations, on islands, in mountains, and are used for applications such as remote monitoring and water pumping [22].

6.9.2 Photovoltaic hybrid system

Hybrid systems usually consist of two or more energy systems used in combination to ensure the required power output, with better efficiency, and lower costs. In practice, a photovoltaic generator is combined with a wind turbine or a fuel generator, or both with an energy storage system. Such a hybrid system has proved to be a good choice for applications which require a continuous supply of high power. A hybrid photovoltaic system optimizes the combined use of several renewables, energy storage, and/or fossil energy sources [22].

6.9.3 Photovoltaic system connected to the grid

The photovoltaic field is directly coupled to the electricity grid using an inverter (DC-AC). Since energy is normally stored in the grid itself, energy storage systems are not necessary unless an autonomous form of energy is needed during power outages. Part of the energy produced is consumed on-site; the surplus energy is injected into the electric network [22].

6.10 Advantages and disadvantages of a PV installation

6.10.1 Advantages

Photovoltaic technology offers a large number of advantages which include [25]:

i. Producing neither gas nor toxic waste as the sun is a clean and renewable source of energy.
ii. Offering high reliability, because it does not have moving parts which makes it particularly suitable for isolated regions such as spacecraft.
iii. Operating cost is very low due to its reduced maintenance and because it does not require neither fuel nor highly specialized personnel.
iv. Modular character allowing simple assembly. It can be sized for applications ranging from milliwatt to megawatt.
v. combining with other energy sources to increase system reliability.
vi. flexibility and can be expanded at any time to meet the user's energy needs.
vii. Requiring almost no maintenance [25].

6.10.2 Disadvantages

i. The manufacturing of the photovoltaic module requires high-cost investments.
ii. For low energy demands in isolated regions, diesel generators are competitive with PV systems.
iii. PV module efficiency is low (the theoretical efficiency for a silicon cell is 28%) [25].

6.11 Energy storage systems

In a photovoltaic installation, energy storage role consists of converting and storing the energy produced by the photovoltaic generator to a different energy form to be used later. The management of solar energy requires considering energy storage according to weather conditions [14, 22, 26, 27]. The storage of electrical energy most often involves some form of intermediate energy (gravity, compression, electrochemical, electrostatic, electromagnetic, kinetic, thermal, etc.) which can then be transformed into electricity. There exist two types of energy storage systems which include short-term storage with a storage time less than 10 min, and Long-term storage with a storage time greater than 10 min.

The selection of the storage system is made according to the fundamental characteristics of this latter which include mass power (in W/kg), mass energy (in Wh/kg), number of operating cycles, cost, and energy efficiency.

6.11.1 Short-term storage systems

6.11.1.1 Superconductor magnetic energy storage

Energy can be stored in the form of a magnetic field created by a current flowing in a superconducting coil. To keep the coil in the superconducting state, it is introduced into a cryostat filled with liquid helium. The energy stored in a magnetic field is expressed by Eq. (6.3):

$$E = \frac{1}{2} \ LI_{SP}^2, \tag{6.3}$$

where L the inductance of the coil and I_{sp} the current passing through it.

The manufacturing and maintenance costs of the system are very high. There are actually very few SMESs in operation. The available systems are between 1 and 10 MW even though research envisages to have systems in the order of 10–100 MW [14, 26, 27].

6.11.1.2 Storage by double-layer capacitors: Supercapacitors

The electric double-layer capacitor is a very high-capacity capacitor; it is commonly called "supercapacitor." The current systems are characterized by their very high-power density (well over 1000 W/kg), longevity (more than 100.00 cycles), and relatively high efficiency (in general over 80%). Because of these characteristics, they are generally well suited to the charge equalization function for a battery or a fuel cell. Indeed, they have a much lower energy density than that of other power sources: only 3–7 Wh/kg. Taking into account weight and space, these systems provide only one or two hundred watt-hours of energy [14].

6.11.1.3 Storage in the form of kinetic energy: Flywheel

Flywheel energy storage system is composed of a moving inertial mass, circular, or non-rotating at a rotation speed Ω. It is constructed of composite material or steel. It makes use of a motor/generator at high rotation speed for the transfer of this energy [14, 26, 27]. This system, suspended on magnetic or ball bearings, operates in a vacuum enclosure to limit losses caused by ventilation and friction. The external enclosure must be very resistant to avoid dispersion of the debris in the event of rupture of the inertia disc during operation.

The kinetic energy stored in this storage system is expressed by Eq. (6.4):

$$E = \frac{1}{2} J \Omega^2. \tag{6.4}$$

where J is the moment of inertia and Ω is the rotational speed.

The operation of the inertial storage system is based on the conversion of energy into a kinetic form, which is then converted to electrical energy when necessary. A flywheel is driven by a reversible electric machine that initially operates as a motor to supply energy to the inertial mass. With the drive system disconnected, the flywheel stores energy in its rotation. Upon request, this latter will be transformed into electrical energy by the generator. Flywheel system is used for the application of electric vehicles [14, 26, 27].

6.11.2 Long-term storage techniques

6.11.2.1 Pumped hydro energy storage

This storage system consists of two large reservoirs located at different heights; water is pumped from the lower reservoir to the upper one. The pumped water returns to the lower basin passing through one or more turbines thus transforming the potential energy of water into mechanical energy. The generators driven by these turbines transform mechanical energy into electrical energy. The energy production of pumped hydro storage for a mass of water, m, at elevation h is governed by Eq. (6.5):

$$E = mgh. \tag{6.5}$$

This storage technique provides up to 1000 MW. However, they are conditioned by geographic, geological, and environmental constraints, as well as the fairly high cost of their construction. This device cannot be installed in the vicinity of consumption centers, and they are difficult to transport. In general, the yield associated with this storage technique is around 75% [14, 27].

6.11.2.2 Energy storage in the form of compressed air

In these installations, the electricity available during off-peak hours is used to compress air with a turbocharger. Compressed air is stored in caverns located at great depths. The weight of the terrain above the cave allows it to resist air pressure. To recover the stored energy, the compressed air is directed to a machine similar to a gas turbine. It is heated in a combustion chamber through a natural gas booster and then expanded in the turbine. The residual heat of the fumes is recovered. In a modern installation, to restore 1 kWh, 0.75 kWh of electricity must be used during the pumping phase and 1.22 kWh of natural gas must be burned during the destocking [14, 26, 27].

6.11.2.3 Energy storage in thermal form

The principle of thermal storage consists in heating a body (sodium, salt, pressurized water, etc.) which does not undergo a change of state during the accumulation phase. The recovered heat is used to produce water vapor. A turbo generator is then driven by this latter.

This system, which might have seemed incongruous a few years ago, is now possible thanks to the development of high-temperature gas turbines (above 1400°C), which allow a conversion efficiency of heat into electricity of about 60% in combined cycles. In addition, at this temperature level, the energy density that can be stored in refractories is considerable. (It takes 500 kWh to heat a ton of refractories to 1400°C.) The refractories necessary to store heat at 1400°C are already widely used in the steel industry. Thermal insulation at this temperature level is also well mastered by this industry. A first analysis also shows that heat losses of such systems are relatively more reduced for large storage capacity. In addition, all the ingredients necessary to develop such energy storage systems are available [14].

6.11.2.4 Energy storage in chemical form

Batteries

When we talk about electricity storage, we immediately think of electrochemical batteries which thus constitute references on the subject. Storing energy in chemical form in electrochemical batteries is by far the most common technique for storing electrical energy [14, 22, 26, 27]. Depending on the battery type, different chemical reactions occur for storing energy in chemical form. Reverse chemical reactions are used for the production of electricity. A battery is an assembly of several cells, independent and complete units having all the electrochemical characteristics necessary for storage. A cell is made up of three basic elements which include a positive electrode (cathode), a negative electrode (anode), and an electrolyte.

The objective of a cell is to create a potential difference between the two electrodes linked to their constituent materials. The electrolyte allows the migration of ions between the anode and the cathode while preventing the transfer of electrons. These will have to pass through an electrical circuit which is the load. Typically, battery technology is defined by a couple of electrode materials. If necessary, the type of electrolyte can also be specified [14, 22, 26, 27].

Fuel cells

A fuel cell is among the technologies envisaged for the future as it engenders numerous research and development work around the world. This technology is evolving quickly as it is driven by the manufacturers' desire to offer products that are economically viable and reliable [14, 26–28].

Fuel cell is an electrochemical device carrying out the reverse operation of the electrolysis of water. While electrolysis dissociates water molecule into hydrogen and oxygen thanks to the passage of a current; the fuel cell makes it possible to bring them together, thus producing electricity, water, and heat. A fuel cell has two electrodes: the anode and the cathode. An oxidation reaction occurs at the anode of the fuel (for example, hydrogen) while, at the cathode, there is a reduction reaction of oxidant (most often oxygen). The electrodes are covered with a catalyst to facilitate redox reactions and are separated by an electrolyte. This latter allows the passage of protons from the anode to the cathode while blocking the electron migration. This latter must take an external circuit (the load) in order to migrate [14, 26–28]. The global reaction occurring within the fuel cell can be written as in the Eq. (6.6):

$$H_2 + \tfrac{1}{2}O_2 \rightarrow H_2O + \text{Electricity} + \text{Heat} \tag{6.6}$$

6.11.3 Comparison of energy storage systems

Some key features of the discussed energy storage systems are as follows. Pumped Hydroelectric storage and compressed air energy storage are large-scale energy storage which require favorable geology. The development of superconductors is still limited to date, which makes them very expensive. Despite the different advantages flywheel systems have compared to other energy storage, their high cost may be an obstacle to their development and use in addition to their sensitivity to shocks. Fuel cells are still prohibitively expensive to invest in; their overall efficiency is very low, and their service life is insufficient. Electrochemical accumulators are characterized by good mass energy. Their use seems interesting as they offer a good cost/performance ratio, high reliability, and good security [14, 27]. Considering all the advantages provided by batteries energy storage, their association with photovoltaic systems is almost imperative in decentralized production systems. Table 6.1 summarizes the advantages and disadvantages of the aforementioned energy storage systems.

6.12 Operation principle of batteries

Two electrodes immersed in an electrolyte constitute an electrochemical chain. Metal blades of different natures can act as electrodes. The electrochemical chain plays the role of a current generator by converting chemical energy into electrical energy. The electrochemical reactions take place at the electrodes. These reactions can generally be described as follows [14, 29]:

At the anode, which is the negative electrode, the oxidation reaction occurs during the discharge mode of the battery as expressed by Eq. (6.7):

$$N_1 \rightarrow N_1^{n+} + ne^- \tag{6.7}$$

where N_1 is the active species of the anode [14, 29].

The released electrons will pass through the external circuit to reach the cathode, which is the positive electrode, at which the reduction reaction takes place as represented in Eq. (6.8) [14, 29]:

$$N_2^{n+} + ne^- \rightarrow N_2 \tag{6.8}$$

where N_2 is the active species of the cathode. The overall redox reaction is as in the Eq. (6.9):

$$N_1 + N_2^{n+} \rightarrow N_1^{n+} + N_2 \tag{6.9}$$

Table 6.1 Comparison between the modes of energy storage.

Technology	Advantages	Disadvantages
Battery	Low cost	Limited lifetime
Compressed air energy storage	High storage capacity	Specific site use of natural gas
Pumped hydro storage	Large storage capacity Low cost	Long construction time Limited favorable site
Flywheel	High power	Low energy density High cost
Superconductor	High power	High cost Low energy density
Supercapacitor	Long lifetime Good efficiency	Low energy density

6.13 Different types of batteries

6.13.1 Lead-acid battery

It was in 1859 that Gaston Planté produced a lead accumulator by forming sheets of pure lead, in sulfuric acid under the influence of an electric current. Lead-acid battery consists of two electrodes (positive and negative) and an electrolyte. The positive electrode is lead dioxide (PbO_2) while the negative one is lead (Pb). The electrolyte, a solution of sulfuric acid (H_2SO_4), allows the flow of ions between the electrodes and creates a current. The density of the electrolyte decreases during the discharge mode. When the battery delivers the current, the active materials will be transformed into sulfate and water [22, 27].

The chemical reaction occurring at the anode:

$$Pb + H_2SO_4 \rightarrow Pb^{2+} + SO_4{}^{2-} + 2H^+ + 2e^- \tag{6.10}$$

The chemical reaction occurring at the cathode:

$$PbO_2 + H_2SO_4 + 2e^- \rightarrow Pb^{2+} + SO_4{}^{2-} + 2OH^{2-} \tag{6.11}$$

Lead-acid battery overall chemical reaction is as shown in Eq. (6.12):

$$PbO_2 + Pb + 2H_2SO_4 \leftrightarrow 2Pb^{2+} + 2SO_4{}^{2-} + 2H_2O \tag{6.12}$$

Reverse chemical reactions occur during the charging mode of the battery. The potential difference between the two electrodes is 2 V. A battery is an assembly of several cells, independent and complete units having all the electrochemical characteristics necessary for storage. Their assembly in a module called battery is only intended to have a higher voltage across the device. Deep discharge of the battery can result in an irreversible loss of capacity. If the battery charging continues for a long period, the voltage can reach a threshold value. This latter causes a decomposition of water (electrolysis) into gaseous oxygen O_2 at the positive electrode and into gaseous dihydrogen H_2 at the negative electrode [22, 27].

Lead-acid batteries have the following advantages:

i. Low cost.
ii. Good performance.
iii. Best compromise in terms of cost/performance/maintenance.
iv. Good lifetime.
v. Favorable electrochemical characteristics.
vi. Wide availability.
vii. Little or no maintenance.
viii. Good resistance to extreme temperatures.

6.13.2 Nickel-cadmium battery (Ni-Cd)

Nickel-Cadmium batteries are made from two electrodes (Nickel and Cadmium hydroxide) immersed in a potash solution. The overall reversible reaction is as expressed by Eq. (6.13):

$$2NiOOH + 2H_2O + Cd \leftrightarrow 2\,Ni(OH)_2 + Cd(OH)_2 \tag{6.13}$$

The positive electrode consists of Nickel hydroxide, while the negative one is made from Cadmium. The electrolyte is based on potash. The voltage varies from 1.15 to 1.45 V per cell with a nominal value of 1.2 V [14, 26, 27]. Although this battery type is less efficient than lead-acid accumulators, Ni-Cd accumulators have a number of advantages. It has very good mechanical resistance (on-board applications); it does not fear frost and endures high temperatures; it supports total discharges, and it does not produce toxic fumes.

6.13.3 Nickel-metal hydride battery (Ni-MH)

Ni-MH batteries differ from Ni-Cd because of the constitution of the negative electrode based on hydrogen adsorbed in a metal. This technology brings interesting improvements as compared to Ni-Cd. The specific energy of this battery type is high. The absence of cadmium makes the reprocessing of the accumulator at its end life much easier. The cost of this type of battery unfortunately remains very high. These batteries were developed with three objectives:

i. To increase capacity per unit of volume.
ii. To favor fast charges.
iii. To eliminate cadmium which is toxic to the environment. This material is already banned in some countries.

Apart from their characteristics, these batteries are not better than NiCd for solar applications. They are even worse when it comes to cold resistance.

6.13.4 Lithium-ion battery (li-ion)

The electrolyte of lithium-ion batteries consists of an organic solution or a solid polymer. The lightness of this metal allows obtaining extremely high-energy density and cell voltage, as well as high power density. The cost of this battery is 10 times higher than that of Pb batteries (600 € /kWh), while they are still commonly used for small power applications [14, 27].

6.13.5 Nickel-zinc battery

In the current situation of environmental issues, the replacement of Cadmium by Zinc would be very welcome. These types of batteries have a high energy density (25% more than the NiCd battery). The cell voltage ranges from 1.6 to 1.8 V. However, their lifetime remains the major problem to be resolved [14].

There exist other types of batteries which have not been mentioned here. A comparison table between the different discussed battery types is presented in Table 6.2 [14, 22, 27].

The hybrid photovoltaic system used in this case study make use of lead-acid batteries as an energy storage system.

6.14 Lead-acid battery

Lead-acid batteries are commonly employed as a storage form in hybrid PV systems. Lead-acid batteries have a good performance, only moderate efficiency, and low energy density; they also have a long lifetime and are less costly as compared to other battery types [22].

The parameters of a lead-acid battery are classified into two groups [22]:

i. The accessible parameters that are represented by visible and measurable physical data from outside the battery such as temperature, voltage, current, impedance, and quantity of the stored energy "capacity." These quantities are of interest to the user and are used to define the characteristics of the accumulators. Among these quantities, there are the external

Table 6.2 Comparison of the technical characteristics of the batteries.

Technology	Lead-acid	Ni-Cd	Ni-MH	Li-Ion
Specific energy (Wh/kg)	35–50	50–60	70–95	60–130
Specific power (W/kg)	80–150	150–400	200–1000	500–4000
Efficiency (%)	>80	75	70	>95
Cycle number	500–1000	800	750–1200	1000
Advantages	Low cost	Reliability Cold performance	Very good energy density	Excellent power and energy density
Disadvantage	Lifetime	Toxicity	Behavior in temperature	High cost

(current and voltage) and the internal quantities (resistance and state of charge).

ii. The inaccessible parameters that are the internal quantities which are related to chemistry and which represent the state of the accumulator such as the quantity of active material, density of the electrolyte, and internal pressure.

6.15 Behaviors of the battery

Batteries have three distinct behaviors:

i. Charging mode: the battery is connected to an energy source and acts as a receiver as it stores energy.

ii. Discharging mode: the battery is connected to a receiver.

iii. Standby mode: the battery is not connected to any power supply or receiver. There is no exchange with the outside environment.

6.16 Battery characteristics and models

Given the relatively complex nature of electrochemical accumulators, it is difficult to develop a general model. The various models existing in literature are of a complex implementation in particular because of the high number of parameters which need to be determined [22, 26]. The factors that affect battery performance include state of charge, storage capacity, temperature, and lifetime.

6.16.1 Storage capacity

The storage capacity expressed in ampere-hours represents the amount of charge that can be obtained during a complete discharge of the battery, initially charged, with a constant current. The nomenclature used to designate the storage capacity obtained for a discharge of n hours is C_n. Generally, a capacity C_5 is equivalent to the quantity of charge obtained during a discharge of 5 h. Similarly, a capacity of $C_{0.5}$ means a quantity of charge measured during a $\frac{1}{2}$ hour discharge [22, 26]. The discharge current (I_n), also called discharge regime, uses the same nomenclature as the capacitance C_n and is defined by Eq. (6.14):

$$I_n = \frac{C_n}{n}. \tag{6.14}$$

6.16.2 Efficiency

6.16.2.1 Faradic yield

The output in ampere-hours (or faradic) η_f is the ratio between the quantity of electricity delivered C_d and the quantity of electricity supplied for the charging C_c of the battery (Ah recovered during the discharging/Ah supplied during the charging). This is expressed by Eq. (6.15) [26]:

$$\eta_f = \frac{C_d}{C_c}. \tag{6.15}$$

6.16.2.2 Energy efficiency

The definition of energy efficiency (η_e) is similar to the definition of faradaic efficiency, except that the quantities involved are energies and not quantities of charge [26]:

$$\eta_e = \frac{E_d}{E_c}. \tag{6.16}$$

6.16.3 Open-circuit voltage

Open-circuit voltage indicates the equilibrium voltage of the battery at rest. It is defined as the difference of the equilibrium potentials between the two electrodes [26].

6.16.4 Lifetime

The performance of batteries decreases throughout their lifetime. The battery end of life is reached when its performance no longer meets the required performance of the intended application [26].

The lifetimes expressed in years are given for floating operation at 25°C. It varies from 2 to 7 years for Pb batteries and from 10 to 25 years for Ni-Cd. The lifetime of batteries decreases by half for each 10°C rise in the temperature of use [14, 26].

6.16.5 Depth of discharge

Depth of discharge is of considerable importance for Pb batteries since they rarely survive a full discharge. However, this parameter does not have much importance for Ni-Cd which can completely discharge. The discharging of a battery is generally limited to 80% of the nominal capacity. For solar applications, the discharge depth hardly exceeds 60%. Accumulators are often oversized in order to increase their lifespan [22, 26].

6.16.6 Self-discharge

The self-discharge rate of an accumulator represents the relative average loss of capacity per month for a given temperature. A battery self-discharge can result from several causes such as a decrease in the electrolyte density. This is mainly due to the presence of small metallic impurities on the electrodes, leading to the formation of micro-batteries and to the generation of weak internal current [26]. Typical values of self-discharge range from 3 to 15% per month for lead acid and up to 30% per month for Ni-Cd, for a temperature of 20°C.

6.16.7 State of charge

Knowing the actual state of charge of a battery is necessary and crucial for its proper use. It enables the user to avoid operating modes which damage the battery such as overcharging. In addition, one can predict whether the battery can ensure its functionality or not, for example, the remaining autonomy of an electric car.

As the state of charge is far from being unequivocally linked to one or few measurable parameters, it is important to work with several electrochemical and electrical properties of the battery, such as open-circuit voltage, loaded voltage, current, internal impedance, temperature, specific gravity of the electrolyte, and conductivity. A good mixing of these parameters by algorithms makes it possible to obtain an estimate of this state dump. It also provides valuable information on the internal state of the battery and its remaining service life [26]. The state of charge of a battery is estimated using three methods which include [26]:

- **a.** Physical methodology: based on the measurement of physical changes (electrolyte concentration) occurring within the battery during the discharging mode.
- **b.** Electrical methods: based on the modification of the electrical parameters (voltage, current) of cells or of the whole battery during the discharge mode.
- **c.** Impedance measurement methods: to have an approximation of the battery internal state battery, this method is based on the superposition of an AC signal over the DC component in order.

6.16.8 Electrical models of the batteries

6.16.8.1 The ideal model

In an ideal model, the battery is simply represented by a voltage source as an equivalent circuit. All other internal parameters are not taken into account [27] (Fig. 6.5).

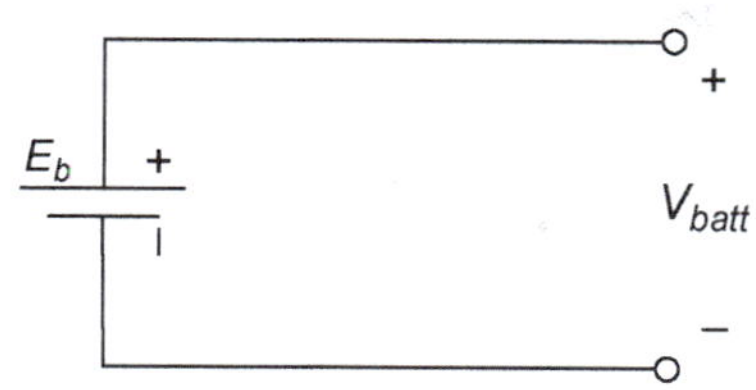

Fig. 6.5 Ideal model of a battery.

6.16.8.2 Simplified model of the battery

A typical model of a battery is shown in Fig. 6.6. It uses the no-load voltage of the battery E_b in series with an internal resistance R_s. This model is widely used and does not take into account variations in the internal impedance of the battery depending on the state of charge and temperature. Such a model is only applicable in certain simulations where the state of charge of the battery is less important [27]. This model takes into account another effect, which is given by the Peukert equation. This latter is a convenient way to characterize the cell behavior and to measure eccentric capacity in mathematical terms. It estimates how the battery available capacity changes according to the rate of discharge [27].

Peukert equation is expressed by Eq. (6.17):

$$C = (I_d)^n T_d. \tag{6.17}$$

Here, I_d represents discharge current in amperes, n is the battery constant (for lead-acid batteries, for example, $n = 1.35$), T_d is the discharge time in hours, and C is the theoretical capacity of the battery in ampere-hours.

Peukert's equation demonstrates that at higher currents, the battery has less energy capacity. The Peukert number n is associated with the battery internal resistance. Higher currents mean more losses and less available capacity. The battery is very efficient if this value is close to 1. The Peukert number is found empirically. For lead-acid batteries, this number is between 1.3 and 1.4.

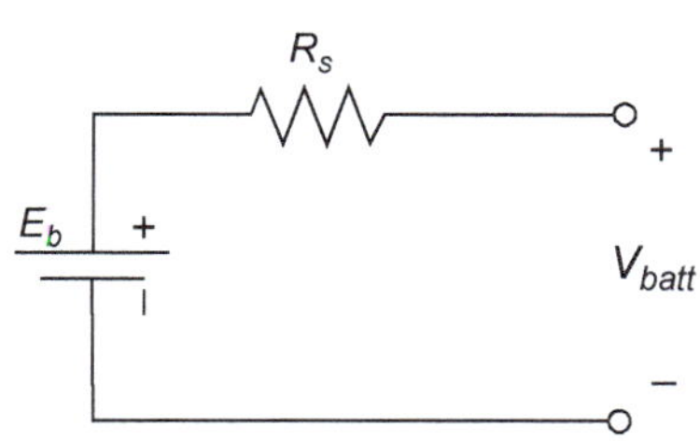

Fig. 6.6 Simplified model of the battery.

6.17 Prediction modeling of a photovoltaic system coupled with a battery energy storage

To study the operation of PV systems with batteries connected to a smart house, a physically based dynamic simulator has been modeled in Python software. The aim of this model is to perform week-long predictions of PV system production and the state of charge of lead-acid batteries. Different PV technologies are used in this study which include monocrystalline, polycrystalline, and amorphous. The objective of the model is to manage the energy needs of the house.

In the first part of this section, the PV model and architecture will be presented. In the second part, system design and prediction modeling are performed. In addition, the prediction results of battery state and the PV output are discussed.

It is essential to perform forecasting of energy production to properly manage energy and meet the household energy demand. This has been triggered by the policy of moving toward smart homes in the future. These latter make use of energy production forecasting as it is considered as a decisive tool for owners to manage energy. In addition, the energy production of a photovoltaic system depends on many factors including solar irradiance, inclination of the panels, technology, as well as temperature. However, the received solar radiation remains the main influential factor. Photovoltaic solar energy is an intermittent energy; its availability varies greatly. Another challenge facing PV systems is that their energy output cannot be stored in large quantities. Therefore, the prediction of energy production remains a key solution to overcome the aforementioned issues. Forecasting of solar irradiation is considered as a very active area of research in recent years, particularly in photovoltaic systems.

6.17.1 Solar radiation

To size photovoltaic systems, it is necessary to define the solar radiation profile of the studied region from the statistical study. Solar radiation propagates in the form of electromagnetic waves. It is an energy supply which is available on a daily basis and which varies according to place, time, and date (Fig. 6.7).

Solar radiation reaches the ground directly and in diffuse form. The global irradiance (G) is the sum of direct normal (I), diffuse (D), and reflected (R) irradiance [30]:

$$G = I + D + R \tag{6.18}$$

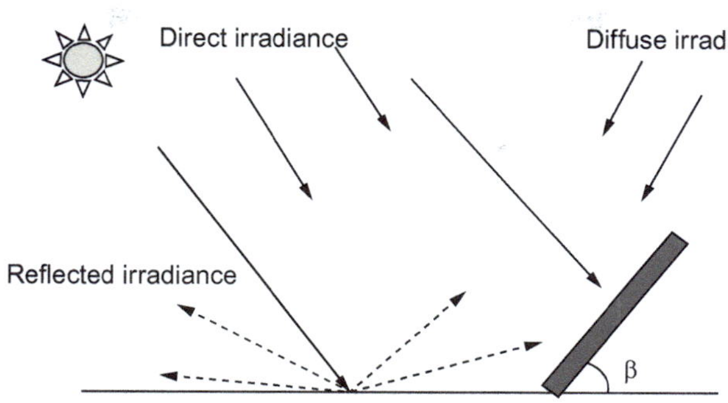

Fig. 6.7 Solar radiation components.

6.17.2 Photovoltaic generator Modeling

Many mathematical models represent photovoltaic generators. In this study, a simplified diode model with six parameters (I_{sc}, V_{oc}, I_{mpp}, V_{mpp}, R_s, R_{sh}) called model with a is used. The proposed model takes into account the influence of temperature and solar radiation. It can be described in a simple way as current source producing a current I_{Ph} which is proportional to the incident light power, and in parallel with a diode. In addition to this, the model takes into consideration the resistive properties of the cell, and the leakage currents which are modeled by a series resistance R_s, and as a shunt resistance R_{sh}, respectively. The generator model is presented in Fig. 6.8.

The series (N_s)/parallel (N_p) association of elementary photovoltaic cells determines the photovoltaic module. If all the PV cells are identical and operate under the same conditions, we obtain the photovoltaic module supplying a current I_M under a voltage V_M, with: $I_M = N_p.I_C$ and $V_M = N_s.V_C$, here I_C and V_C are the current and voltage of cell, respectively.

From the equivalent electrical circuit presented in Fig. 6.8, the expression of the current of the I_M module is given by Eq. (6.19) [31–33]:

$$I_M = I_{ph} - I_d - I_{sh}. \tag{6.19}$$

The output current of the photovoltaic system can be presented as a function of the output voltage of the PV system as in the Eq. (6.20):

$$I_M = I_{sc} - I_0\left(e^{\frac{q(V_M + I_M R_s)}{KT}} - 1\right) - \left(\frac{V_M + I_M R_s}{R_{sh}}\right). \tag{6.20}$$

where I_M, V_M, I_{sc}, I_0, R_s, R_{sh}, K, T, *and* q are the module's current, module's voltage, the light-generated current, reverse saturation current, series resistance, shunt resistance, resistance, the Boltzmann constant, the temperature, and the electronic charge, respectively. Note that Eq. (6.20) is applicable

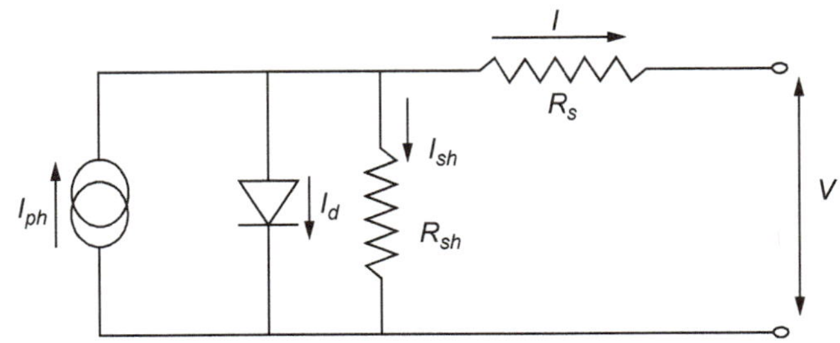

Fig. 6.8 Equivalent circuit of the photovoltaic generator.

only under standard test conditions ($G = 1000\,\text{W/m}^2$ and $T = 25°\text{C}$) [31–33]:

In an operating condition, the output current I_M can be written as Eq. (6.21):

$$I_M = I_{sc}\left[1 - K_1\left(e^{K_2 V_M^m} - 1\right)\right]. \tag{6.21}$$

The coefficients K_1 and K_2 are given by Eqs. (6.22)–(6.23):

$$K_1 = 0.01175, \tag{6.22}$$

$$K_2 = \frac{K_4}{(V_{oc})^m}. \tag{6.23}$$

The coefficients m, K_3, and K_4, and m are given by Eqs. (6.24)–(6.26):

$$m = \frac{\ln\left(\frac{K_3}{K_4}\right)}{\ln\left(\frac{V_{mpp}}{V_{oc}}\right)}, \tag{6.24}$$

$$K_3 = \ln\left[\frac{\left(I_{sc}(1 + K_1) - I_{mpp}\right)}{K_1 I_{sc}}\right], \tag{6.25}$$

$$K_4 = \ln\left(\frac{(K_1 + 1)}{K_1}\right), \tag{6.26}$$

where I_{mpp}, V_{mpp}, I_{sc}, *and* V_{oc} represent current at maximum power, voltage at maximum power, the short-circuit current, and the open-circuit voltage, respectively [34].

When temperature and insolation change, the current and the voltage are given by Eqs. (6.28)–(6.29), respectively:

$$\Delta T_c = T_c - T_{STC}(°\text{C}), \tag{6.27}$$

$$\Delta \mathrm{I}_M = \alpha_{sc}\left(\frac{G}{G_{STC}}\right)\Delta \mathrm{T}_c + \left(\frac{G}{G_{STC}} - 1\right) I_{sc,STC} \quad (\mathrm{A}), \tag{6.28}$$

$$\Delta \mathrm{V}_M = -\beta_{oc}\Delta \mathrm{T}_c - R_s \Delta \mathrm{I}_M \quad (\mathrm{V}), \tag{6.29}$$

where T_c, G represent the temperature and radiation under any conditions, respectively. T_{STC}, G_{STC} are temperature and radiation under standard conditions, respectively. $\Delta \mathrm{T}_c$, $\Delta \mathrm{I}_M$ are the variation of temperature, the variation of the current relative to the radiation and the temperature, respectively. $\Delta \mathrm{V}_M$ is the variation of voltage relative to insolation and temperature, and βoc is the increment coefficient of open-circuit voltage when the temperature increases by one degree Celsius (V/°C) (under standard operating conditions) [34].

The new values of the voltage and current will be given by Eqs. (6.30)–(6.31):

$$\mathrm{I} = I_{STC} + \Delta \mathrm{I}_M \quad (\mathrm{A}), \tag{6.30}$$

$$\mathrm{V} = V_{STC} + \Delta \mathrm{V}_M \quad (\mathrm{V}). \tag{6.31}$$

The prediction of PV production and solar irradiation has been achieved through the development of a dynamic simulator. This latter is based on physical resources representing the energy production of three studied PV systems. The model is written as a set of Python modules corresponding to the components of the PV system. These modules are interconnected by a unique set of data structures used to denote the PV production profile over a one-week forecast period. Solar irradiance Incident on the PV panels is forecasted by the developed model based on the site location and the time of the year, combined with cloud cover forecasts. To obtain the weather forecast data, Dark Sky Application Programing Interface has been used. This latter is considered as one of the most accurate forecasting sources [35, 36]. The generated data are in the form of a json file.

The forecasted temperature and cloud cover data of the site are automatically evaluated and interpolated every morning at midnight, over a one-week forecast period. These latter are then combined with the predicted irradiance to obtain the solar irradiance profile theoretically using Eqs. (6.32)–(6.36) [37–39]:

$$DNI = G_{SC} * 0.73^{AM^{0.678}} \quad \left(\frac{W}{m^2}\right), \tag{6.32}$$

$$Predicted\ DNI = (1 - Cloud\ Cover) * DNI \quad \left(\frac{W}{m^2}\right), \tag{6.33}$$

$$Predicted\ DI = 0.2 * predicted\ DNI \quad \left(\frac{W}{m^2}\right), \tag{6.34}$$

$$Predicted\ GI = Predicted\ DN + Predicted\ DI \quad \left(\frac{W}{m^2}\right), \tag{6.35}$$

$$AM = \frac{1}{Cos\,(Zenith\ angle)}. \tag{6.36}$$

where G_{SC} is the solar constant and is equal to 1366 W/m^2; AM is the air mass. Eq. (6.32) shows that approximately 73% of the incident solar radiation on the atmosphere is transferred to the Earth. In addition, the inconsistency in the atmospheric layers is also taken into consideration by Eq. (6.32). This is represented by the power term of 0.678 [38, 40]. To obtain this latter factor, the observed data have been fit to an empirical distribution. 0.2 in Eq. (6.34) denotes the ratio of (DI) to (DNI) on a clear day.

Fig. 6.9 illustrates the methodology used presenting the input and output data.

6.17.3 Presentation of the studied system

The PV system is be composed of photovoltaic modules, electrochemical storage devices (batteries), inverter DC/AC, photovoltaic charge controller, and supervision module. Fig. 6.10 presents Block diagram of the PV system used in this case study.

6.17.3.1 Energy converter (DC/AC)

The modeled PV system includes an inverter which converts DC voltage and current to AC voltage and current. A distinction is therefore made between the continuous part denoted DC (Direct Current), which is upstream of the inverter, and the alternative part denoted AC (Alternating Current), which is downstream of the inverter.

Currently, most of the electrical appliances on the market are adapted to the AC voltage. Manufacturers do not always offer revised equivalents for DC voltage. As a result, the presence of an inverter becomes essential to provide the user with a wide range of choice for electrical equipment.

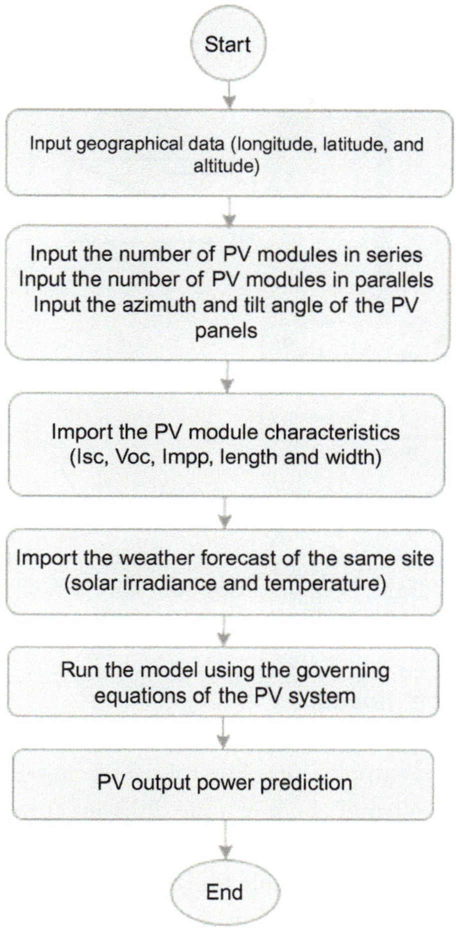

Fig. 6.9 Approach used to predict PV output power.

6.17.3.2 Photovoltaic charge controller

Batteries are sensitive elements which must be handled with care to avoid operational and durability problems described in the previous section. The PV charge controller occupies an essential and indispensable role within a photovoltaic installation. It must ensure a crucial function which is the management of the charge and the discharge of the batteries. If the battery is not charged properly, damage of this latter may occur. The photovoltaic regulator must protect the batteries against risks of degradation during the charging process. The state of charge is checked by measuring the voltage

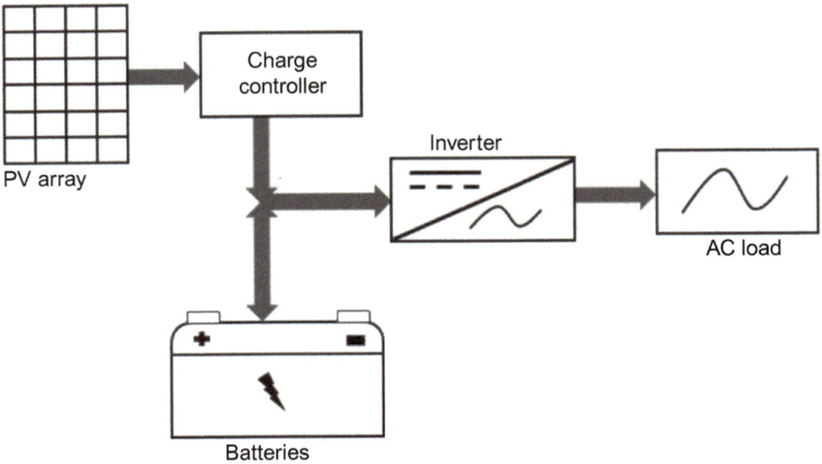

Fig. 6.10 Block diagram of the PV system studied.

of the battery bank. If the voltage drops below a predefined level, the photovoltaic regulator disconnects the battery bank from the electrical devices.

6.17.3.3 Supervision module

In a photovoltaic system, it is essential to optimally manage the energy flow between the different components. The role of the supervision module is to control the energy exchanges between the different components of the system and to regulate the storage charge/discharge process [41]. Therefore, the supervision module is responsible for managing the energy coming from the PV panels and the storage system according to the load demand. The energy management system is based on two scenarios:

1. PV panels provide energy to the load and batteries if the solar energy is sufficient.
2. PV panels and batteries provide energy to the load if the solar energy is insufficient.

When the current is negative, the use of batteries is required. On the other hand, when the current is positive, the battery charging is allowed using Eq. (6.37) (Fig. 6.11):

$$P = P_{PV} - P_{Load}. \tag{6.37}$$

6.17.4 Design and dimensioning of a photovoltaic system

In general, the design and sizing should follow the following steps:

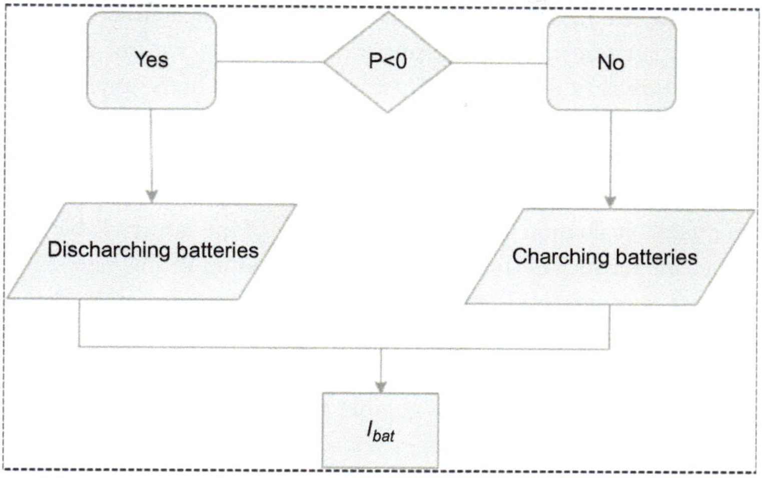

Fig. 6.11 Diagram of management of the production energy.

1. Definition of the electrical need.
2. Evaluation of local solar irradiance.
3. Dimensioning of the PV array.
4. Sizing of the batteries.
5. Selection of the solar charge controller and the inverter.
6. Sizing of the wiring system.

6.17.4.1 Identification of electrical needs

The identification of the electrical needs is an important preliminary work which impacts the dimensioning of the photovoltaic field and the battery park. An underestimation of the electrical requirement will cause malfunctions such as an electrical cutoff, or premature aging of the batteries, while an overestimation will significantly increase the installation cost of the plant. This part draws up an inventory of appliances and provides a method for calculating electrical needs. It is first necessary to carry out an inventory of equipment which consume electrical energy. These are diverse and varied. A non-exhaustive inventory is offered in this work.

Identifying electrical needs is based on the calculation of the daily consumed electrical energy. These needs are typically expressed in Wh/day (or kWh/day). The approach used in this work is as follows:

First, electrical devices that will be powered by the stand-alone photovoltaic system are identified. For each of these devices, the nominal operating power must be determined. For this, one can rely on site measurements directly, or use the information provided in the technical/data sheets of the devices. As a last resort, if no information is available, we can approximate the electrical power of the device. Then, an estimation of the appliance's daily usage time should be made. That is, it is essential to know the habits of the users. The product of the electric power by the usage time of the appliance provides the daily energy consumed. Finally, the sum of the calculated daily energy provides an overall evaluation of the electrical needs of the house. Table 6.3 presents the characteristics of the different appliances used in this case study.

Between summer and winter, daily energy consumption varies by around 1%. This is due to the fact that lighting is the only consumption item that depends on the season, and thanks to the very energy-efficient LED technology, its consumption is negligible compared to other items. In this case, we can consider that the energy consumption is constant over the seasons, around 19.3795 kWh/day.

6.17.4.2 Evaluation of the local irradiation and dimensioning of the PV array

The photovoltaic field is dimensioned based on the daily electric energy which should be produced by the photovoltaic field (E_{elec}). This latter must be at least equal to the daily energy consumed by electric devices (E_{need}):

$$E_{elec} \geq E_{need}. \tag{6.38}$$

The electrical energy (E_{elec}) is calculated by Eq. (6.39):

$$E_{elec} = \frac{P_p * E_i * PR}{P_i}, \tag{6.39}$$

where E_{elec} is the daily electrical energy potentially generated by the PV installation (kWh/day). P_p is the peak power of the photovoltaic array (kW_p). P_i is the radiative power under standard test conditions (kW/m^2). E_i is the daily solar irradiation received by a unit surface ($1\,m^2$) of the photovoltaic array, taking into account its orientation and inclination (kWh/m^2/day). PR is the performance ratio of the photovoltaic installation which depends on the ventilation of the modules and the MPPT. In this case study, PV modules are well-ventilated modules and an inverter with MPPT is used. Therefore, the PR is equal to 0.75. E_{need} is the daily energy consumed by electrical appliances (kWh/day).

Eq. (6.39) can be written as

$$\frac{P_p * E_i * PR}{P_i} \geq E_{need}. \tag{6.40}$$

Table 6.3 Nominal power of the electrical appliances used in the house.

Electrical appliance	Nominal power (W)	Daily use time (h/day)	Daily energy consumption (Wh)
Lamps	10*7=70	4	280
TV	100	5	500
Refrigerator	300	12	3600
Coking stove	2000	3	6000
Clothes washer + dryer	3000	2	6000
Computer	65	3	195
Oven	3380	0.5	1690
Coffee maker	600	0.08	48
Wi-fi box	6	12	72
Vacuum cleaner	350	0.17	59.5
Iron	1000	0.14	140
Extractor hood	139	3	417
Hair dryer	1800	0.2	360
Blender	600	0.03	18
Total			19,379.5

The peak power P_p of the photovoltaic installation is expressed as Eq. (6.41):

$$P_p \geq \frac{E_{need} * P_i}{E_i * PR}. \tag{6.41}$$

E_i depends on the geographical location, the orientation, and the tilt of the photovoltaic array. Therefore, it is necessary to obtain solar irradiation data for the studied region. In this case study, Ifrane city in Morocco has been used. The daily solar irradiation of this latter is presented in Fig. 6.12.

The monthly peak power P_p of this installation is then calculated using Eq. (6.41). Fig. 6.13 illustrates the variation of P_p monthly. It is shown that almost a peak power of 4 kW_p is sufficient to cover the electrical needs of the house during the months of April, May, June, July, and August. On the other hand, during the month of January in which solar irradiance reaches its lowest value, the peak power required is equal to 6 kW_p. Thus, if we wish to cover 100% of the electrical needs, whatever the month of the year, it is recommended to use the most unfavorable case as a reference case, that is, to size the photovoltaic field according to this latter. In this case, the reference case is equivalent to 6 kW_p.

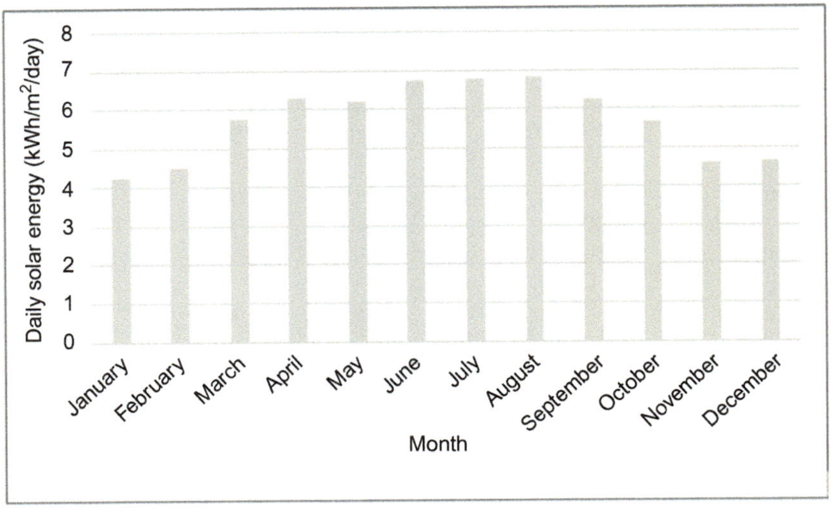

Fig. 6.12 Average daily solar energy of each month in the studied region.

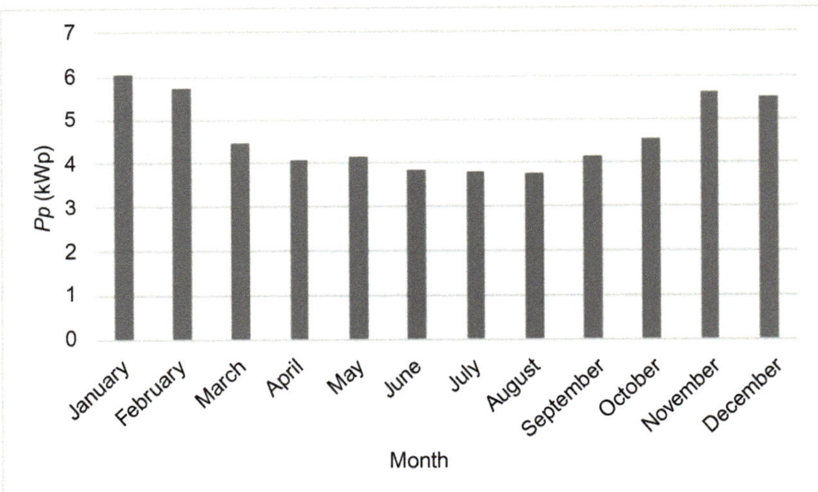

Fig. 6.13 Average daily peak power need for the house for each month.

6.17.4.3 Sizing of the battery bank

The dimensioning of the battery consists of making the following two technical choices:

i. Battery voltage.
ii. Battery capacity.

The calculation of the battery bank voltage is performed with an aim to minimize the voltage drop in the cables and to avoid overheating of the cables. These latter behave in practice like an electrical resistance which induces a voltage drop between the two ends. The voltage drop is expressed by Eq. (6.42):

$$\varepsilon = \frac{\rho * L * I}{S * U_{Bat}}. \tag{6.42}$$

The voltage drops (ε) should not be too large between the place where the voltage is generated (the battery bank) and the place in which it is delivered (regulator). However, this voltage drop depends on the following parameters which include:

i. The voltage (U) from the cables, in this case, the voltage of the U_{Bat} battery bank.
ii. The resistivity of the conductive material (ρ) (copper or aluminum) under the operating temperature conditions, expressed in Ω (mm^2/m).
iii. The cable length (L), expressed in m.
iv. The electrical power passing through the cables (P), expressed in W.
v. The section of cables (S), expressed in mm^2.

The battery voltage is calculated using Eq. (6.43):

$$U_{Bat} = \frac{\rho * 2 * L * I}{S * \varepsilon}. \tag{6.43}$$

The factor 2 in Eq. (6.43) makes it possible to take into account the outward and return distances of the cable. The greater the length L, the higher the voltage of the battery bank. Close proximity between the battery bank and the inverter is more favorable. Note that the regulator should not be located less than 50 cm from the battery bank for safety reasons. By increasing the section of the cables, the voltage of the battery bank can be decreased.

When calculating the battery voltage, the most unfavorable configuration must be taken into account. That is, all electrical equipment of the house operates simultaneously. In this study, the electric power that the batteries must supply is equal to $P = 6000$ W. In order to ensure an optimal operation of the battery bank, the voltage drop between the battery bank and the inverter must be at most 1%, i.e., $\varepsilon = 0.01$, and should have a typical length of $L = 3$ m. The cables used in this study is made of copper with $\rho_0 = 0.01851\ \Omega.\text{mm}^2/\text{m}$. Taking into account the effect of the temperature on the cable, the resistivity becomes:

$$\rho = \rho_0 \times 1.25 = 0.02314\ \Omega.\text{mm}^2/\text{m}.$$

And Eq. (6.42) becomes:

$$S = \frac{\rho * 2 * L * P}{U_{Bat}{}^2 * \varepsilon} = \frac{0.02314 * 2 * 3 * 6000}{U_{Bat}{}^2 * 0.01} = \frac{83304}{U_{Bat}{}^2}.$$

The cable section of the cables must not exceed 50 mm^2 that is S $\leq$ 50 mm^2. This results in $U_{Bat} \geq$ 40.8 V [(83,304) / ($U_{Bat}{}^2$) $\leq$ 50 mm^2].

Therefore, a battery bank voltage equal to 48 V is used. The section of the cable placed between the battery bank and the inverter will be equal to $S = 50\,\text{mm}^2$. It is important to check that the 50 mm^2 cable can withstand the current flowing through it. Indeed, under a voltage of 48 V and a power of 6000 W, the resulting current rises to $I_{Bat} = \frac{P_{Bat}}{U_{Bat}} = \frac{6000}{48} = 125\,A$. The admissible current of a 50 mm^2 copper cable is 194 A, which is much higher than the calculated 125 A.

The second parameter to be determined when sizing the battery bank is its capacity. Capacity is an indicator used to assess the amount of energy stored in a battery. Usually, it is expressed in Ah. The capacity of the battery bank is calculated using the Eq. (6.44):

$$C_{T_d} \geq \frac{N_d * E_{need}}{D_P * K_T}. \tag{6.44}$$

C_{T_d} is battery capacity associated with discharge time (Ah), N_d is the reserve autonomy (day), D_P is the maximum depth of discharge, E_{need} is the Daily energy consumed (kWh/day) and is equal to $E_{need} = \frac{I_{need}}{U_{Bat}}$, K_T is the capacity temperature coefficient $K_T = N_d * 24$ (hour).

The daily energy requirement is 19.3795 kWh/day. This value should be expressed in Ah/day. With a voltage of 48 V, the daily requirement is:

$$(19379.5\,\text{Wh/day})/(48\,\text{V}) = 404\,\text{Ah/day}$$

Solar batteries are still very expensive. Just 1-day autonomy has been chosen because the house will be connected to the grid at low voltage. This means that the batteries can power, independently of the photovoltaic field, the electrical equipment defined in Table 6.3 for 1 day. Deep discharges are detrimental for the lifespan of lead-acid batteries. Maximum depth of discharge of 90% has been used, i.e., $D_P = 90\%$. In this study, the batteries will operate at $-10°$C in winter, and it is necessary to apply the corrective coefficient $K_T(\text{C}) = 0.85$, according to Table 6.4 (using manufacturer's data) [42]:

Table 6.4 Different corrective coefficient corresponds to battery operating temperature.

Battery operating temperature (°C)	−20	−10	0	10	20	30	40	50
Corrective coefficient (applied to C_{24})	0.8	0.85	0.9	0.95	1	1.04	1.1	1.13

By applying Eq. (6.44), the capacity of the battery is calculated:

$$C_{1*24} \geq \frac{1*404}{0.9*0.85} Ah = 528\, Ah.$$

Thus, the C_{24} capacity of the battery must be at least 528 Ah. The value of the nominal capacity C_{10} can be deduced using the relation: $C_{10} \approx 71\% \times C_{24}$, that is equal to 375 Ah. Therefore, the case makes use of four lead batteries of 12 V and 100 Ah connected in series.

6.17.4.4 Selection of the solar charge controller and inverters

The photovoltaic regulator must be compatible with the other components of the PV installation such as PV panels and the battery bank.

This study makes use of a PV charge controller with the MPPT function. The is latter has been chosen because of several aspects including:

i. Wiring the modules will be simple. With an MPPT regulator, there are fewer constraints on the voltage of the photovoltaic field.

ii. The efficiency of the photovoltaic installation is improved by around 15%. This is relevant on the one hand because the power of the photovoltaic field exceeds 1 kW_p (in this case study, the power is 6 kW_p). On the other hand, because the site temperature is very low in winter (in cold climate, the voltage of the photovoltaic modules increases, which impacts the system efficiency in the absence of an MPPT system).

The dimensioning of the regulator is carried out according to 4 majors' criteria including:

1. The maximum power of the photovoltaic field must be less than the nominal power of the regulator.

2. The open-circuit voltage of the photovoltaic field (increased by the safety factor) must be lower than the maximum voltage admissible by the inverter.
3. The output current of the photovoltaic field must not exceed the maximum current admissible by the regulator.
4. The regulator must be compatible with the voltage of the battery bank.

The selection of the inverter is based on three criteria which include:

1. Compatibility in power: The input power of the inverter, on the DC side, should be less than the maximum power allowed by the inverter. The inverter must at all times request the maximum power from the photovoltaic group to which it is connected. The search for the maximum power point is carried out by the Maximum Power Point Tracking (MPPT) system integrated upstream of the inverter.
2. Voltage compatibility: An inverter is characterized by a maximum admissible input voltage U_{max}, and the voltage delivered by the modules should be less than U_{max}. It is also necessary to ensure that the voltage delivered by the photovoltaic group is included in the MPPT voltage range of the inverter to which it is connected. If this is not the case, there will be no damage to the inverter, but only a loss of power.
3. Current compatibility: When the input current of the inverter on the DC side is higher than the maximum current admissible by the inverter, it continues to operate but supplies just the power corresponding to its maximum current.

From these three criteria, the sizing of the inverters will determine the PV modules wiring. In this case study, a PV system of approximatively of 6 kW_p is needed to cover the electrical needs of the load.

The PV installation of the house consists of three technologies which include monocrystalline, polycrystalline, and amorphous of power of 2.04 kW_p, 2.04 kW_p, and 1.86 kW_p, respectively. The first two systems consist of 8 photovoltaic panels each connected in series. The third PV system is amorphous; it is composed of two parallel strings compromising of 6 PV panels connected in series. The PV systems are oriented south with a 32° tilt and an azimuth of 0°. Each technology is connected to the SMA inverter of 2 kW. The solar modules do not have any cleaning schedule. Table 6.5 illustrates the specifications techniques of PV panels and inverter used. This case study is located in Ifrane, Morocco. The geographical data of this site are presented in Table 6.6.

Table 6.5 PV system parameter under standard test conditions.

Technologies	mc-Si	pc-Si	a-Si
PV characteristics			
Rated power P_{mpp} (Wp)	255	255	155
Open-circuit voltage V_{oc} (V)	37.8	38	85.5
Operating voltage V_{mpp} (V)	31.4	30.9	65.2
Short-circuit current I_{sc} (A)	8.66	8.88	2.54
Operating current I_{mpp} (A)	8.15	8.32	2.38
Efficiency (%)	15.51	15.21	9.9
PV system length (mm)	1675		1412
PV system width (mm)	1001		1112
PV system height (mm)	31	33	35
PV system weight (kg)	21.2	18	19.5
Inverter characteristics			
Maximum DC power (kW)	2.1		
Maximum DC voltage (V)	700		
Nominal AC power (kW)	2.0		
MPPT DC voltage range (V)	175–560		
Maximum efficiency (%)	96		

Table 6.6 Geographic site data for Ifrane to do the prediction.

Number of days	Latitude (°)	Longitude (°)	Altitude (m)
7 days	33.5	−5.12	1665

6.17.5 Simulation results and discussions

The prediction period used in this study is 7 days from January 31 till April 5, 2020. Fig. 6.14A and B illustrate the predicted solar radiation based on temperature and cloud cover for the whole studied week and for the first day of that week. Dark Sky API has been used to obtain the cloud cover and temperature used in this model. Table 6.7 illustrates the summary of the PV/batteries sytem studied.

Fig. 6.15 presents the PV output power prediction based on previous solar radiation predictions over the investigated one-week period of monocrystalline, polycrystalline, and amorphous technology, respectively.

The simulation results show that the weather conditions of the studied case scenarios vary. The minimum temperature is obtained during the first 3 days with a high average cloud cover. However, the minimum cloud

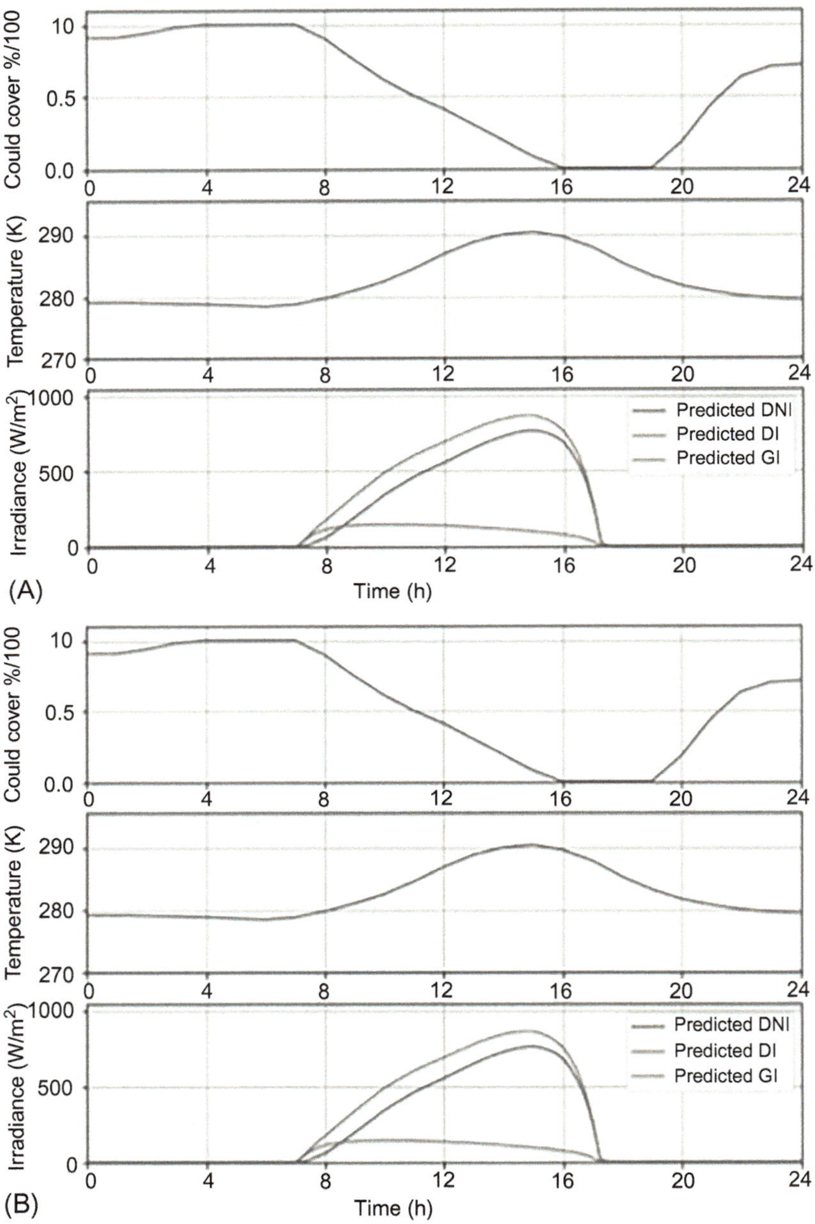

Fig. 6.14 Solar irradiance prediction. (A) cloud cover forecast fraction (top), the predicted outdoor temperature (middle), and the forecasted direct normal (DNI), diffuse (DI), and global (GI) irradiance components (bottom) over a period of 1 week; (B) prediction of solar irradiance, cloud cover, and temperature for the first day of the week.

Table 6.7 Summary of the PV system studied.

Daily electrical needs: E_{need} (kWh)	19.3795
Average daily solar radiation for the worst month: E_i (kWh/m^2/day)	4.2
Battery DC voltage: U_{batt} (V)	48
Number of days of autonomy: N_d	1
Battery discharge depth: D_p	0.9
PV array power (kW$_p$)	6
Nominal batteries capacity (Ah)	400

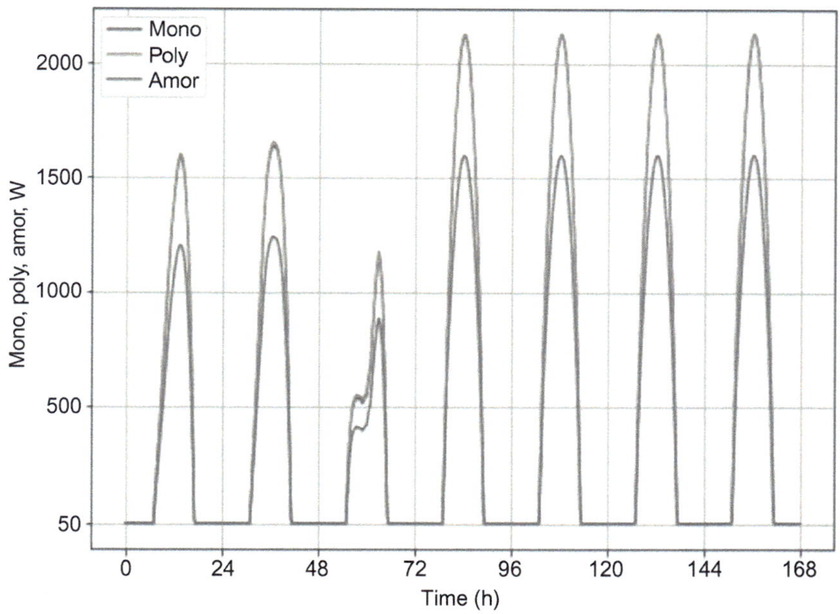

Fig. 6.15 The predicted power production of the three investigated PV technologies over a period of 1 week.

cover equivalent to approximately 0% is recorded during the last 4 days with increasing temperature, and a very high forecasted DNI, and GI as well as a low DI. As such, the weather conditions for the first day are illustrated in Fig. 6.14B. It can be seen that the highest predicted DNI, and GI are achieved between 2 pm and 4 pm, while the highest predicted DI is obtained at 9 am. The studied indices including global solar radiation, temperature, and cloud cover have an important impact on the PV output power prediction which is illustrated in Fig. 6.15. Due to the low predicted temperature and solar irradiance during the first 3 days, the forecasted PV power production of the three investigated types is low as compared to the last 4 days.

Figs. 6.16 and 6.17 summarize the prediction of the net electrical power of the house over 1 week and for the first day of the same week, respectively. Figs. 6.16A and 6.17A illustrate the predicted power production as negative and consumption as positive over 1 week, and for the first day of the same week, respectively. Figs. 6.16B and 6.17B illustrate the predicted primary battery charging level over 1 week and for the first day of the same week, respectively. Figs. 6.16C and 6.17C show the contest performance measure, epsilon, which indicates the fraction of power consumption that is produced by the PV to that consumed by the battery system over 1 week and for the first day of the same week, respectively. The results demonstrate that the PV array provides nearly all the energy needed over this week to meet expected house energy demands.

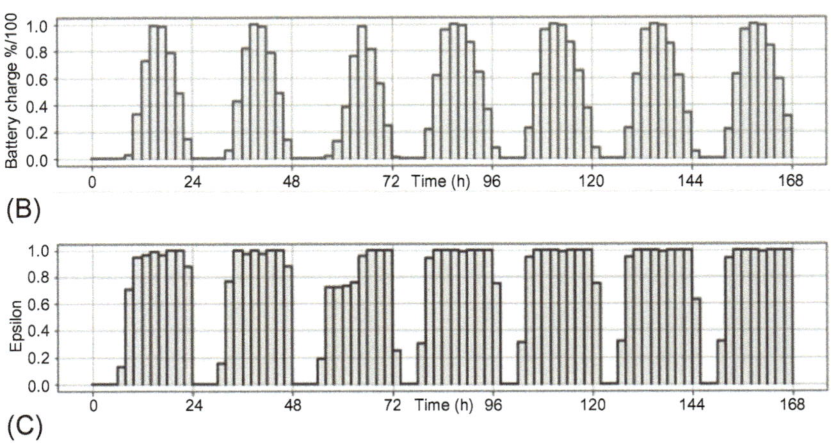

Fig. 6.16 Net electrical power prediction of the house over 1 week. (A) predicted net output power production and consumption; (B) predicted state of battery charge; (C) the contest performance measure epsilon.

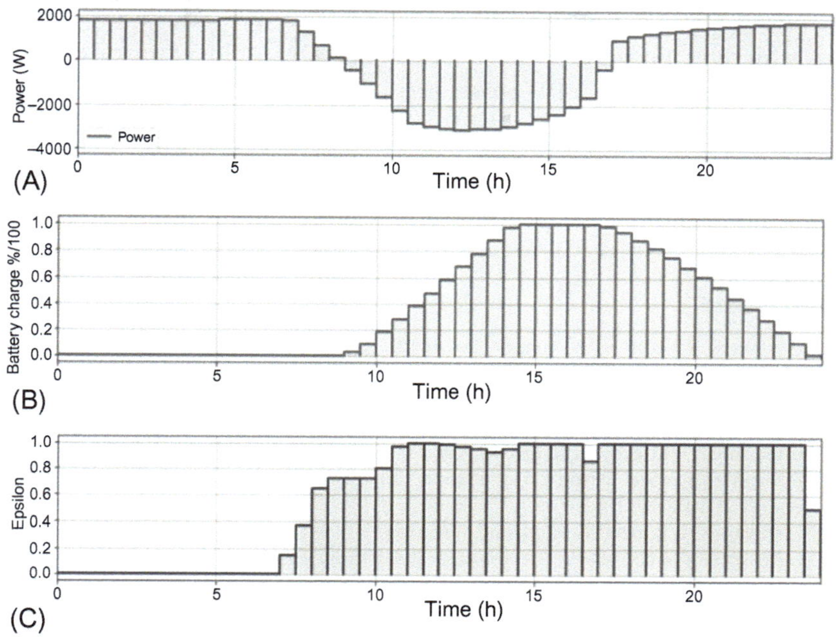

Fig. 6.17 Net electrical power prediction of the house for the first day of study week. (A) Predicted net output power production and consumption; (B) predicted state of battery charge; (C) the contest performance measure epsilon.

6.18 Conclusions

A study of a photovoltaic/battery hybrid system connected to the electric network with an aim to ensure continuous energy provision has been performed in this chapter. This system operates at its optimum power with the use of a PV charge controller and an inverter.

The first part of this chapter discusses the principle of photovoltaic conversion, and some generalities about PV technologies and energy storage, as well as the characteristics and the operating principle of battery energy storage systems.

The following part was devoted to the presentation of the proposed PV model. The design and sizing of the photovoltaic system operating in MPPT and coupled to lead-acid batteries have been performed to meet the power demand of the load.

The last part presents a case study of a house equipped with a hybrid photovoltaic/battery system connected to the grid. The sun's irradiance is predicted by the developed model based on the site location and the time of the year. Dark Sky API has been utilized to obtain the daily cloud-cover forecasts which are used to determine the instantaneous incident solar irradiance during the day. Finally, a PV power prediction model was carried out based on solar irradiance forecast and PV technologies characteristics. The predicted PV power output results demonstrate that the hybrid system provides nearly all the electrical needs of the house.

References

[1] Brockway PE, Owen A, Brand-Correa LI, Hardt L. Estimation of global final-stage energy-return-on-investment for fossil fuels with comparison to renewable energy sources. Nat Energy 2019; https://doi.org/10.1038/s41560-019-0425-z.

[2] Hanif I, Faraz Raza SM, Gago-de-Santos P, Abbas Q. Fossil fuels, foreign direct investment, and economic growth have triggered CO2 emissions in emerging Asian economies: some empirical evidence. Energy 2019; https://doi.org/10.1016/j.energy.2019.01.011.

[3] Handayani K, Krozer Y, Filatova T. From fossil fuels to renewables: an analysis of long-term scenarios considering technological learning. Energy Policy 2019; https://doi.org/10.1016/j.enpol.2018.11.045.

[4] York R, Bell SE. Energy transitions or additions?: why a transition from fossil fuels requires more than the growth of renewable energy. In: Energy research and social science. 2019; https://doi.org/10.1016/j.erss.2019.01.008.

[5] Sampaio PGV, González MOA. Photovoltaic solar energy: conceptual framework. In: Renewable and sustainable energy reviews. 2017; https://doi.org/10.1016/j.rser.2017.02.081.

[6] Hernandez JA, Velasco D, Trujillo CL. Analysis of the effect of the implementation of photovoltaic systems like option of distributed generation in Colombia. In: Renewable and sustainable energy reviews. 2011; https://doi.org/10.1016/j.rser.2011.02.003.

[7] Nair NKC, Garimella N. Battery energy storage systems: assessment for small-scale renewable energy integration. Energ Buildings 2010; https://doi.org/10.1016/j.enbuild.2010.07.002.

[8] Vieira FM, Moura PS, de Almeida AT. Energy storage system for self-consumption of photovoltaic energy in residential zero energy buildings. Renew Energy 2017; https://doi.org/10.1016/j.renene.2016.11.048.

[9] Goel S, Sharma R. Performance evaluation of stand alone, grid connected and hybrid renewable energy systems for rural application: a comparative review. In: Renewable and sustainable energy reviews. 2017; https://doi.org/10.1016/j.rser.2017.05.200.

[10] Cai B, Liu Y, Ma Y, Huang L, Liu Z. A framework for the reliability evaluation of grid-connected photovoltaic systems in the presence of intermittent faults. Energy 2015; https://doi.org/10.1016/j.energy.2015.10.068.

[11] Shivashankar S, Mekhilef S, Mokhlis H, Karimi M. Mitigating methods of power fluctuation of photovoltaic (PV) sources—a review. In: Renewable and sustainable energy reviews. 2016; https://doi.org/10.1016/j.rser.2016.01.059.

[12] Barelli L, Bidini G, Bonucci F, Castellini L, Castellini S, Ottaviano A, Pelosi D, Zuccari A. Dynamic analysis of a hybrid energy storage system (H-ESS) coupled to a photovoltaic (PV) plant. Energies 2018; https://doi.org/10.3390/en11020396.

[13] International Renewable Energy Agency (IRENA). Electricity storage and renewables: costs and markets to 2030. In: Electricity-storage-and-renewables-costs-and-markets. 2017.

[14] Ould Amrouche S, Rekioua D, Rekioua T, Bacha S. Overview of energy storage in renewable energy systems. Int J Hydrog Energy 2016; https://doi.org/10.1016/j.ijhydene.2016.06.243.
[15] Cleveland CJ, Morris C. Section 15 - Photovoltaics. In: Handbook of energy. 2014; https://doi.org/10.1016/B978-0-12-417013-1.00015-7.
[16] Jestin Y. Down-shifting of the incident light for photovoltaic applications. In: Comprehensive renewable energy. 2012; https://doi.org/10.1016/B978-0-08-087872-0.00131-1.
[17] McEvoy A, Markvart T, Castañer L. Solar cells. In: Solar cells. 2013; https://doi.org/10.1016/C2011-0-07789-X.
[18] Riverola A, Vossier A, Chemisana D. Fundamentals of solar cells. In: Nanomaterials for solar cell applications. 2019; https://doi.org/10.1016/B978-0-12-813337-8.00001-1.
[19] Sangster AJ. Solar photovoltaics. Green Energ Technol 2014; https://doi.org/10.1007/978-3-319-08512-8_7.
[20] Perraki V, Kounavis P. Effect of temperature and radiation on the parameters of photovoltaic modules. J Renew Sustain Energy 2016; https://doi.org/10.1063/1.4939561.
[21] Bayod-Rújula AA. Solar photovoltaics (PV). In: Solar hydrogen production: processes, systems and technologies. 2019; https://doi.org/10.1016/B978-0-12-814853-2.00008-4.
[22] Attari K, Elyaakoubi A, Asselman A. Performance analysis and investigation of a grid-connected photovoltaic installation in Morocco. Energy Rep 2016; https://doi.org/10.1016/j.egyr.2016.10.004.
[23] National Renewable Energy Laboratory. Perovskite efficiency chart. National Renewable Energy Laboratory; 2019.
[24] Bhatia SC. Solar photovoltaic systems. In: Advanced renewable energy systems. 2014; https://doi.org/10.1016/b978-1-78242-269-3.50005-x.
[25] Bhatia SC. Energy resources and their utilisation. In: Advanced renewable energy systems. 2014; https://doi.org/10.1016/b978-1-78242-269-3.50001-2.
[26] Salameh Z. Chapter 4 – Energy storage. In: Renewable energy system design. 2014; https://doi.org/10.1016/B978-0-12-374991-8.00004-0.
[27] Williamson SS, Cassani PA, Lukic S, Blunier B. Energy storage. In: Power electronics handbook. 2011; https://doi.org/10.1016/B978-0-12-382036-5.00046-X.
[28] Mokrani Z, Rekioua D, Rekioua T. Modeling, control and power management of hybrid photovoltaic fuel cells with battery bank supplying electric vehicle. Int J Hydrog Energy 2014; https://doi.org/10.1016/j.ijhydene.2014.03.215.
[29] Pistoia G. Chapter 1 - Basic battery concepts. In: Batteries for portable devices. 2005; https://doi.org/10.1016/B978-044451672-5/50001-6.
[30] Ioan S, Calin S. Solar radiation 2.1. 13–28. https://doi.org/10.1016/B978-0-12-811662-3.00002-5.
[31] Koohi-Kamali S, Rahim NA, Mokhlis H, Tyagi VV. Photovoltaic electricity generator dynamic modeling methods for smart grid applications: a review. In: Renewable and sustainable energy reviews. 2016; https://doi.org/10.1016/j.rser.2015.12.137.
[32] Lamba R, Kaushik SC. Modeling and performance analysis of a concentrated photovoltaic-thermoelectric hybrid power generation system. Energy Convers Manag 2016; https://doi.org/10.1016/j.enconman.2016.02.061.
[33] Xiong G, Zhang J, Shi D, He Y. Parameter extraction of solar photovoltaic models using an improved whale optimization algorithm. Energy Convers Manag 2018; https://doi.org/10.1016/j.enconman.2018.08.053.
[34] Chowdhury S, Chowdhury SP, Taylor GA, Song YH. Mathematical modelling and performance evaluation of a stand-alone polycrystalline PV plant with MPPT facility, In: IEEE power and energy society 2008 General meeting: conversion and delivery of

electrical energy in the 21st centuryPES; 2008; https://doi.org/10.1109/PES.2008.4596376.
[35] Dark Sky - 7, Meknès-Tafilalet. (n.d.). Retrieved May 5, 2020, from https://darksky.net/forecast/33.538,-5.108/us12/en
[36] Segarra EL, Du H, Ruiz GR, Bandera CF. Methodology for the quantification of the impact of weather forecasts in predictive simulation models. Energies 2019;12(7):1–16. https://doi.org/10.3390/en12071309.
[37] Laue EG. The measurement of solar spectral irradiance at different terrestrial elevations. Sol Energy 1970; https://doi.org/10.1016/0038-092X(70)90006-X.
[38] Meinel AB, Meinel MP, Glaser PE. Applied solar energy: an introduction. Phys Today 1977; https://doi.org/10.1063/1.3037372.
[39] PVEducation. (n.d.). Retrieved May 6, 2020, from https://www.pveducation.org/
[40] Precup R-E, Kamal T, Hassan SZ. Solar photovoltaic power plants: advanced control and optimization techniques. In: Power systems. 2019; https://doi.org/10.1155/2017/1041375.
[41] Zaouche F, Rekioua D, Gaubert J-P, Mokrani Z. Supervision and control strategy for photovoltaic generators with battery storage. Int J Hydrog Energy 2017;42(30):19536–55. https://doi.org/10.1016/j.ijhydene.2017.06.107.
[42] Masters GM. Renewable and efficient electric power systems. In: Renewable and efficient electric power systems. 2004; https://doi.org/10.1002/0471668826.

CHAPTER 7

Comparison of objective functions on the sizing of hybrid PV and wind energy systems with and without energy storage systems

Loiy Al-Ghussain[a] and Onur Taylan[b,c]

[a]Mechanical Engineering Department, University of Kentucky, Lexington, KY, United States
[b]Mechanical Engineering Program, Middle East Technical University, Northern Cyprus Campus, Mersin, Turkey
[c]Center for Solar Energy Research and Applications (GÜNAM), Middle East Technical University, Ankara, Turkey

Nomenclature

A_{CO_2}	annual avoided CO_2 emissions (ton)
AH	number of autonomous operating hours of the RES
C_C	RES capital cost (USD)
CF	annual capacity factor (%)
C_M	annual maintenance cost (USD)
D	electrical demand (kWh)
D_{EX}	excess energy (kWh)
D_G	demand covered by the utility grid (kWh)
D_M	maximum hourly demand (kWh)
D_{RES}	demand cover by the RES at time t (kWh)
DC	ESS depth of charge limit (kWh)
DS_F	annual demand supply fraction (%)
E_C	electricity cost (USD/kWh)
E_{PV}	electrical energy production of a PV power plant (kWh)
E_{RES}	electrical energy production of by a RES (kWh)
E_W	electrical energy production of a wind turbine (kWh)
GT	local grid tariff (USD/kWh)
H	hub height (m)
H_g	reference or ground height (m/s)
I	global insolation on horizontal (Wh/m^2)
$I_{b,n}$	direct normal or beam insolation (Wh/m^2)
$I_{b,t}$	direct or beam insolation on a tilted surface (Wh/m^2)
I_{CO2}	CO_2 intensity of electricity (kg/kWh)
I_d	diffuse insolation on horizontal (Wh/m^2)
$I_{d,t}$	diffuse insolation on a tilted surface (Wh/m^2)
I_R	reference insolation at standard conditions (Wh/m^2)
I_T	global insolation on a tilted surface (Wh/m^2)

Hybrid Energy System Models
https://doi.org/10.1016/B978-0-12-821403-9.00004-4

K shape parameter of the Weibull distribution for wind speed
L_{loc} longitude of the location (degrees)
L_{st} standard meridian for the local time zone (degrees)
LS lifespan of the RES (years)
M_A surface area of PV panel (m^2)
N_m number of PV modules
N_w number of wind turbines
n day number
P_e average electrical power produced by the wind turbine (kW)
$P_{e,R}$ rated wind turbine electrical power (kW)
P_{PV} PV rated power (kW)
R_F annual renewable energy fraction (%)
Rv annual net revenues (USD)
R_1 first-year net revenues (USD)
r annual discount rate (%)
SC storage capacity of electrical energy (kWh)
SE^t stored electrical energy at time t (kWh)
SE_A^t available electrical energy in the ESS at time t (kWh)
T_A ambient temperature (°C)
T_{PV} PV module's temperature (°C)
$T_{R,NOC}$ reference PV cell temperature at nominal operating conditions (°C)
$T_{R,STC}$ PV cell temperature at standard test conditions (°C)
t_z local time zone (h)
t time in a year (h)
t_s solar time (h)
t_{std} standard time (h)
u_C cut-in wind speed (m/s)
u_F furling wind speed (m/s)
u_g wind speed at ground level (m/s)
u_H wind speed at hub height (m/s)
u_R rated wind speed (m/s)
$\overline{u}$ mean wind speed at hub height (m/s)

Greek letters

α wind shear coefficient
β tilt angle of the PV modules (degrees)
β_R PV cell temperature coefficient (1/°C)
γ surface azimuth angle (degrees)
γ_s solar azimuth angle (degrees)
δ declination angle (degrees)
η_l losses in the PV system (%)
η_r ESS round-trip efficiency (%)
η_{PV} PV module efficiency (%)
$\eta_{PV,R}$ PV reference module efficiency (%)
θ incident angle (degrees)
θ_z zenith angle (degrees)

σ	wind speed standard deviation (m/s)
ϕ	latitude angle (degrees)
ω	hour angle (degrees)

Subscripts and superscripts

l	long term
s	short term

Acronyms and abbreviations

DoD	depth of discharge
ESS	energy storage system
HESS	hybrid ESS (battery/PHS)
LCOE	levelized cost of electricity (USD/kWh)
METU NCC	Middle East Technical University Northern Cyprus Campus
NOCT	nominal operating PV cell temperature (°C)
NPV	net present value (USD)
PBP	simple payback period (years)
PHS	pumped hydro storage
PV	photovoltaic
RES	renewable energy system
TMY	typical meteorological year
WTS	wind turbine system
ZBB	zinc-bromine battery

7.1 Introduction

Increasing energy demand and shortage of nonrenewable fuel resources [1–3] worlwide force industries to continuously research improvements to their energy-consuming processes [4, 5] or to look for optimal quality configurations [6] reducing required mending energy, or innovate unique configurations to improve system's performance and energy usage [78]. Renewable energy systems (RESs) are also part of the solution. Among all the available RESs, solar including photovoltaic (PV) and CSP [79,81] and wind energy systems stand out as attractive and alternative options for energy generation with a considerable growth rate in the global market [7]. The technical and economic feasibility of such systems has been extensively studied in the literature [8–18], where a large number of studies have presented several optimization techniques to come up with the optimal capacities or sizes of the components with either maximum economic and technical benefits or both [12,14,19–23,82]. Furthermore, integrating energy storage systems (ESSs) to RESs not only increases the reliability and autonomy of RESs but also increases the matching between energy generation profiles and the demand [21,24,80]. Therefore, tremendous effort has been invested in developing optimal sizing methodologies for PV/ESS and wind/ESS [9, 13–16, 20, 25–32].

The hybridization between PV system and wind turbine systems (WTSs) improves the techno-economic feasibility of the system by improving the matching between the energy generation and the demand, thus decreasing the dependency on the grid. Several studies have worked on techno-economic feasibilities of hybrid PV + WTS, where the major conclusion of the studies was the vitality of the optimal sizes of both PV and WTS on the techno-economic feasibility of the system [17, 21, 33–37]. Therefore, the studies in the literature focused on developing single- or multiobjective optimization techniques [33–35, 38–40].

Although the hybridization of PV and WTS enables a better matching between the electricity supply and the demand, the hybrid PV + WTS system still cannot cover the whole demand and feasibly achieve full autonomy. The integration of ESS with hybrid PV + WTS system increases the system's ability to meet more demands by reallocating the excess energy to match the electricity demand during the deficiency periods, which could enhance the system's techno-economic feasibility where the optimization of the system's capacities including the ESS capacity is vital for the techno-economic feasibility [22, 41–44]. In the literature, several studies considered the integration of ESS with hybrid PV + WTS systems, such as batteries [34, 35, 44–47], pumped hydro storage (PHS) [21, 48, 49], compressed air storage system [3, 50], and fuel cells as energy storage [24, 51–53].

The optimization of the system's capacities is vital, where the optimization methodologies presented in the literature generally aim at one objective, which is either maximizing the technical feasibility regardless of the economic one [44] or maximizing the economic benefits alone [22, 41–43, 54]. However, some studies considered the maximization of the technical and economic feasibilities at the same time [21, 25, 34–36, 54–57]. For instance, Al-Ghussain et al. presented in several studies [24, 34–36, 47] a new optimization methodology for different RES configurations with or without ESSs, where the optimization methodology was based on increasing the RES share with the cost of electricity less than or equal to the local electricity tariff, which allowed the maximum utilization of the renewable resources in an economically feasible way. Moreover, Jamshidi and Askarzadeh [25] developed a multiobjective optimization scheme to find the capacities of a hybrid PV-fuel cell/diesel generator system to meet the electrical energy demand of an off-grid community in Kerman, Iran. The objectives of the algorithm were minimizing the loss of power supply probability and the total net present cost of the system, while Zhang et al. [54] developed an algorithm to find the optimal stand-alone different PV and WTS configurations with battery or hydrogen tank as ESS to meet the residential electricity demand in Iran based on minimizing the life cycle cost of the system.

Therefore, this study aims to find optimal components' sizes or capacities using three different sizing objectives for different combinations of PV and WTS configurations with and without ESSs. The optimization objectives are (i) maximization of the renewable energy share without any economic constraint, (ii) minimization of the cost of electricity without any technical constraint, and (iii) maximization of the renewable energy share with the cost of electricity being less than or equal to the grid tariff. The results are presented in terms of optimal components' sizes, demand met by the system, cost of energy, net present value (NPV), and annual avoided CO_2 emissions due to implementation of the system. Middle East Technical University Northern Cyprus Campus is used to demonstrate the proposed methodology in this study, as the meteorological and demand data of the campus are available.

7.2 Theory and methodology

7.2.1 Solar geometry model

Several angles are required to calculate the total radiation incident on the PV module to estimate the electrical energy generation by the PV modules. The angle of incidence is one of these essential angles, which is the angle between the surface normal and the direct solar radiation. The angle of incidence depends on the positions of the sun, earth, and object's surface. The positions of these elements must be geometrically found to find the angle of incidence. The position of the sun is described by two angles at any location: the zenith angle and the solar azimuth angle. Both angles require solar time (t_s) per Eq. (7.1), which converts local time to the solar time [58]:

$$t_s = t_{std} + 4 \times (L_{st} - L_{loc}) + E, \tag{7.1}$$

where L_{st} is estimated using Eq. (7.2), L_{loc} using Eq. (7.3), while the constant E can be estimated using Eq. (7.4) [58]:

$$L_{st} = \begin{cases} -t_z \times 15, & t_z \leq 0 \\ 360 - t_z \times 15, & t_z > 0 \end{cases}, \tag{7.2}$$

$$L_{loc} = \begin{cases} L_{loc}, & \text{if } L_{loc} \text{ in the West,} \\ 360 - L_{loc}, & \text{if } L_{loc} \text{ in the East,} \end{cases} \tag{7.3}$$

$$E = 229.2 \times (0.000075 + 0.001868 \times \cos B - 0.032077$$

$$\times \sin B - 0.014615 \times \cos(2 \times B) - 0.04089 \times \sin(2 \times B)), \tag{7.4}$$

where B is a constant and can be estimated using Eq. (7.5) [58],

$$B = (n - 1) \times \frac{360}{365}. \tag{7.5}$$

The solar time is used in the angular form which will be used in the estimation of the incidence angle using Eq. (7.6), where the angular form is known as the hour angle (ω) [58]:

$$\omega = (t_s - 12) \times 15. \tag{7.6}$$

The zenith angle, which can be calculated using Eq. (7.7), is the angle between the normal direction of a horizontal surface and the direct insolation, whereas the solar azimuth angle, which can be calculated using Eq. (7.8), is the angle between the horizontal component of the direct insolation and the South direction [58]:

$$\cos\theta_z = \cos\phi\cos\delta\cos\omega + \sin\phi\sin\delta, \tag{7.7}$$

$$\gamma_s = \text{sign}(\omega) \times \left| \cos^{-1}\left(\frac{\cos\theta_z \sin\phi - \sin\delta}{\sin\theta_z \cos\phi} \right) \right|, \tag{7.8}$$

where δ, named the declination angle, represents the angle between the earth's equator at solar noon and the direct insolation and can be estimated using Eq. (7.9) [58]:

$$\delta = 23.45 \times \sin\left(360 \times \frac{284 + n}{365}\right). \tag{7.9}$$

To estimate the solar insolation on a tilted PV module, the orientation of the module should be determined geometrically, which involves the surface tilt angle (β) and the surface azimuth angle (γ). The surface azimuth angle is the angle between the South direction and the horizontal component of the surface normal, whereas the surface tilt angle is the inclination of the module's surface with respect to the horizontal surface [58]. Finally, after the estimation of the solar geometry angles, the incident angle is estimated for any hour angle using Eq. (7.10). The surface tilt and azimuth angles need to be optimized to maximize the solar insolation on the PV module, which would maximize the electrical energy output. For METU NCC, the optimal tilt and azimuth angles are 25 and 0 degrees (PV modules oriented toward the south), respectively, to maximize the annual energy production [59]:

$$\cos\theta = \cos\theta_z \cos\beta + \sin\theta_z \sin\beta \cos(\gamma_s - \gamma). \tag{7.10}$$

7.2.2 Solar resources model

The total solar insolation, I_T, on any tilted surface has three main components: (i) direct, (ii) diffuse, and (iii) reflected insolation. The reflected insolation is commonly neglected in PV applications as the amount of reflected insolation is small compared to the other components. Eq. (7.11) can be used for calculating the total insolation on a tilted surface [58]:

$$I_T = I_{b,t} + I_{d,t}, \tag{7.11}$$

where the direct (beam) insolation on a tilted surface, $I_{b,t}$, can be estimated using Eq. (7.12), whereas the diffuse one, $I_{d,t}$, can be estimated using Eq. (7.13) [58]. Note that the diffuse component is estimated based on the isotropic sky model, which assumes that the diffuse radiation has the same magnitude from all directions above the surface [60]:

$$I_{b,t} = I_{b,n} \times \cos\theta, \tag{7.12}$$

$$I_{d,t} = I_d \times \left(\frac{1 + \cos\beta}{2}\right). \tag{7.13}$$

In this study, the hourly typical meteorological year (TMY) dataset was used in the simulations, where TMY dataset on a horizontal surface at METU NCC was generated using Meteonorm v7.1 software.

7.2.3 Energy estimation from a PV system

After the amount of insolation on a tilted surface is estimated, the energy production from a PV module can be estimated. The efficiency of a PV module can be estimated using Eq. (7.14), which takes into account the effects of ambient temperature and solar insolation on the modules:

$$\eta_{PV} = \eta_{PV,R} \times [1 - \beta_R \times (T_{PV} - T_{R,\mathrm{STC}})], \tag{7.14}$$

Table 7.1 Technical specifications of the PV modules [61].

Parameter	Value
P (W)	285
$\eta_{PV,\ R}$ (%)	17.41
β_R (1/°C)	0.0041
$T_{R,\ STC}$ (°C)	25
$T_{R,\ NOCT}$ (°C)	20
NOCT (°C)	45
I_R (W/m^2)	800
A_m (m^2)	1.63

where T_{PV} can be estimated using Eq. (7.15),

$$T_{PV} = T_A + (\text{NOCT} - T_{R,\text{NOCT}}) \times \frac{I_T}{I_R}. \tag{7.15}$$

In this study, the hourly TMY dataset for the ambient temperature at METU NCC was generated using Meteonorm v7.1. In addition, specifications of the PV module of CS6K-285M from Canadian Solar were used [61], which are tabulated in Table 7.1. Based on the estimations for the total insolation on the PV module and the efficiency of the PV module, the electrical energy generation of the PV power plant can be calculated using Eq. (7.16):

$$E_{PV} = \eta_{PV} \times I_T \times \text{M}_\text{A} \times \text{N}_m \times \eta_l, \tag{7.16}$$

where η_l is the efficiency for the balance of the system, which includes energy losses in inverters, wiring, shading, and due to dust on modules, and in this study, η_l was assumed to be 0.85 [62–64].

7.2.4 Energy estimation from a wind turbine

The electricity production of a wind turbine highly depends on the wind speed at its hub height (H), which can be estimated through extrapolation from the wind speed measurements at the ground level (u_g at H_g) using the power law as in Eq. (7.17), where α is referred to as the wind shear coefficient. In this study, α was taken as 1/7 [36]:

$$u_H = u_g \times \left(\frac{H}{H_g}\right)^{\alpha}. \tag{7.17}$$

After obtaining the wind speed at the hub height, the electrical power generation of the wind turbine can be estimated using Eq. (7.18):

$$P_e = \begin{cases} 0 & \text{for} \quad u_H \langle u_C \text{ or } u_H \rangle u_F, \\ P_{e,R} \times \dfrac{(u_C)^K - (u_H)^K}{(u_C)^K - (u_R)^K} & \text{for} \quad u_C \leq u_H \leq u_R, \\ P_{e,R} & \text{for} \quad u_R < u_H \leq u_F, \end{cases} \tag{7.18}$$

where, based on the Justus theory, K can be calculated as [65]

$$k = \left(\frac{\sigma}{\overline{u}}\right)^{-1.086} \quad \text{for } 1 \leq k \leq 10. \tag{7.19}$$

Finally, the electrical power energy generation by the wind turbine(s) can be estimated as

$$E_W = P_e \times N_w, \tag{7.20}$$

where N_w is the number of wind turbines.

In this study, Meteonorm v7.1 software was used to obtain hourly wind speeds at ground level. Based on the TMY data, $\overline{u}$ is equal to 4.1 m/s and σ is equal to 3.27 m/s at hub height in METU NCC. The wind turbine with a rated power of 2 MW from the GAMESA company (G114-2.0) [66] was used whose specifications are tabulated in Table 7.2.

7.2.5 Electrical energy demand model

The temporal variations of the solar and wind resources, as well as the electrical energy demand, create a challenge for the harmonization between the energy generation of RESs and the demand. This challenge is tackled by introducing energy storage systems (ESSs) to the RESs, which enhances the harmonization between the demand and the energy supply by ESS-integrated RES. This study investigates four ESS integration scenarios; (1) no ESS integration into RES; any deficiency in the demand is met by the grid, and any excess energy is dumped for free into the grid, (2) only battery integration into RES, (3) only PHS integration into RES, and (4) hybrid ESS (battery + PHS) integration into RES. In Scenarios (2)–(4), ESS is only charged by the excess energy from the RES.

Table 7.2 Wind turbine technical specifications [66].

Parameter	Value
P_e (MW)	2
H (m)	140
u_C (m/s)	2
u_R (m/s)	9
u_F (m/s)	21

Fig. 7.1 shows the energy flowchart used in this study. The proposed modeling scheme compares the demand and energy production from the RES. In the case of the presence of ESS, the modeling scheme also checks the energy level in the ESS. If the production is greater than the demand, the excess energy is provided to the utility company for free, if there is no EES or if the ESS is full. If the ESS is not full, the excess energy is stored in the ESS until it is full. Additionally, in the case of hybrid ESS, the excess energy is used to charge the batteries first, and if the excess energy is more than the required amount to fully charge the batteries, the remaining excess is stored in the PHS. If the PHS is also full, the remaining excess energy is provided to the grid for free. On the other hand, if the demand is greater than the production, the deficit energy is provided by the grid at the local tariff in case there is no ESS or the energy stored in the ESS is at the recommended depth of charge limit (DC). If there is sufficient energy to meet the deficit, the ESS is used to meet the deficit. In the case of the hybrid ESS, the batteries are used to supply the deficit for 4 minutes, which is the lag time of PHS. After the first 4 minutes, the PHS will supply the deficit energy if it has sufficient stored energy. If the amount of deficit cannot be supplied by the ESS, the system will use the power grid to meet the deficit at the local tariff.

In this study, zinc-bromine battery bank (ZBB) is considered as the short-term ESS, where it is assumed that such battery banks are available in any size, while PHS is considered as the long-term ESS. The round-trip efficiency (η_r) and the depth of discharge (DoD) are the only ESS specifications used in the simulations of this study, and both ESSs are assumed to have a constant lifespan [59]. Table 7.3 presents the values of the technical specifications of the ZBB and PHS used in this study.

In PHS, the turbines and the water gates require about 4 minutes to produce electricity [59]. This duration is referred to as the lag time of PHS. Therefore, in the case of an energy deficit in the RES, the energy cannot be met by the PHS, and so either grid connection is required or another ESS can be introduced to meet the deficit during this lag time. In this study, the ZBB is proposed as a solution during the lag time to increase the autonomy and the renewable energy share of the system. Since hourly values are used in this study, the demand is assumed to be constant during an hour, and the size of the ZBB (SC) can be estimated using Eq. (7.21):

$$SC^s = \frac{D_{\max}}{\mathrm{DoD}_s \times \eta_{r,s} \times N_{lag}}, \tag{7.21}$$

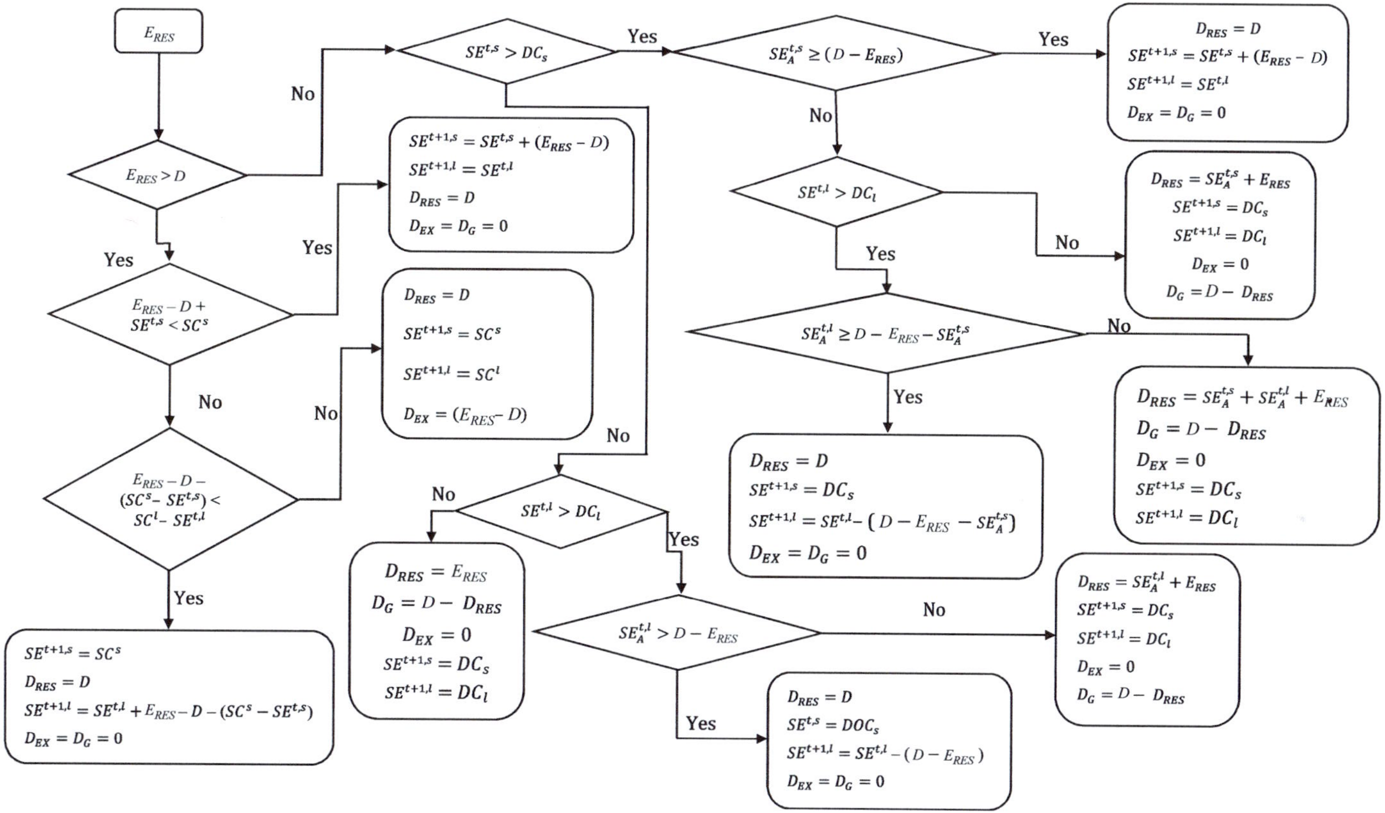

Fig. 7.1 The RES energy flow chart.

Table 7.3 Technical specifications for ZBB and PHS used in this study [59].

Parameter	Battery	PHS
η_r (%)	72	85
DoD (%)	80	89
Discharge time (h)	1	8

where $D_{\max}$ is the maximum hourly demand, the subscript s stands for the short-term storage, which is the ZBB in this study, and N_{lag} is the number of lag times in an hour, which is 15 in this study, i.e., 60 minutes divided by the lag time of the PHS.

7.2.6 Technical and economic assessment of the RES

Both technical and economic feasibility of any future energy project has crucial importance for the governments and investors. The technical feasibility can be represented by the harmonization between the demand and the energy supplied by the RES as well as the autonomy of the system. The RES fraction (R_F), which is the renewable energy share (D_{RES}) of the total demand (D) in a year, is used to inspect the matching between the demand profile and the profile of the energy supplied by the proposed RES. In addition, the demand supply fraction (DSF) quantifies the autonomy of the RES in terms of the number of hours, and it is defined as the fraction of the number of hours in a year where the RES totally meets the demand. R_F and DSF are calculated as [59, 64]

$$R_F = \sum_{i=1}^{8760} \frac{D_{RES}}{D}, \tag{7.22}$$

$$DSF = \frac{AH}{8760}, \tag{7.23}$$

where the numerator AH shows the number of autonomous hours in a year and the denominator is the total number of hours in a year.

Furthermore, the annual capacity factor (CF) is used to assess the technical performance of an energy system, and it can be calculated using Eq. (7.24):

$$CF = \frac{\sum_{i=1}^{8760} E_{RES}}{(P_{e,R} \times N_w + P_{PV}) \times 8760}. \tag{7.24}$$

Several parameters have been reported in the literature to quantify the economic feasibility of the energy systems, where the levelized cost of electricity (LCOE) is commonly used in this regard. LCOE represents the cost of electricity generated by the energy system, which can be estimated using Eq. (7.25). In Eq. (7.25), instead of the generated energy by RES, the demand covered by the RES is used to include the economic effect of temporal mismatch between the demand and the energy supply by RES. Thus, Eq. (7.25) overestimates the LCOE by using D_{RES} instead of E_{RES}:

$$\mathrm{LCOE} = \frac{C_C + \sum_{i=1}^{\mathrm{LS}} \frac{C_M}{(1+r)^i}}{\sum_{i=1}^{\mathrm{LS}} \frac{\sum_{i=1}^{8760} D_{RES}}{(1+r)^i}}. \tag{7.25}$$

In this study, it is assumed that all systems have the same lifetime and the capital costs include the component cost, cost of the balance of the system, installation, and labor and land cost. Table 7.4 presents the economic values used in this study.

At the time of this study, the electrical metering scheme in Northern Cyprus is unidirectional, such that the excess energy produced by the RES is given to the utility company for free [59]. That is why D_{RES} is used instead of E_{RES} in Eq. (7.25) to calculate LCOE in this study. Additionally, the deficit energy, that is, the amount of electrical energy demand that cannot be met by the proposed RES, is supplied by the grid at the local

Table 7.4 The economical parameters of the RES and ESS used in this study, in addition to the grid tariff and the annual discount rate for METU NCC.

Parameter	Value	Ref.
Capital cost of the PV system (USD/kW)	1533	[67]
Annual maintenance cost of the PV system (USD/kW)	24.68	[68]
Capital cost of WTS (USD/kW)	1516	[67]
Annual maintenance cost of WTS (USD/kW)	39.53	[69]
Capital cost of ZBB (USD/kWh)	250	[70]
Capital cost of PHS (USD/kWh)	68	[59, 70]
System lifetime (year)	20	[59, 71, 72]
Local electricity/grid tariff (USD/kWh)	0.175	[64]
Annual discount rate (%)	9	[73]

electricity grid tariff (GT). Most of the studies in the literature present the LCOE; however, it omits the renewable energy integration of the RES as LCOE and GT are two different economic parameters. To incorporate LCOE and GT in the same economic metric, electricity cost (E_C) is defined, which is mainly the weighted average of LCOE and GT based on the energy produced by RES and energy received from the grid, respectively. E_C can be calculated using Eq. (7.26):

$$E_C = \frac{\sum_{i=1}^{8760} D_{RES} \times \text{LCOE} + \sum_{i=1}^{8760} D_G \times \text{GT}}{D}. \quad (7.26)$$

Another essential economic measure is the net present value (NPV), which indicates the investment's worth considering the discounted values of cash inflows and outflows in each year over the lifetime of the system to the present. Higher NPV shows better economic feasibility of the system. The NPV of the RES is calculated as [74]

$$\text{NPV} = \sum_{i=1}^{\text{LS}} \frac{Rv}{(1+r)^i} - C_C. \quad (7.27)$$

Finally, the simple payback period (PBP), as another economic parameter, is used to present the number of years to recover the initial investment of the system as follows:

$$\text{PBP} = \frac{C_C}{Rv_1}. \quad (7.28)$$

7.2.7 Optimization procedure

Finding the sizes of each component in the proposed RES involves a nonlinear relation between the sizes and the technical results, such as E_{RES} and D_{RES}, and the economic parameters, such as LCOE and E_C. Therefore, a nonlinear optimization algorithm, the generalized reduced gradient (GRG) algorithm, is used to find the optimal component sizes in this study. In the hybrid ESS scenario, the capacity of the battery is found using Eq. (7.21), and so, it is excluded from the optimization. The capacities of the components in twelve RES configurations are optimized, where investigated scenarios are tabulated in Table 7.5. As aforementioned, this study aims to compare different sizing objectives, where a comparison between three common objectives was presented in this study, as shown in Table 7.6.

Table 7.5 The RES configuration considered in this study.

Scenario #	Configuration	Scenario #	Configuration	Scenario #	Configuration
1	PV	5	Wind	9	PV/Wind
2	PV/ZBB	6	Wind/ZBB	10	PV/Wind-ZBB
3	PV/PHS	7	Wind/PHS	11	PV/Wind-PHS
4	PV/HESS	8	Wind/HESS	12	PV/Wind-HESS

Table 7.6 The optimization objectives and constraints analyzed in this study.

Scenario #	Optimization objective	Optimization constraint
A	Maximum RES fraction	–
B	Minimum electricity cost	–
C	Maximum RES fraction	E_C being less than or equal to the grid tariff

7.2.8 Environmental assessment of the RES

RESs run on clean energy resources where the deployment and the use of RESs instead of conventional power generation systems that run on fossil fuel assist significantly in the mitigation of greenhouse gases, especially CO_2 [75]. Therefore, it is important to estimate the avoided amount of CO_2 emissions as a result of the installation of RESs. The annual avoided CO_2 emissions can be estimated using Eq. (7.29):

$$A_{CO_2} = I_{CO_2} \times \sum_{i=1}^{8760} E_{RES}, \tag{7.29}$$

where I_{CO_2} is estimated to be 0.584 kg/kWh for Northern Cyprus [76].

7.2.9 Case study: Middle East Technical University Northern Cyprus Campus

Middle East Technical University Northern Cyprus Campus (METU NCC) is used as the location to demonstrate the proposed methodology

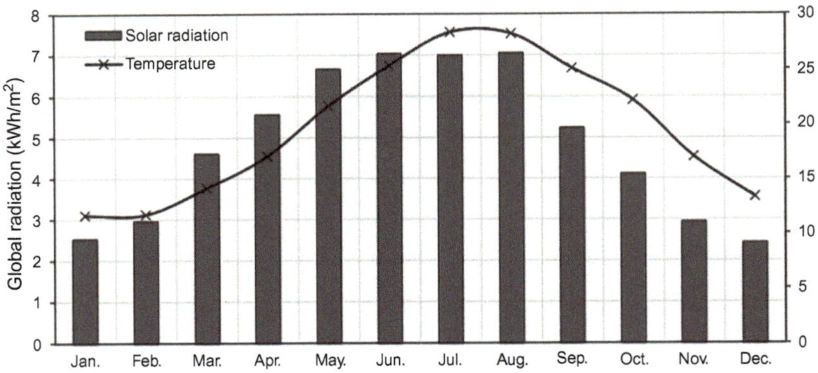

Fig. 7.2 The average daily global insolation on the horizontal surface as well as the average monthly temperature in METU NCC.

in this study. The campus is located in Guzelyurt, Northern Cyprus, which has a relatively sunny, hot climate with up to 340 sunny days throughout the year [64]. Fig. 7.2 shows the average daily GHI and ambient temperature at METU NCC. Moreover, with an average wind speed of 2.81 m/s, METU NCC is classified as moderately windy based on the European Wind Classifications [77]. Fig. 7.3A shows the average wind speeds and their corresponding wind energy values for each month of the year, whereas Fig. 7.3B shows the histogram of wind speeds at 10-m height from the ground level in METU NCC. Furthermore, the location of METU NCC is approximately 12 km away from the Mediterranean sea with about 100 m elevation difference, which makes the campus suitable for the PHS. To demonstrate the proposed optimization methodology to size the components of a hybrid RES with ESS, METU NCC is used as a case study in this work. For that purpose, the hourly demand data for the year of 2015 and hourly TMY dataset for METU NCC were obtained from Kib-Tek, the local electricity authority in Northern Cyprus and Meteonorm v7.1, respectively. Fig. 7.4 shows the daily electricity consumption data of METU NCC that is used in this study.

7.3 Results and discussion

The optimization of RESs is vital to ensure the full utilization of renewable energy resources in a feasible way, either economically or technically or both. The optimization of the RESs to achieve the maximum technical feasibility without economic considerations results in oversizing the RES,

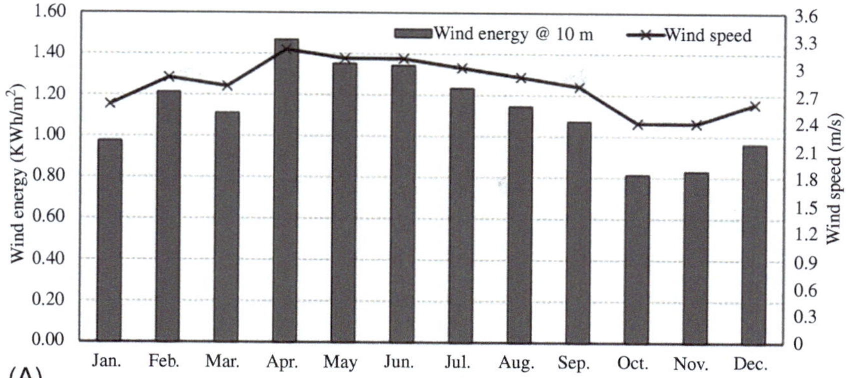

(A)

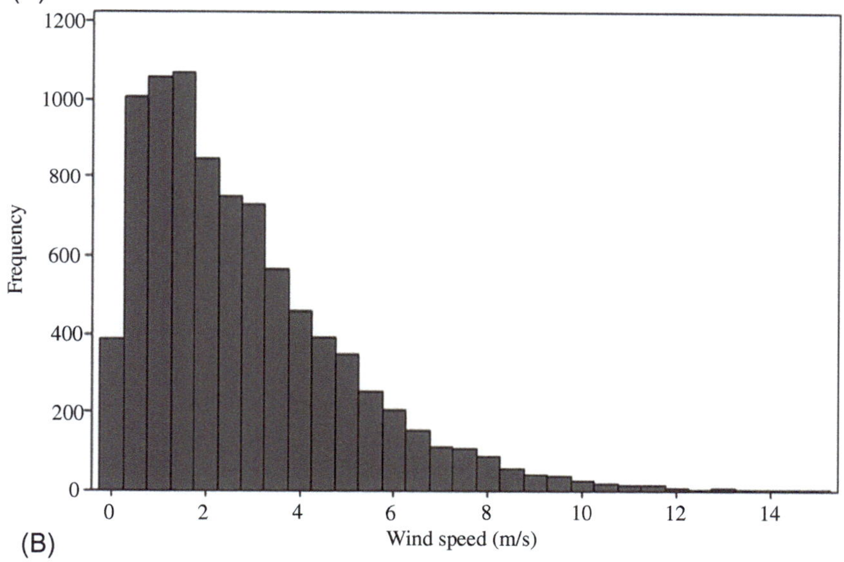

(B)

Fig. 7.3 Wind Energy in METU NCC: (A) the average hourly wind speed and wind energy at 10m and (B) the frequency of the wind speeds at 10m in METU NCC.

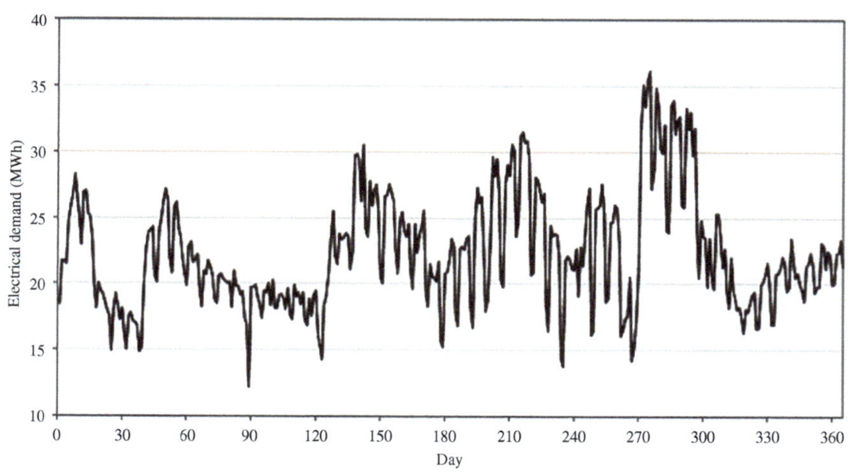

Fig. 7.4 METU NCC daily electrical demand in 2015.

while sizing the RES based on achieving the maximum economic feasibility only with technical consideration results in undersizing of the RES. Therefore, a comparison between the technical and economic parameters of optimized RESs using different optimization scenarios is presented in the following sections with and without ESS.

7.3.1 Configurations with PV system

7.3.1.1 PV system without energy storage (Scenario 1)

As a base case, Scenario 1 investigates the technical and economic feasibility of PV systems without any ESS. The optimization of the PV system based on maximizing the technical feasibility, in this study, maximizing the RES fraction, results in oversizing the PV system without economic feasibility. The resultant optimal PV capacity to achieve the maximum RES fraction is very large compared to the demand due to the mismatch between the energy generation from the PV system and the demand due to the natural fluctuation of the solar resources throughout the day and the year. Table 7.7 shows the optimal PV capacities in METU NCC using the three optimization objectives and constraints shown in Table 7.6.

Notice in Table 7.7 that the optimal PV capacity based on maximizing the RES fraction, as Scenario A, is almost 1000 times larger than the peak demand of METU NCC, which was about 1.2 MW. This capacity resulted in a very high cost of electricity and negative NPV. Moreover, notice that even with the huge PV capacity in Scenario 1A, the RES fraction and the *DSF* were around 50% where it is impossible to meet the demand during the night using the solar resources without the energy storage system. On the other hand, notice that Scenario 1B had capacity close to the peak demand; however, due to the mismatch, the system was able to cover only 1/4th of the demand in less than 10% of the time throughout the year with the cost of electricity of 0.1651 USD/kWh, while Scenario 1C resulted in almost twice

Table 7.7 Optimal PV capacities using the three optimization scenarios in METU NCC.

Parameter	Scenario A	Scenario B	Scenario C
PV capacity (kW)	1.55×10^6	1465	2681
R_F (%)	53.7	25.5	36.2
DSF (%)	52.0	8.26	23.9
E_C (USD/kWh)	36.76	0.1651	0.1750
NPV (M USD)	−2719.72	0.74	0.00
PBP (years)	63.0	6.9	9.1
A_{CO_2} (tons)	1,362,421	1287	2356

the PV capacity in Scenario 1B, and the size in Scenario 1C covered almost 1/3th of the demand and almost 1/4th of the time in an economically feasible way with the cost of electricity being equal to the grid tariff.

7.3.1.2 PV system with zinc-bromine battery (Scenario 2)

The integration of ESS would increase the probability of meeting the demand during the night using the solar resources, which would alter the economic and technical feasibility of the system. However, this change depends on several technical and economic parameters of the ESS, and Scenario 2 investigates the technical and economic feasibility of PV systems with zinc-bromine system (ZBB) as a short-term ESS. The excess energy from the PV system is stored in the ESS during the daylight and used during the night to meet the deficit. Table 7.8 shows the optimal PV and ZBB capacities in METU NCC using the three optimization objectives and constraints shown in Table 7.6.

Notice in Table 7.8 that the optimal PV capacity based on maximizing the RES fraction was significantly larger than the peak demand and even larger than the PV capacity in Scenario 1A. However, the PV system in Scenario 2A covered almost 95% of the demand and 92% of the time, where the integration of ZBB allowed the excess energy to be stored and used at night to meet the deficit. The PV capacity in Scenario 2B was the same in Scenario 1B, where the lowest cost of electricity can be achieved without having the ZBB. The higher cost of the ZBB compared to the PV system is the main reason for the optimization in Scenario 2B to discard it with no significant contribution to the revenues with small capacities due to the efficiency and the rated DoD, which decreases the amount of energy that can be retrieved from the ZBB. On the other hand, the PV capacity in Scenario 2C decreased compared to Scenario 1C, while the RES fraction and the *DSF* increased

Table 7.8 Optimal PV and ZBB capacities using the three optimization scenarios in METU NCC.

Parameter	Scenario A	Scenario B	Scenario C
PV capacity (kW)	6.58×10^6	1465	2437
ZBB (kWh)	12,677	0	1647
R_F (%)	94.9	25.5	39.3
DSF (%)	92.6	8.26	29.4
E_C (USD/kWh)	155.8	0.1651	0.1750
NPV (M USD)	−11,563	0.74	0.41
PBP (years)	62.66	6.88	8.30
A_{CO_2} (tons)	5,782,283	1287	2141

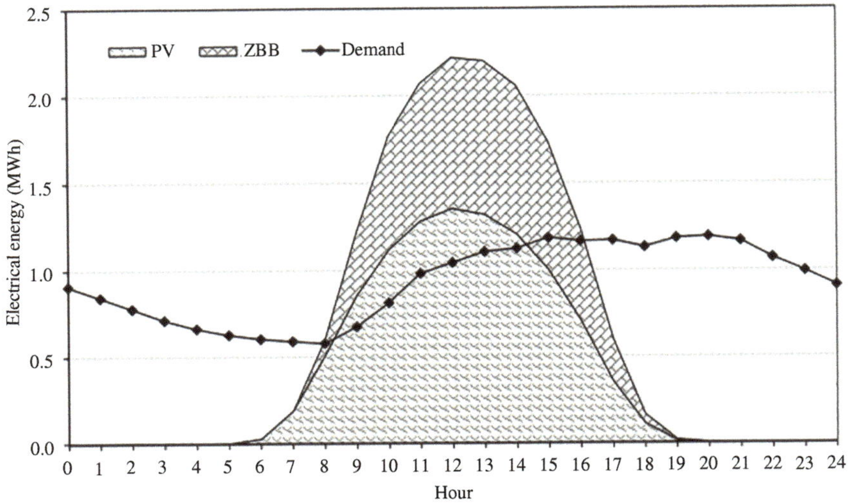

Fig. 7.5 The average hourly energy generation from the PV and the average hourly energy available in the ZBB as well as the average hourly demand of METU NCC.

where the integration of the ZBB caused the increase in the cost of the electricity. The PV capacity decreased to satisfy the constraint and the objective of Scenario C. The increase in the RES fraction and the *DSF* was due to the contribution of the ZBB in meeting the deficit during the night as shown in Fig. 7.5.

7.3.1.3 PV system with pumped-hydro storage (Scenario 3)

Scenario 3 considers the integration of PHS as a long-term ESS into PV systems. The integration of PHS could significantly contribute to increasing the RES fraction, where the low cost of this ESS as well as its high efficiency makes it an attractive option. However, one of the PHS' shortcomings is the lag time, which prevents the contribution of it in increasing the autonomy of the system. Table 7.9 shows the optimal PV and PHS capacities in METU NCC using the three optimization objectives and constraints shown in Table 7.6.

Notice in Table 7.9 that the PV capacity in Scenario 3A was similar to the one in Scenario 2A with lower ESS capacity, where the efficiency and the rated DoD of the PHS are larger than the ZBB, which allows more energy to be retrieved from the PHS, i.e., the 11 MWh PHS contributes similar to the 12 MWh ZBB. As a result, the same RES fraction was obtained. However, Scenario 3A had lower *DSF* compared to Scenario 2A due to the lag time of

Table 7.9 Optimal PV and PHS capacities using the three optimization scenarios in METU NCC.

Parameter	Scenario A	Scenario B	Scenario C
PV capacity (kW)	6.58×10^6	1712	5819
PHS (MWh)	11.5	0.93	20.7
R_F (%)	94.9	30.3	89.5
DSF (%)	52.16	11.62	37.32
E_C (USD/kWh)	155.7	0.1633	0.1750
NPV (M USD)	−11,561	0.87	0.00
PBP (years)	62.6	6.9	9.1
A_{CO_2} (tons)	5,782,283	1504	5112

the PHS, where the *DSF* in Scenario 3A is almost the same as in Scenario 1A, which is the PV system without ESS. Moreover, notice in Scenario 3B that the PV capacity achieves the lowest cost of electricity increased and became 1.7 MW with almost 1 MWh PHS, unlike Scenario 2B where the exclusion of ZBB was the only way to achieve the minimum cost of electricity. This result is mainly due to the different specific costs of ZBB and PHS. In addition, the integration of the PHS decreased the minimum cost of electricity that can be observed by comparing Scenarios 2B and 3B. Furthermore, notice in Scenario 3C that the PV capacity almost doubled compared to Scenarios 1C and 2C, while the PHS size was about ten times more than ZBB size in Scenario 2C. This larger size of ESS enabled almost 100% increase in the RES fraction where more energy is being stored and used during the night.

7.3.1.4 PV system with hybrid energy storage system (Scenario 4)

As aforementioned, PHS has a lag time of 4 minutes, which eliminates its role in increasing the autonomy of RES. The integration of the ZBB to supply the demand during these 4 minutes of the PHS could increase the autonomy of the system. Scenario 4 checks the technical and economic feasibility of hybrid ESS for PV systems, and Table 7.10 shows the optimal PV, PHS, and ZBB capacities in METU NCC using the three optimization objectives and constraints.

Notice in Table 7.10 that the PV capacity in Scenario 4A decreased almost 20 times compared to Scenarios 2A and 3A, while the RES fraction remained constant. This result shows that the integration of PHS and ZBB helped in achieving the maximum RES fraction with less PV capacity by storing the excess during the daylight and using it to meet the deficit during

Table 7.10 Optimal PV and HESS capacities using the three optimization scenarios in METU NCC.

Parameter	Scenario A	Scenario B	Scenario C
PV capacity (kW)	302.4×10^3	2408	5297
ZBB (kWh)	1358	1358	1358
PHS (MWh)	26.6	2.83	17.4
R_F (%)	94.9	42.9	85.9
DSF (%)	99.9	32.6	83.3
E_C (USD/kWh)	7.186	0.1686	0.1750
NPV (M USD)	−520.9	0.81	0.34
PBP (years)	77.1	7.7	8.8
A_{CO_2} (tons)	1,425,135	2115	4653

the night. However, the PV capacity is still very large compared to METU NCC peak demand with no economic feasibility of such capacity. Moreover, notice in Scenario 4B that the PV capacity increased by more than 50% compared to Scenarios 2B and 3B, where the minimum cost of electricity increased due to the relatively higher cost of ZBB. In the hybrid ESS scenarios, as aforementioned in Section 7.2.5, the ZBB was sized to meet the deficit during the lag time of the PHS, and so the size of ZBB was excluded from the optimization. On the other hand, in Scenario 4C compared to Scenario 3B, the PV capacity decreased due to the integration of ZBB. The decrease in PV capacity decreased the RES fraction, but the autonomy of the system was increased by almost 100% due to the integration of ZBB where the system is capable of meeting more demand during the day.

7.3.2 Configurations with wind turbines

7.3.2.1 Wind turbines without energy storage (Scenario 5)

Similar to PV systems, the optimization of the WTS was based on maximizing the technical and economic feasibility. In this study, maximizing the RES fraction results in oversizing the WTS without economic feasibility. The resultant optimal WTS capacity to achieve the maximum RES fraction is very large compared to the peak demand due to the mismatch between the energy generation from the WTS and the demand. Table 7.11 shows the optimal WTS capacities in METU NCC using the three optimization objectives and constraints shown in Table 7.6.

Notice in Table 7.11 that even though the capacity of the WTS is very large compared to the peak demand, it is less than the optimal capacities of PV system in all Scenarios A in Section 7.3.1. The reason is the wind

Table 7.11 Optimal WTS capacities using the three optimization scenarios in METU NCC.

Parameter	Scenario A	Scenario B	Scenario C
WTS capacity (MW)	438	2	2
R_F (%)	71.1	44.5	44.5
DSF (%)	69.0	26.5	26.5
E_C (USD/kWh)	11.11	0.1477	0.1477
NPV (M USD)	−812.8	2.03	2.03
PBP (years)	40.7	5.5	5.5
A_{CO_2} (tons)	647,064	2955	2955

resource profile and the performance of WTS, as the wind energy is available during the night, unlike the solar resources, and therefore, smaller system capacity is needed to achieve the maximum RES fraction in Scenario 5. Fig. 7.6 shows the average electricity generation throughout the day by 2 MW PV and 2 MW WTS to illustrate the differences between PV and WTS energy generations throughout the day. Moreover, notice in Scenario 5B that the minimum cost of electricity was lower the PV scenarios of the same optimization objective, even though the WTS capacity was larger. Lower cost in WTS was obtained because WTS produces more energy compared to the PV with a capacity factor of 28.9% compared to 17.7%

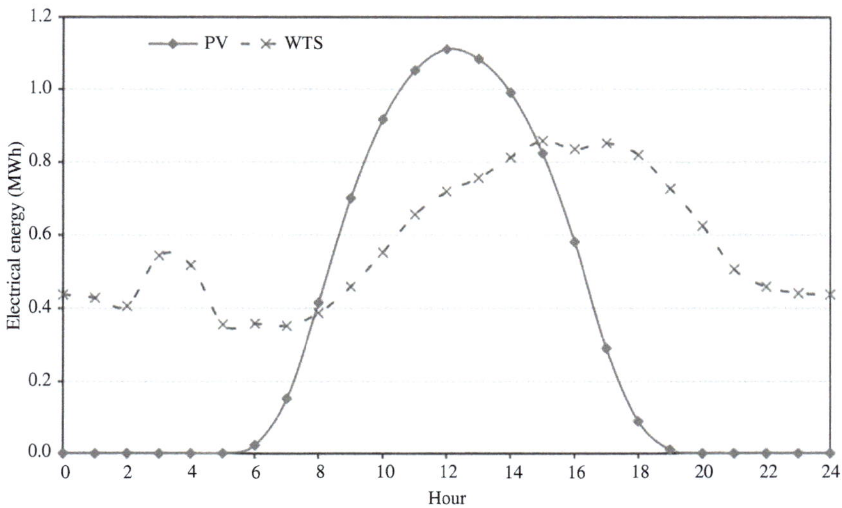

Fig. 7.6 The average hourly energy generation from 2 MW PV and 2 MW WTS in METU NCC.

for the PV. In addition, notice that Scenarios 5B and 5C had the same results where this is due to the rated capacity of each wind turbine used, as it is not possible to have a fraction number of wind turbines, unlike the PV capacity.

7.3.2.2 Wind turbines with zinc-bromine battery (Scenario 6)

Scenario 6 investigates the feasibility of WTS with ZBB, and Table 7.12 shows the optimized results of this scenario. The integration of ZBB increased the RES fraction in all the optimization scenarios when compared to WTS without any ESS in Scenario 5, where ZBB assisted in reallocating the excess energy. So, ZBB enhanced the matching between the energy generation from WTS and the demand. Notice in Table 7.12 that the WTS capacity in Scenario 6A increased almost four times compared to Scenario 5A, where Scenario 6A required the huge capacity of the ZBB, which enabled more energy to be stored to satisfy the optimization objective which is maximizing the RES fraction in Scenarios A. Moreover, the RES fraction increased by 0.5% in Scenario 6B when compared to the one in Scenario 5B where the contribution of ZBB was smaller as a result of the losses in the energy stored in the battery due to the rated DoD and the round trip-efficiency of the ZBB. Furthermore, notice in Scenario 6C that the integration of ZBB with the same WTS capacity increased the RES fraction and *DSF* by about 8.2% and 14%, respectively, while maintaining its economic feasibility.

7.3.2.3 Wind turbines with pumped-hydro storage (Scenario 7)

The integration of the PHS into WTS is investigated in Scenario 7. PHS increases the RES fraction; however, it does not contribute to autonomy due to the 4-min lag time. This result can be observed in Table 7.13 by comparing *DSF*. For instance, *DSF* decreased from about 100% to 69% when

Table 7.12 Optimal WTS and ZBB capacities using the three optimization scenarios in METU NCC.

Parameter	Scenario A	Scenario B	Scenario C
WTS capacity (MW)	1792	2	2
ZBB (kWh)	46,635	113.3	6322.7
R_F (%)	99.9	45.0	53.2
DSF (%)	99.9	27.5	41.5
E_C (USD/kWh)	45.56	0.1476	0.1750
NPV (M USD)	−3362	2.07	1.58
PBP (years)	39.3	5.5	6.8
A_{CO_2} (tons)	2,647,349	2955	2955

Table 7.13 Optimal WTS and PHS capacities using the three optimization scenarios in METU NCC.

Parameter	Scenario A	Scenario B	Scenario C
WTS capacity (MW)	852	2	4
PHS (MWh)	42.5	8.66	0.47
R_F (%)	99.9	55.2	58.0
DSF (%)	69.0	26.5	43.2
E_C (USD/kWh)	21.55	0.1369	0.1750
NPV (M USD)	−1589.0	2.83	0.00
PBP (years)	40.1	5.1	9.1
A_{CO_2} (tons)	1,258,673	2955	5909

PHS replaced ZBB, i.e., when Scenarios 6A and 7A are compared. Table 7.13 shows that, in Scenario 7A, the WTS capacity that achieves the maximum RES fraction decreased by almost half when compared to Scenario 6A. This result is because the amount of stored energy that can be retrieved from the PHS is more than the amount of energy that can be retrieved from the ZBB, as the round-trip efficiency and the rated DoD of the PHS are greater than the ZBB. Moreover, notice in Scenario 7B that the minimum cost of electricity is the lowest achieved so far compared to all the previous scenarios where relatively low cost and higher efficiency of the PHS compared with ZBB were the main reasons for getting lower cost of electricity. Finally, the integration of PHS in Scenario 7C made 4 MW WTS feasible where the net revenues gained from this integration allowed the WTS capital cost to be doubled. Therefore, higher RES fraction and *DSF* could be obtained with smaller PHS when compared to Scenario 7B.

7.3.2.4 Wind turbines with hybrid energy storage system (Scenario 8)

Scenario 8 checks the feasibility of the integration of a hybrid ZBB/PHS ESS (HESS) into WTS. Table 7.14 shows the optimal WTS and HESS capacities in METU NCC using the three optimization objectives and constraints shown in Table 7.6. Hybrid ESS increased the autonomy of the system and reduced the optimal WTS capacity required to achieve the maximum RES fraction, where instead of increasing the WTS capacity to meet the demand during the PHS deficiency period, the energy stored in the ZBB was used to meet the demand; thus, the autonomy increased. For instance, when Scenarios 7A and 8A are compared, *DSF* increased

Table 7.14 Optimal WTS and HESS capacities using the three optimization scenarios in METU NCC.

Parameter	Scenario A	Scenario B	Scenario C
WTS capacity (MW)	618	2	4
ZBB (kWh)	1358.1	1358.1	1358.1
PHS (MWh)	40.53	7.44	1.24
R_F (%)	99.9	55.1	63.6
DSF (%)	99.9	44.2	55.5
E_C (USD/kWh)	15.65	0.145	0.1750
NPV (M USD)	−1145	2.57	0.34
PBP (years)	40.9	5.5	8.7
A_{CO_2} (tons)	912,981	2955	5909

by about 30%. Similarly, *DSF* increased from 26.5% to 44.2% when ZBB was introduced to WTS/PHS system, while RES fraction was about the same at 55%, i.e., Scenario 7B is compared with Scenario 8B. Moreover, in Scenario 8C, the system was able to meet almost 64% of the demand and 55% of the time throughout the year in an economically feasible way, where the integration between the WTS and HESS enhanced the matching between the generation and demand profiles.

7.3.3 Configurations with PV/wind turbine hybrid system

7.3.3.1 PV/wind turbine hybrid system without energy storage (Scenario 9)

The hybridization between solar and wind energy systems enhances the matching profile between the energy generation and the demand where both systems work in a synergistic and corporative way. Table 7.15 shows

Table 7.15 Optimal PV + WTS capacities using the three optimization scenarios in METU NCC.

Parameter	Scenario A	Scenario B	Scenario C
PV capacity (kW)	570.60×10^3	150	2496
WTS capacity (MW)	288	2	2
R_F (%)	83.0	46.3	62.6
DSF (%)	81.9	27.5	46.5
E_C (USD/kWh)	20.80	0.1480	0.1750
NPV (k USD)	−1533.0	2.01	0.00
PBP (years)	54.0	5.7	9.1
A_{CO_2} (tons)	926,765	3086	5147

the optimal PV and WTS capacities in METU NCC without any ESS. In Scenario 9A, the total PV and WTS capacities are smaller than the capacities of PV in Scenario 1A and WTS in Scenario 5A. The RES fraction and *DSF* are higher, where the synergistic performance between the PV and WTS is the reason behind this. Furthermore, in Scenario 9B, the electricity cost is less than the cost in Scenario 1B, PV without ESS, and higher than Scenario 5B, WTS without ESS. This result is mainly because the integration of WTS reduced the cost of electricity of the hybrid system due to the higher capacity factor of WTS compared to PV systems, even though the specific cost of WTS is higher than the one of PV. Finally, the hybrid PV + WTS could cover almost 62% of the demand and *DSF* was about 46% due to the synergistic performance of the two systems as shown in Fig. 7.7.

7.3.3.2 PV/wind turbine hybrid system with zinc-bromine battery (Scenario 10)

The integration of ZBB into hybrid PV + WTS is investigated as Scenario 10. ZBB would increase the synergistic performance of the hybrid RES by covering more demands where without any ESS, the hybrid system can cover up only up to 83% of the demand and 82% of the time. On the other hand, adding ZBB could increase the ability of the hybrid RES to cover almost all the demand throughout the year, as shown in Table 7.16, however, in an infeasible way as negative NPV and high cost

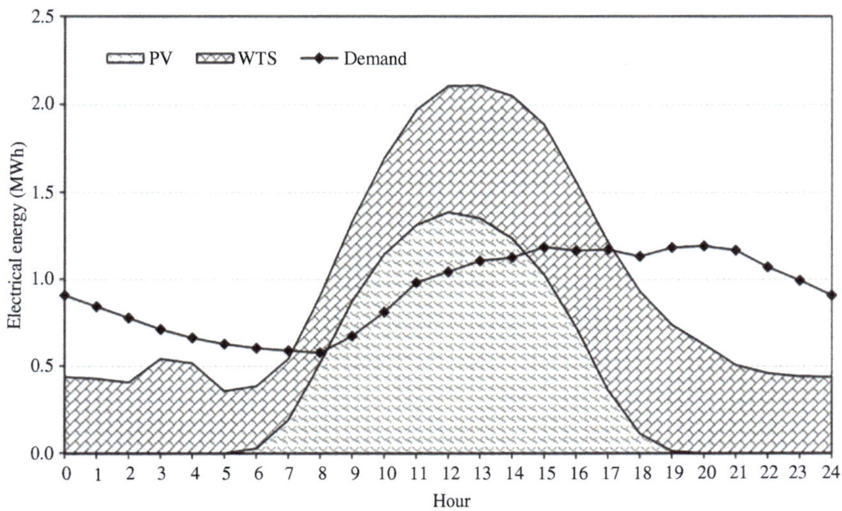

Fig. 7.7 The average hourly energy generation from PV and WTS as well as the average hourly demand of METU NCC.

Table 7.16 Optimal PV + WTS and ZBB capacities using the three optimization scenarios in METU NCC.

Parameter	Scenario A	Scenario B	Scenario C
PV capacity (kW)	6754	196	2046
WTS capacity (MW)	172	2	2
ZBB (kWh)	26,000	199.3	4011.3
R_F (%)	99.9	47.7	71.9
DSF (%)	99.9	29.7	62.1
E_C (USD/kWh)	4.677	0.1480	0.1750
NPV (k USD)	−328.2	2.06	1.00
PBP (years)	50.1	5.7	8.0
A_{CO_2} (tons)	260,032	3127	4752

of electricity denote. Notice in Table 7.16 that, Scenario 10A, the PV and WTS capacities were considerably less than Scenario 9A while achieving higher RES fraction and higher autonomy, which reduced the cost of electricity. This reduction was due to the utilization of the energy stored in ZBB when ESS was unable to meet the demand instead of trying to cover the deficit by increasing the PV and WTS capacities. Moreover, in Scenario 10B when compared to Scenario 9B, the PV capacity slightly increased with the same cost of electricity due to the contribution of the ZBB. Finally, notice in Scenario 10C that even though the PV capacity was decreased by about 20%, the RES fraction was increased by almost 15% due to the reallocation of the excess energy using the ZBB when compared to the results of Scenario 9C.

7.3.3.3 PV/wind turbine hybrid system with pumped-hydro storage (Scenario 11)

Scenario 11 includes the feasibility analysis of a hybrid PV + WTS with PHS as an ESS. The integration of PHS increases the RES fraction of PV + WTS; however, it does not contribute directly to the autonomy of it due to the 4-min lag time, as shown in Table 7.17 compared to the ZBB and no ESS scenarios. Table 7.17 shows that, in Scenario 11A, the optimal WTS and PV capacities, which achieved the maximum RES fraction, were smaller than the respective capacities in Scenario 10A, whereas the ESS capacity increased by about 5.2 MWh. The reason behind this result is that the amount of stored energy that can be retrieved from the PHS is larger than the amount of energy that can be retrieved from the ZBB as the round-trip efficiency and the rated DoD of the PHS are higher than the ZBB. Moreover, in Scenario 11B, the minimum cost of electricity is lower than

Table 7.17 Optimal PV + WTS and PHS capacities using the three optimization scenarios in METU NCC.

Parameter	Scenario A	Scenario B	Scenario C
PV capacity (kW)	4475	310	777
WTS capacity (MW)	104	2	4
PHS (MWh)	31.2	9.8	53.4
R_F (%)	99.9	60.1	96.1
DSF (%)	75.7	28.5	46.8
E_C (USD/kWh)	2.76	0.1366	0.1750
NPV (k USD)	−192.18	2.86	0.00
PBP (years)	59.6	5.4	9.1
A_{CO_2} (tons)	157,572	3227	6592

Scenario 7B, i.e., WTS with PHS, where the synergistic performance of PV and WTS increased the net revenues of the system along with the reallocation of excess energy by the PHS. Finally, notice that the integration of PHS in Scenario 11C made an additional wind turbine feasible similar to Scenario 7C. With about 777 kW PV, the system was able to cover about 96% of the demand and about 47% of the time in a feasible way by enhancing the match between the demand and PV + WTS energy generation.

7.3.3.4 PV/wind turbine hybrid system with hybrid energy storage system (Scenario 12)

As aforementioned, the hybridization of PV and WTS would increase the matching between the demand and energy generation profiles where adding a hybrid ZBB/PHS with no lag time can assist in achieving a very high demand coverage rate and very high autonomy rates as shown in Table 7.18. Notice that the RES capacities in Scenario 12A are considerably smaller than all the other A scenarios with almost the same maximum RES fraction and *DSF* due to the performance of the hybrid RES. Finally, the system in Scenario 12C was able to cover almost 95% of the demand and 93% of the time in a feasible way with only 3.3 MW PV and 2 MW WTS, while Scenario 12A required 4.7 MW PV and 80 MW WTS to cover 99.9% of the demand and 99.9% of the time, which resulted in an infeasible system. An increase of almost 80 MW RES capacity was required to increase the RES fraction by almost 5% only. Therefore, it can be concluded that the hybridization between PV and WTS, as well as the integration of hybrid PHS/ZBB, is economically and technically feasible if Scenario C is used as shown in Fig. 7.8.

Table 7.18 Optimal PV+WTS and HESS capacities using the three optimization scenarios in METU NCC.

Parameter	Scenario A	Scenario B	Scenario C
PV capacity (kW)	4670	348	3319
WTS capacity (MW)	80	2	2
ZBB (kWh)	1358.1	1358.1	1358.1
PHS (MWh)	31.1	8.1	30.0
R_F (%)	99.9	60.3	94.6
DSF (%)	99.9	48.5	93.2
E_C (USD/kWh)	2.168	0.1447	0.1750
NPV (k USD)	−147.8	2.59	0.34
PBP (years)	70.6	5.8	8.8
A_{CO_2} (tons)	122,288	3260	5870.56

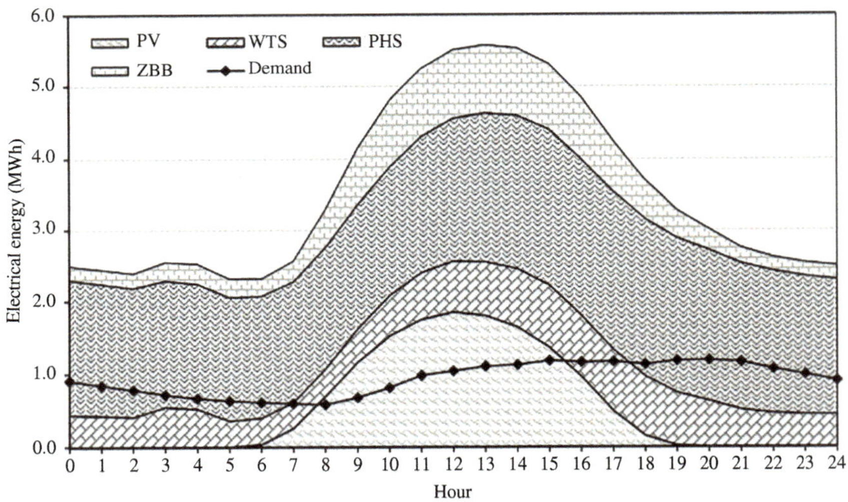

Fig. 7.8 The average hourly energy generation from PV and WTS, as well as the energy available in PHS and ZBB, in addition to the average hourly demand of METU NCC.

7.3.4 Comparison of the RES scenarios

The investigated scenarios compare not only PV vs. WTS or ZBB vs. PHS but also three different optimization schemes. Fig. 7.9 shows a comparison between the technical and economic parameters of the RES scenarios in METU NCC. The maximization of the RES fraction without any economic constraint, as Scenarios A, resulted in an economically infeasible system with enormous installed RES capacities compared to the peak demand

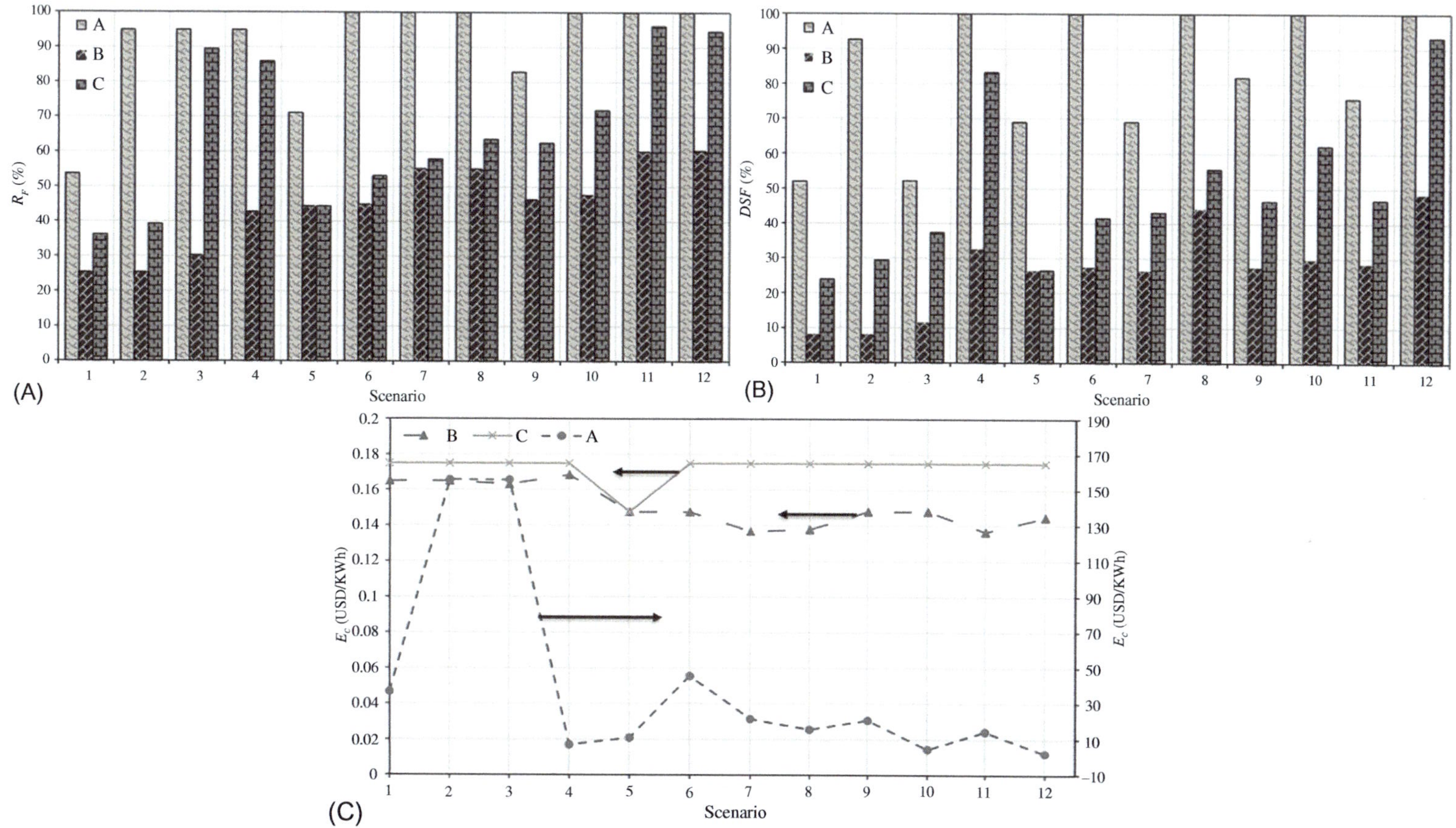

Fig. 7.9 The technical and economic parameters of the investigated RES scenarios: (a) the RES fraction, (b) the demand supply fraction, and (c) the cost of electricity.

of METU NCC, while the minimization of the cost of electricity without any technical constraints, as Scenarios B, resulted in undersizing of the RES with small RES fractions and *DSF* values. Alternatively, the maximization of the RES fraction with cost of electricity being less than or equal to the local grid tariff, as Scenarios C, achieved both the economic and technical feasibilities of the RES. These results show that studies that considered RES and ESS component sizes using only technical or economic optimization objectives miss the opportunity of better utilization of the RES and ESS. A similar conclusion can be drawn for the studies that do not consider hybrid RES or ESS.

Notice in Fig. 7.9a that the RES configurations in Scenarios A had RES fraction higher than 90% if they had ESS. The energy generation from the RES did not match the demand, although the hybridization between PV and WTS enhanced the matching; however, there is a need for energy reallocation using ESS. On the other hand, the RES configurations based on Scenario B resulted in RES fractions below 60%, while RES configurations based on Scenario C had RES fraction between 35% and 95%. Among all configurations, the hybrid PV + WTS-HESS had the largest RES fraction in all optimization objectives.

Similar to the RES fraction, notice in Fig. 7.9b that the maximum *DSF* values were achieved in the RES configurations of Scenario A, where the RES configurations of Scenario B achieved the lowest *DSF*. While the hybrid PV + WTS configuration from Scenario C was able to achieve *DSF* close to Scenario A in an economically feasible way, as shown in Fig. 7.9c. Fig. 7.9c shows that all the RES configurations in Scenario A were economically infeasible with electricity much larger than the grid tariff (0.175 USD/kWh). On the other hand, the optimal configurations resulted from Scenario C were able to achieve high RES fraction as well as high *DSF* without compromising the economic feasibility of the system. The optimization of the RESs using Scenario A resulted in oversizing of the system components while trying to cover more demands where almost 99% of the energy generated is being dumped into the grid for free, as shown in Fig. 7.10, which significantly increased the cost of electricity with almost no dependency on the grid. On the other hand, the dependency on the grid was relatively high in the case of the optimal RES configurations of Scenario B, where almost all the generated energy by RES was used to cover the demand, while the majority of RES configurations of Scenario C had less dependency on the grid compared to the configurations of Scenario B with almost the same dumped energy ratios and smaller demand covered to energy generation ratios.

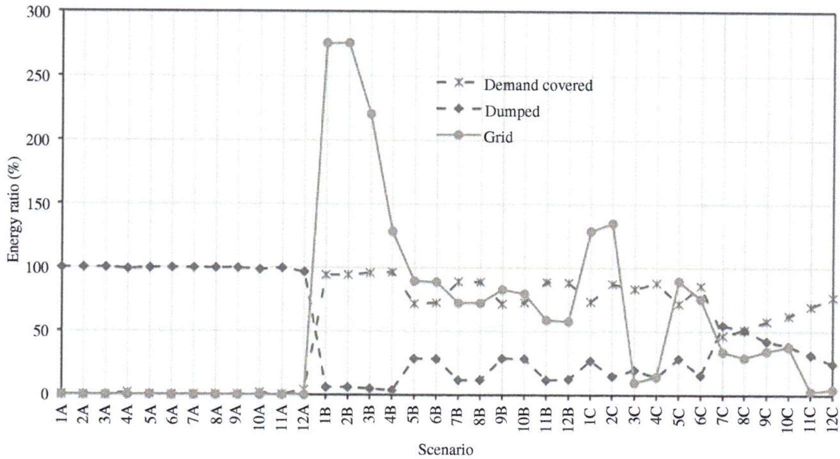

Fig. 7.10 The ratio of demand covered, the energy dumped into the grid, and the energy supplied by the grid to the energy generated by the RES for each investigated RES scenario.

7.4 Conclusions

This study considered different sizing methodologies using techno-economic analysis of a hybrid PV + WTS with and without ESS options, namely, ZBB, PHS, and hybrid ZBB/PHS. The sizing methodologies include different optimization schemes to maximize the renewable energy fraction without any economic consideration, to minimize the cost of electricity without any technical considerations, and to maximize the renewable energy fraction in which the cost of electricity is less than or equal to the grid tariff. Middle East Technical University Northern Cyprus Campus is used to implement these sizing methodologies in this study; however, the methodologies are general so that they can be applied to other locations where meteorological and demand data are available.

The results showed that Scenarios A, in which the sizes of the components were determined to maximize the renewable energy fraction without any economic consideration, gave huge capacities for all configurations, as the temporal mismatch between the renewable resources and demand was tried to be minimized by increasing the capacities of RES. Although high renewable energy fractions and demand side fractions were obtained, the cost of electricity was higher than the grid tariff, and net present values were negative in these scenarios. Therefore, this optimization methodology is not suggested to minimize the mismatch and to finalize the sizes of RES and ESS components. In Scenarios B, the component sizes were determined to minimize the cost of electricity. The results of these scenarios were feasible, which means that the cost of electricity was lower than the grid tariff, and net present values were positive. This methodology can be suggested from the technical and economic perspectives; however, another methodology, Scenario C, was offered to enhance the environmental benefits of RES. Scenario C considered the optimal sizes of RES and ESS when the RES fraction was maximum at the cost of electricity being equal to the grid tariff. The component sizes in Scenarios C are larger than the ones in Scenarios B at a penalty of higher electricity cost; however, larger sizes meant that more CO_2 emissions were avoided due to higher utilization of RES. Therefore, the last optimization scheme is suggested if one-way tariff is utilized. In the countries, where net-metering or two-way tariff is applicable, sizes between Scenarios B and C would be suggested depending on the grid tariff and the power purchase tariff by the utility company.

PV and WTS are also compared in this study as RES options. As the solar resources are higher than wind resources in METU NCC and the specific cost of wind turbines is higher than PV modules, PV systems seemed to be advantageous than WTS. Another advantage of PV systems was found to be that wind turbines can only be installed in integer increments, 2 MW in this study, whereas the PV system increments can be done module by module, 285 W in this study. However, energy production from WTS was distributed throughout the day, whereas energy production from PV system was possible only during the day. Due to the intermittency of solar resources, a hybrid PV + WTS system was suggested as a technically and economically feasible option in METU NCC.

Additionally, ZBB and PHS are compared in this study as ESS options. ZBB can be considered as a short-term, smaller storage option, whereas PHS can be considered as a long-term, larger storage option. When specific costs of these ESSs are compared, PHS is cheaper than ZBB; however, PHS has a lag time during which it cannot provide electrical energy. Therefore, it is suggested in this study that ZBB can compensate for the lag time of PHS in a way that higher autonomy can be achieved. In this study, 47% higher autonomy was achieved when ZBB was added to PHS for hybrid RES in Scenario C.

As a result, optimum PV and WTS sizes were suggested to be 3.3 and 2 MW, respectively, with 1.4 MWh of ZBB and 30 MWh of PHS. This configuration could provide 95% of the annual energy of the campus and could fully meet the demand at 93% of the hours in a year under Scenario C. Higher percentages yielded in infeasible results, as in Scenario A. However, in Scenario B, higher NPV can be obtained by reducing the PV capacity to 348 kW and PHS capacity to 8.1 MWh, while keeping WTS and ZBB sizes the same. In this scenario, 60% of the demand can be met, and the demand can be fully met during 93% of the time.

References

[1] Najjar YSH, Abubaker AM. Exergy analysis of a novel inlet air cooling system with gas turbine engines using cascaded waste-heat recovery. Int J Exergy 2017;22(2):183–204.
[2] Najjar YSH, Abubaker AM, El-Khalil AFS. Novel inlet air cooling with gas turbine engines using cascaded waste-heat recovery for green sustainable energy. Energy 2015;93:770–85.
[3] Najjar YSH, Abubaker AM. Using novel compressed-air energy storage systems as a green strategy in sustainable power generation—a review. Int J Energy Res 2016;40 (12):1595–610.
[4] Wilson JE, Grib SW, Ahmad AD, Renfro MW, Adams SA, Salaimeh AA. Study of near-cup droplet breakup of an automotive electrostatic rotary bell (ESRB) atomizer using high-speed shadowgraph imaging. Coatings 2018;8:174.
[5] Ahmad AD, Abubaker AM, Salaimeh AA, Akafuah NK. Schlieren visualization of shaping air during operation of an electrostatic rotary bell sprayer : impact of shaping air on droplet atomization and transport. Coatings 2018;8:279.
[6] Darwish Ahmad A, et al. Spatial positioning and operating parameters of a rotary bell sprayer: 3D mapping of droplet size distributions. Fluids 2019;4(3).
[7] Gielen D, Boshell F, Saygin D, Bazilian MD, Wagner N, Gorini R. The role of renewable energy in the global energy transformation. Energy Strat Rev 2019;24:38–50.
[8] Al-assad R, Ayadi O. Techno-economic assessment of grid connected photovoltaic systems in Jordan. In: The 8th international renewable energy congress (IREC 2017). IEEE; 2017. p. 1–5. July.
[9] Li C-H, Zhu X-J, Cao G-Y, Sui S, Hu M-R. Dynamic modeling and sizing optimization of stand-alone photovoltaic power systems using hybrid energy storage technology. Renew Energy 2009;34(3):815–26.
[10] Al-Salaymeh A, Al-Hamamre Z, Sharaf F, Abdelkader MR. Technical and economical assessment of the utilization of photovoltaic systems in residential buildings: the case of Jordan. Energy Convers Manage 2010;51(8):1719–26.
[11] Al-Ghussain L, Abujubbeh M, Fahrioglu M. Assessment of PV investments in Northern Cyprus. In: 16th International conference on clean energy (ICCE-2018); May 2018. p. 9–11.
[12] Jaber S, Hawa AA. Optimal design of PV system in passive residential building in Mediterranean climate. Jordan J Mech Indust Eng 2016;10(1):39–49.
[13] Castronuovo ED, Lopes JAP. Optimal operation and hydro storage sizing of a wind-hydro power plant. Int J Electr Power Energy Syst 2004;26(10):771–8.
[14] Papaefthymiou SV, Papathanassiou SA. Optimum sizing of wind-pumped-storage hybrid power stations in island systems. Renew Energy 2014;64:187–96.
[15] Barton JP, Infield DG. Energy storage and its use with intermittent renewable energy. IEEE Trans Energy Convers 2004;19(2):441–8.
[16] Kaldellis JK, Kavadias KA. Optimal wind-hydro solution for Aegean Sea islands' electricity-demand fulfilment. Appl Energy 2001;70(4):333–54.
[17] Abujubbeh M, Marazanye VT, Qadir Z, Fahrioglu M, Batunlu C. Techno-economic feasibility analysis of grid-tied PV-wind hybrid system to meet a typical household demand: case study—Amman, Jordan. In: 2019 1st Global power, energy and communication conference; 2019. p. 418–23.

[18] Abujubbeh M, Fahrioglu M. Determining maximum allowable PV penetration level in transmission networks: case analysis—Northern Cyprus power system. In: Proc. 2019 IEEE 1st glob. power, energy commun. conf. (GPECOM 2019); 2019. p. 292–7.
[19] Askari IB, Ameri M. Techno-economic feasibility analysis of stand-alone renewable energy systems (PV/bat, wind/bat and hybrid PV/wind/bat) in Kerman, Iran. Energy Sources Part B: Econ Plann Policy 2012;7(1):45–60.
[20] Kosmadakis IE, Elmasides C, Eleftheriou D, Tsagarakis KP. A techno-economic analysis of a PV-battery system in Greece. Energies 2019;12(7).
[21] Al-Ghussain L, Taylan O, Baker DK. An investigation of optimum PV and wind energy system capacities for alternate short and long-term energy storage sizing methodologies. Int J Energy Res 2018;1–15.
[22] Le HT. Sizing energy storage systems for wind power firming: an analytical approach and a cost-benefit analysis. In: Power & energy society general meeting, No. II; 2008. p. 1–8.
[23] Campos-Gaona D, Madariaga A, Zafar J, Anaya-Lara O, Burt G. Techno-economic analysis of energy storage system for wind farms: the UK perspective. In: 2018 International conference on smart energy systems and technology (SEST 2018)—proceedings; 2018. p. 1–6.
[24] Al-Ghussain L, Samu R, Taylan O, Fahrioglu M. Techno-economic analysis of photovoltaic-hydrogen fuel cell/pumped hydro storage system for micro grid applications: case study in Cyprus. In: 2018 International conference on photovoltaic science and technologies (PVCon); 2018. p. 1–6.
[25] Jamshidi M, Askarzadeh A. Techno-economic analysis and size optimization of an off-grid hybrid photovoltaic, fuel cell and diesel generator system. Sustain Cities Soc 2019;44:310–20.
[26] Kaldellis JK, Zafirakis D, Kondili E. Optimum sizing of photovoltaic-energy storage systems for autonomous small islands. Int J Electr Power Energy Syst 2010;32(1):24–36.
[27] Kaldellis JK, Zafirakis D, Kaldelli EL, Kavadias K. Cost benefit analysis of a photovoltaic-energy storage electrification solution for remote islands. Renew Energy 2009;34(5):1299–311.
[28] Cherif A, Jraidi M, Dhouib A. A battery ageing model used in stand alone PV systems. J Power Sources 2002;112(1):49–53.
[29] Khatib T, Mohamed A, Sopian K, Mahmoud M. A new approach for optimal sizing of standalone photovoltaic systems. Int J Photoenergy 2012;2012.
[30] Uddin K, Gough R, Radcliffe J, Marco J, Jennings P. Techno-economic analysis of the viability of residential photovoltaic systems using lithium-ion batteries for energy storage in the United Kingdom. Appl Energy 2017;206:12–21.
[31] Bagalini V, Zhao BY, Wang RZ, Desideri U. Solar PV-battery-electric grid-based energy system for residential applications: system configuration and viability. Research 2019;2019:1–17.
[32] Mat Isa N, Wei Tan C, Yatim AHM. A techno-economic assessment of grid connected photovoltaic system for hospital building in Malaysia. IOP Conf Ser Mater Sci Eng 2017;217(1).
[33] El-Tous Y, Al-Battat S, Abdel HS. Hybrid wind-PV grid connected power station case study: Al-Tafila, Jordan. Int J Energy Environ 2012;3(4):605–16.
[34] Al-Ghussain L, Taylan O, Fahrioglu M. Sizing of a photovoltaic-wind oil shale hybrid system: case analysis in Jordan. J Sol Energy Eng Incl Wind Energy Build Energy Conservs 2018;140:1–12.
[35] Al-Ghussain L, Ahmed H, Haneef F. Optimization of hybrid PV-wind system: case study Al-Tafilah cement factory, Jordan. Sustain Energy Technol Assess 2018;30:24–36.

[36] Al-Ghussain L, Taylan O. Sizing methodology of a PV/wind hybrid system: case study in Cyprus. Environ Prog Sustain Energy 2019;38(3):e13052.
[37] Yadav AK, Malik H, Bin Arif MS. Techno economic feasibility analysis of different combination of PV-wind-diesel-battery hybrid system. Elsevier Ltd.; 2018.
[38] Menniti D, Pinnarelli A, Sorrentino N. A method to improve microgrid reliability by optimal sizing PV/wind plants and storage systems. In: IET conference publications, April 2016; 2009. p. 816.
[39] Castaneda M, Fernandez LM, Sanchez H, Cano A, Jurado F. Sizing methods for stand-alone hybrid systems based on renewable energies and hydrogen. In: 2012 16th IEEE Mediterranean electrotechnical conference; 2012. p. 832–5.
[40] Aiad M, Badran A, Shihabi S. Optimal selection of hybrid PV/wind systems for Jordanian conditions. In: Global conference on renewables and energy efficiency for desert regions (GCREEDER); September 2013.
[41] Chen SX, Gooi HB, Wang MQ. Sizing of energy storage for microgrids. IEEE Trans Smart Grid 2012;3(1):142–51.
[42] Bahramirad S, Reder W, Khodaei A. Reliability-constrained optimal sizing of energy storage system in a microgrid. IEEE Trans Smart Grid 2012;3(4):2056–62.
[43] Borowy BS, Salameh ZM. Methodology for optimally sizing the combination of a battery bank and PV array in a wind/PV hybrid system. IEEE Trans Energy Convers 1996;11(2).
[44] Kellogg WD, Nehrir MH, Venkataramanan G, Gerez V. Generation unit sizing and cost analysis for stand-alone wind, photovoltaic, and hybrid wind/PV systems. IEEE Trans Energy Convers 1998;13(1):70–5.
[45] Singh G, Baredar P, Singh A, Kurup D. Optimal sizing and location of PV, wind and battery storage for electrification to an island: a case study of Kavaratti, Lakshadweep. J Energy Storage 2017;12:78–86.
[46] Maleki A, Askarzadeh A. Optimal sizing of a PV/wind/diesel system with battery storage for electrification to an off-grid remote region: a case study of Rafsanjan, Iran. Sustain Energy Technol Assess 2014;7:147–53.
[47] Al-Ghussain L, Samu R, Fahrioglu M. Techno-economic feasibility of PV/wind-battery storage: case analysis in Zimbabwe. In: 16th International conference on clean energy (ICCE-2018); May 2018. p. 9–11.
[48] Bekele G, Tadesse G. Feasibility study of small hydro/PV/wind hybrid system for off-grid rural electrification in Ethiopia. Appl Energy 2012;97:5–15.
[49] Notton G, Mistrushi D, Stoyanov L, Berberi P. Operation of a photovoltaic-wind plant with a hydro pumping-storage for electricity peak-shaving in an island context. Sol Energy 2017;157:20–34.
[50] Sadeghi S, Askari IB. Prefeasibility techno-economic assessment of a hybrid power plant with photovoltaic, fuel cell and Compressed Air Energy Storage (CAES). Energy 2019;409–24.
[51] Hosseinalizadeh R, Shakouri GH, Amalnick MS, Taghipour P. Economic sizing of a hybrid (PV–WT–FC) renewable energy system (HRES) for stand-alone usages by an optimization-simulation model: case study of Iran. Renew Sustain Energy Rev 2016;54:139–50.
[52] Samal S, Hota PK. Design and analysis of solar PV-fuel cell and wind energy based microgrid system for power quality improvement. Cogent Eng 2017;4(1):1–21.
[53] Al-Ghussain L, Samu R, Taylan O, Fahrioglu M. Sizing renewable energy systems with energy storage systems in microgrids for maximum cost-efficient utilization of renewable energy resources. Sustain Cities Soc 2020;102059.
[54] Zhang W, Maleki A, Rosen MA, Liu J. Optimization with a simulated annealing algorithm of a hybrid system for renewable energy including battery and hydrogen storage. Energy 2018;163:191–207.

[55] Sanajaoba Singh S, Fernandez E. Modeling, size optimization and sensitivity analysis of a remote hybrid renewable energy system. vol. 143. Elsevier Ltd.; 2018.
[56] Eriksson ELV, Gray EMA. Optimization of renewable hybrid energy systems—a multi-objective approach. Renew Energy 2019;971–99.
[57] Sawle Y, Gupta SC, Bohre AK. Socio-techno-economic design of hybrid renewable energy system using optimization techniques. Renew Energy 2018;119:459–72.
[58] Duffie JA, Beckman WA. Solar engineering of thermal processes. 3rd ed. Hoboken, NJ: Wiley; 2006.
[59] Sadati SMS. Assessment of renewable energy based micro-grids for small communities. Middle East Technical University Northern Cyprus Campus; 2016.
[60] Loutzenhiser PG, Manz H, Felsmann C, Strachan PA, Frank T, Maxwell GM. Empirical validation of models to compute solar irradiance on inclined surfaces for building energy simulation. Sol Energy 2007;81(2):254–67.
[61] Solar C. Photovoltaic panels datasheet. Canada: Ontario; 2017.
[62] Reich NH, Mueller B, Armbruster A, Van Sark WGJHM, Kiefer K, Reise C. Performance ratio revisited: is PR>90% reaslistic? Prog Photovolt Res Appl 2012;20 (6):717–26.
[63] Ueda Y, Kurokawa K, Kitamura K, Yokota M, Akanuma K, Sugihara H. Performance analysis of various system configurations on grid-connected residential PV systems. Sol Energy Mater Sol Cells 2009;93(6–7):945–9.
[64] Sadati SMS, Jahani E, Taylan O. Technical and economic analyses for sizing PV power plant with storage system for METU NCC. In: ASME international mechanical engineering congress and exposition; 2015.
[65] Manwell JF, McGowan JG, Rogers AL. Wind energy explained: theory, design and application. 2nd ed. Chichester, UK: Wiley; 2009.
[66] Gamesa. Greater energy produced from low and medium wind sites. Zamudio, Spain: Gamesa; 2014.
[67] Fichter T, Trieb F, Moser M, Kern J. Optimized integration of renewable energies into existing power plant portfolios. Energy Procedia 2013;49:1858–68.
[68] Sangster AJ. Solar photovoltaics. Green Energy Technol 2014;194(4):145–72.
[69] Breeze P. Wind power. Power Gener Technol 2014;1(5):223–42.
[70] Zakeri B, Syri S. Electrical energy storage systems: a comparative life cycle cost analysis. Renew Sustain Energy Rev 2015;42:569–96.
[71] Yang H, Wei Z, Chengzhi L. Optimal design and techno-economic analysis of a hybrid solar-wind power generation system. Appl Energy 2009;86(2):163–9.
[72] Liu G, Rasul MG, Amanullah MTO, Khan MMK. Feasibility study of stand-alone PV-wind-biomass hybrid energy system in Australia. In: Asia-Pacific power energy eng. conf. (APPEEC); 2011.
[73] Steinbach J, Staniaszek D. Discount rates in energy system analysis [Discussion paper]; 2015.
[74] Net present value—The Environmental Literacy Council.
[75] Al-Ghussain L. Global warming: review on driving forces and mitigation. Environ Prog Sustain Energy 2018;38(1):13–21.
[76] Breu F, Guggenbichler S, Wollmann J. Carbon dioxide intensities of fuels and electricity for regions and countries; 2003.
[77] Radics K, Bartholy J. Estimating and modelling the wind resource of Hungary. Renew Sustain Energy Rev 2008;12(3):874–82.
[78] Abubaker AM, Najjar YSH, Ahmad AD. A uniquely finned tube heat exchanger design of a condenser for heavy-duty air conditioning systems. Int J Air-Conditioning Refrig 2020;28(01):2050004.
[79] Darwish Ahmad A, Abubaker AM, Najjar YSH, Manaserh YMA. Power boosting of a combined cycle power plant in Jordan: an integration of hybrid inlet cooling & solar systems. Energy Convers Manag 2020;214(January):112894.

[80] Al-Ghussain L, Samu R, Taylan O, Fahrioglu M. Sizing renewable energy systems with energy storage systems in microgrids for maximum cost-efficient utilization of renewable energy resources. Sustain Cities Soc 2020;55:102059.
[81] Al-Ghussain L, Al-Oran O, Lezsovits F. Statistical estimation of hourly diffuse radiation intensity of Budapest City. Environ Prog Sustain Energy 2020;, e13464.
[82] Al-Ghussain L, Samu R, Fahrioghlu M, Taylan O. Techno-economic comparative analysis of renewable energy systems: case study in Zimbabwe. Inventions 2020;5(3):27.

CHAPTER 8

Economic dispatch of large-scale integrated heat and power systems by application of a novel harmony search approach

Amir Fakhim-Babaei[a], Morteza Nazari-Heris[b], Behnam Mohammadi-Ivatloo[a], and Somayeh Asadi[b]
[a]Faculty of Electrical and Computer Engineering, University of Tabriz, Tabriz, Iran
[b]Department of Architectural Engineering, Pennsylvania State University, University Park, PA, United States

8.1 Introduction

Combined heat and power (CHP) plants that recover wasted heat during power generation by thermal plants generate electrical and thermal energy simultaneously. The CHP plants can increase the efficiency of supplying heat and power demands by 90% and, such plants can decrease the pollution by almost rate of 13%–18% [1, 2]. Several research studies have thus considered the application of cogeneration units as efficient concepts to obtain maximum efficiency of the network and achieve more cost savings to supply the demands of industrial sections and microgrids. The optimal production problem of cogeneration plants aims to obtain minimum operation cost of supplying heat and power demands. The CHP economic dispatch (CHPED) aims at providing load demand with optimal generation cost taking into account the constraints of network and generation plants. The CHPED is a nonlinear nonconvex problem, which should be solved taking into account valve impact of thermal plants, limitations of minimum and maximum generation of heat and power by the units, and heat and power demand balance [3]. The valve-point loading impact is investigated as a nonconvex term to the function of the problem that should be handled by using a powerful method [4]. Moreover, heat and power generated by cogeneration plants are in bidirectional dependence, making the CHPED problem more complex [5]. In other words, the power and heat generation of such plants is restricted to a limited zone called the feasible operating region (FOR).

Hybrid Energy System Models
https://doi.org/10.1016/B978-0-12-821403-9.00006-8

The solution of CHPED problem is obtained in several studies by proposing different mathematical and heuristic methods. Several mathematical concepts are used for solving the CHPED problem consisting of nonlinear programming [6], Tabu search [7], Lagrangian relaxation [8], and benders decomposition [9]. Moradi-Dalvand et al. have proposed a two-stage mathematical programming in Ref. [10] to deal with the sinusoidal term of the operation cost of thermal plants in the CHPED problem, where the introduced model provides a convex FOR for the CHP plants and deals with nondifferentiable term of valve-point loading influence of thermal plants in different stages. A branch and bound-based optimization method has been applied in Ref. [1] to study daily scheduling of CHP plants to supply power and heat demands considering load uncertainty. Lin et al. [11] have proposed a decentralized solution to the problem based on a robust model, which aims to increase the efficiency of the solution method and minimize the communication burden using benders decomposition.

However, some challenging constraints of the CHPED problem, such as FOR of the CHP plants and nonlinear cost function of the CHP and thermal plants, highlight the need for a strong solution methodology to obtain optimal generation scheduling of the units. On the other hand, heuristic concepts have been commonly employed for solving CHPED; these concepts consist of combined harmony search (HS) and the particle swarm procedure [12], the teaching-learning-based optimization concept [13], the brain storm method [14], the cuckoo technique [15], a mixture of HS and the Nelder-Mead procedure [16], social cognitive optimization [17], a genetic algorithm [18], the whale optimization procedure [19], a group search method [3] and the crisscross concept [20]. The cuckoo search concept is utilized in Ref. [21] using a penalty function to improve the capability of the approach in solving CHPED. Geem et al. proposed a new method in Ref. [22], where two competing objectives are considered for solving the CHPED. In Ref. [23], the CHPED problem is studied using the firefly algorithm (FA) to address the valve impact, spinning reserve, and limitations of the units ramping. The aim of implementation of heuristic procedures to solve the CHPED is generally obtaining a better solution in terms of production cost or increasing the convergence speed of the optimization method in maintaining the solution. In addition, Feng et al. implemented a combination of HS and Nelder-Mead methods in Ref. [16] to take advantage of both optimization methods and increase the robustness of the solution approach.

The harmony search (HS) concept is defined as a strong optimization technique for solving optimization problems, especially those relating to energy systems. HS, which was first introduced by Geem et al. [24], is a heuristic optimization technique has been widely implemented to solve power system problems such as hydrothermal scheduling [25], allocation [26], power system design [27], and dispatch [28, 29]. The authors have employed HS method to solve the CHPED problem considering the high performance of this method in solving optimization problems and its wide applications in engineering problems. An improved HS (IHS) method is introduced in this research in order to improve the strength of the concept. The IHS procedure is applied on two large test systems to provide optimal scheduling of the plants and investigate the performance of the proposed IHS concept. The investigations show the ability of the method in achieving the minimum operation cost of the system in less computational time.

The organization of this chapter is as follows. The formulation is provided in Section 8.2, and the introduced IHS optimization technique is reported in Section 8.3. The case studies and simulation results are provided in Section 8.4, and the research study is concluded in Section 8.5.

8.2 Problem formulation

The CHPED aims to obtain optimal power and heat production for minimizing the demand supply cost, which contains the costs of cogeneration plants, thermal plants, and boilers [30].

8.2.1 Cogeneration units

The production cost of cogeneration plants can be calculated using the quadratic function of the produced power and heat as follows:

$$C_j\left(P_j^c, H_j^c\right) = a_j\left(P_j^c\right)^2 + b_j P_j^c + c_j + d_j\left(H_j^c\right)^2 + e_j H_j^c + f_j H_j^c P_j^c \tag{8.1}$$

where $C_j(P_j^c, H_j^c)$ is the generation cost of the jth cogeneration unit. The power and heat generated by jth CHP units are indicated as P_j^c and H_j^c, respectively. Coefficients of jth CHP are defined by a_j, b_j, c_j, d_j, e_j, and f_j. The limitation of power and heat generated by CHP plants should be limited as follows:

$$P_j^{\min}\left(H_j^c\right) \leq P_j^c \leq P_j^{\max}\left(H_j^c\right) \tag{8.2}$$

$$H_j^{\min}\left(P_j^c\right) \leq H_j^c \leq H_j^{\max}\left(P_j^c\right) \tag{8.3}$$

where the lower and upper amounts of power production of the jth cogeneration plant are indicated by $P_j^{\min}$ and $P_j^{\max}$, respectively. In addition, $H_j^{\min}$ and $H_j^{\max}$ are the lower and upper bounds of heat production of the jth cogeneration unit, respectively.

8.2.2 Thermal plants

The generation cost of thermal plants is a function of generated power calculated as follows:

$$C_i(P_i^p) = a_i(P_i^p)^2 + b_i P_i^p + c_i \tag{8.4}$$

where P_i^p and $C_i(P_i^p)$ are the electrical energy and generation cost of the ith thermal plant, respectively. The coefficients of the ith thermal unit are a_i, b_i, and c_i. The cost considering the valve effect of such units is calculated as follows:

$$C_i(P_i^p) = a_i(P_i^p)^2 + b_i P_i^p + c_i + \left| d_i \sin\left(e_i\left(P_i^{min} - P_i^p\right)\right)\right| \tag{8.5}$$

where $P_i^{\min}$ is the lower bound of power production of ith thermal units. In addition, limitation of power generation of thermal units is as follows:

$$P_i^{min} \leq P_i^p \leq P_i^{max} \tag{8.6}$$

where the maximum capacity of the thermal plant i is defined by $P_i^{\max}$.

8.2.3 Boilers

The generation cost of boilers can be stated as follows:

$$C_k\left(H_k^h\right) = a_k\left(H_k^h\right)^2 + b_k H_k^h + c_k \tag{8.7}$$

where H_k^h and $C_k(H_k^h)$ indicate the heat produced and generation cost of boiler k, respectively. a_k, b_k, and c_k are the coefficients of the kth boiler.

8.2.4 Load balance

The electrical energy generated by cogeneration plants and thermal units provide the load demand as follows:

$$\sum_{i=1}^{N_p} P_i^p + \sum_{j=1}^{N_c} P_j^c = P_d \tag{8.8}$$

where P_d is the electrical load demand P_d. In addition, the heat generated by CHP plants and boilers should satisfy the heat demand as follows:

$$\sum_{j=1}^{N_c} H_j^c + \sum_{k=1}^{N_h} H_k^h = H_d \tag{8.9}$$

where H_d is the heat load demand of the system H_d.

8.2.5 Objective function

The optimization procedure should provide a minimum generation cost of cogeneration plants, thermal units, and boilers, calculated as follows:

$$\min \sum_{i=1}^{N_p} C_i(P_i^p) + \sum_{j=1}^{N_c} C_j\left(P_j^c, H_j^c\right) + \sum_{k=1}^{N_h} C_k\left(H_k^h\right) \tag{8.10}$$

where the number of cogeneration units, thermal plants, and boilers are defined by N_c, N_p and N_h, respectively.

8.3 The proposed improved HS

The HS procedure is a strong technique, which is found based on a musician's efforts in improvement of harmonies using the own experiments. The main advantages of HS are fewer parameters and easy application on optimization problems. The main optimization levels of the HS concept can be introduced as follows.

A. *The initialization of algorithm parameters*

Initialization of data related to the parameters of the optimization procedure is the first step of the HS concept.

B. *Harmony memory (HM)*

HM is obtained by producing random vectors stated as follows:

$$\mathrm{HM} = \begin{bmatrix} y_{1,1} & y_{1,2} & \cdots & y_{1,N-1} & y_{1,N} \\ y_{2,1} & y_{2,2} & \cdots & y_{2,N-1} & y_{2,N} \\ \vdots & \vdots & \vdots & \vdots & \vdots \\ y_{HMS-1,1} & y_{HMS-1,2} & \cdots & y_{HMS-1,N-1} & y_{HMS-1,N} \\ y_{HMS,1} & y_{HMS,2} & \cdots & y_{HMS,N-1} & y_{HMS,N} \end{bmatrix} \tag{8.11}$$

C. *Improvisation of new harmony (NH)*

Two main operators of the procedure include *HMCR* and *PAR*, which are employed in the process as follows:

(1) *Harmony memory consideration rate (HMCR)*

HMCR is the probability of selecting *NH* from the *HM*.

$$X_i^t = \left\{ \begin{array}{ll} x_i^t \in \{x_i^1, x_i^2, \ldots, x_i^{HMS}\} & \text{with probability } HMCR \\ x_i^t \in X_i & \text{with probability } (1 - HMCR) \end{array} \right\} \quad (8.12)$$

(2) *Pitch adjustment rate (PAR)*

PAR is employed in the process of optimization for checking whether *NH* requires a pitch adjusting level.

$$x_i^t = \left\{ \begin{array}{ll} Yes & p = PAR \\ No & p = (1 - PAR) \end{array} \right\} \quad (8.13)$$

$$x_i^t = x_i^t \pm \text{rand}()*bw \quad (8.14)$$

D. *Update for new harmony*

After integrating *HM* and *NH*, harmonies are sorted according to the fitness function. The worst harmonies will be removed from the *HM*.

E. *Termination procedure of the method*

The process will be finished once the maximum number of iterations has been carried out.

In this study, the IHS procedure is presented. The traditional HS concept selects *NH* randomly from *HM*. On the other hand, the chance of selection of memories with lower fitness is higher in the introduced IHS method. The proposed framework can be performed using the following method:

$$X_i^t = \left\{ \begin{array}{ll} x_i^t = x_i^{\left[\text{rand}()^a * HMS\right]+1} & \text{with probability } HMCR \\ x_i^t = X_i & \text{with probability } (1 - HMCR) \end{array} \right. \quad (8.15)$$

where $a > 1$. The proposed model is effective in obtaining an optimal solution with minimum cost and faster convergence. *HMCR*, *PAR*, and *BW* are calculated by a similar concept. An exponential function is utilized in order to improve the quality of the method in the optimization process:

$$\begin{aligned} HMCR &= HMCR_{start} + (HMCR_{end} - HMCR_{start})*(i/\max Iter)^b \\ PAR &= PAR_{start} + (PAR_{end} - PARR_{start})*(i/\max Iter)^c \\ BW &= BW_{start} + (BW_{end} - BW_{start})*(i/\max Iter)^d \end{aligned} \quad (8.16)$$

Fig. 8.1 illustrates *HMCR* through 1000 iterations for $HMCR_{start} = 0.95$, $HMCR_{end} = 0.99$ considering 0.5, 1, and 2 values for *b*.

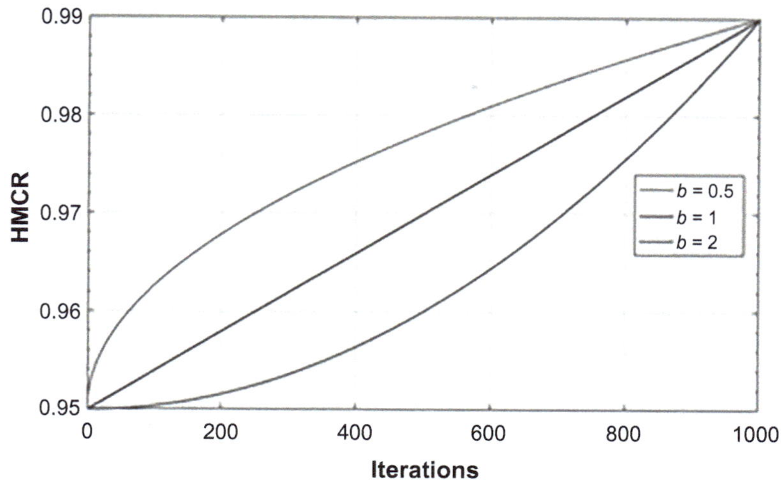

Fig. 8.1 Change of *HMCR*.

8.4 Case study and simulation results

The IHS concept was implemented to provide an optimal solution for cogeneration systems. The studied systems are 84 unit systems and 96 plant system. The first case study contains 40 thermal plants, 24 cogeneration plants, and 20 boilers [5]. The second one includes 52 thermal units, 24 CHP plants, and 20 boilers. Cost coefficients of the CHP units, thermal plants, and boilers, and feasible operating zones of the cogeneration plants, are provided in Tables 8.1–8.4, respectively. The same characteristics for the second case study are provided in Tables 8.5–8.8.

The parameters used for the proposed IHS algorithm are provided in Table 8.9.

Table 8.10 provides an optimal solution of the electrical and thermal energy supply for the thermal plants, cogeneration plants, and boilers for case study 1. The sum of electrical energy supplied by cogeneration plants and thermal plants meets the electrical demand of the cogeneration network (12,700 MW), and the sum of heat generated by cogeneration plants and boilers supplies the heat demand of the network (5000 MWth).

Table 8.11 compares the obtained generation cost of the studied test system 1 using the proposed IHS method and the recent publication utilizing whale optimization algorithm (WOA) [17]. WOA has provided $290,123.97424 generation cost for the studied system; however, the proposed IHS procedure has reported a generation cost of $287,982.8803, which suggests that the method is able to reduce the production cost by

Table 8.1 Parameters of thermal plants of case study 1.

Unit	a_i	b_i	c_i	d_i	e_i	P^{min}	P^{max}
1, 2	0.0069	6.73	94.705	100	0.084	36	114
3	0.02028	7.07	309.54	100	0.084	60	120
4	0.00942	8.18	369.03	150	0.063	80	190
5	0.0114	5.35	148.89	120	0.077	47	97
6	0.01142	8.05	222.33	100	0.084	68	140
7	0.00357	8.03	287.71	200	0.042	110	300
8	0.00492	6.99	391.98	200	0.042	135	300
9	0.00573	6.6	455.76	200	0.042	135	300
10	0.00605	12.9	722.82	200	0.042	130	300
11	0.00515	12.9	635.2	200	0.042	94	375
12	0.00569	12.8	654.69	200	0.042	94	375
13	0.00421	12.5	913.4	300	0.035	125	500
14	0.00752	8.84	1760.4	300	0.035	125	500
15, 16	0.00708	9.15	1728.3	300	0.035	125	500
17	0.00313	7.97	647.85	300	0.035	220	500
18	0.00313	7.95	649.69	300	0.035	220	500
19	0.00313	7.97	647.83	300	0.035	242	550
20	0.00313	7.97	647.81	300	0.035	242	550
21, 22	0.00298	6.63	785.96	300	0.035	254	550
23, 24	0.00284	6.66	794.53	300	0.035	254	550
25, 26	0.00277	7.1	801.32	300	0.035	254	550
27, 28, 29	0.52124	3.33	1055.1	120	0.077	10	150
30	0.0114	5.35	148.89	120	0.077	47	97
31, 32, 33	0.0016	6.43	222.92	150	0.063	60	190
34	0.0001	8.95	107.87	200	0.042	90	200
35, 36	0.0001	8.62	116.58	200	0.042	90	200
37, 38, 39	0.0161	5.88	307.45	80	0.098	25	110
40	0.00313	7.97	647.83	300	0.035	242	550

Table 8.2 Parameters of CHP plants of case study 1.

Unit	a_j	b_j	c_j	d_j	e_j	f_j
41, 42, 43, 44, 49, 50, 51, 52	0.0345	14.5	2650	0.030	4.200	0.031
45, 46, 47, 48, 53, 54, 55, 56	0.0435	36.0	1250	0.027	0.600	0.011
57, 58, 59, 60	0.1035	34.5	2650	0.025	2.203	0.051
61, 62, 63, 64	0.072	20.0	1565	0.020	2.340	0.040

Table 8.3 FOR of cogeneration plants of case study 1.

Unit	Marginal points
41, 42, 43, 44, 49, 50, 51, 52	[98.8, 0], [81, 104.8], [215, 180], [247, 0]
45, 46, 47, 48, 53, 54, 55, 56	[44, 0], [44, 15.9], [40, 75], [110.2, 135.6], [125.8, 32.4], [125.8, 0]
57, 58, 59, 60	[20, 0], [10, 40], [45, 55], [60, 0]
61, 62, 63, 64	[35, 0], [35, 20], [90, 45], [90, 25], [105, 0]

Table 8.4 Parameters of boilers of case study 1.

Plant	a_k	b_k	c_k	H^{min}	H^{max}
65, 66, 67, 68	0.038	2.0109	950	0	2695.20
69, 70, 71, 72	0.038	2.0109	950	0	60
77, 78, 79, 80, 81, 82, 83, 84	0.052	3.0651	480	0	120

Table 8.5 Parameters of thermal plants of case study 2.

Unit	a_i	b_i	c_i	d_i	e_i	P^{min}	P^{max}
1, 14, 27, 40	0.00028	8.1	550	300	0.035	0	680
2, 3, 15, 16, 28, 29, 41, 42	0.00056	8.1	309	200	0.042	0	360
4, 5, 6, 7, 8, 9, 17, 18, 19, 20, 21, 22, 30, 31, 32, 33, 34, 35, 43, 44, 45, 46, 47, 48	0.00324	7.74	240	150	0.063	60	180
10, 11, 23, 24, 36, 37, 49, 50	0.00284	8.6	126	100	0.084	40	120
12, 13, 25, 26, 38, 39, 51, 52	0.00284	8.6	126	100	0.084	55	120

Table 8.6 Parameters of CHP plants of case study 2.

Unit	a_j	b_j	c_j	d_j	e_j	f_j
53, 55, 59, 61, 65, 67, 71, 73	0.0345	14.5	2650	0.030	4.200	0.031
54, 56, 60, 62, 66, 68, 72, 74	0.0435	36.0	1250	0.027	0.600	0.011
57, 63, 69, 75	0.1035	34.5	2650	0.025	2.203	0.051
58, 64, 70, 76	0.072	20.0	1565	0.020	2.340	0.040

Table 8.7 FOR of cogeneration plants of case study 2.

Unit	Marginal points
53, 55, 59, 61, 65, 67, 71, 73	[98.8, 0], [81, 104.8], [215, 180], [247, 0]
54, 56, 60, 62, 66, 68, 72, 74	[44, 0], [44, 15.9], [40, 75], [110.2, 135.6], [125.8, 32.4], [125.8, 0]
57, 63, 69, 75	[20, 0], [10, 40], [45, 55], [60, 0]
58, 64, 70, 76	[35, 0], [35, 20], [90, 45], [90, 25], [105, 0]

Table 8.8 Parameters of boilers of case study 2.

Plant	a_k	b_k	c_k	H^{min}	H^{max}
77, 82, 87, 92	0.038	2.0109	950	0	2695.20
78, 79, 83, 94, 88, 89, 93, 94	0.038	2.0109	950	0	60
80, 81, 85, 86, 90, 91, 95, 96	0.052	3.0651	480	0	120

Table 8.9 Parameters of the optimization procedure.

Parameter	Quantity	Parameter	Quantity
$PAR_{initial}$	0.001	bw_{inital}	0.45
PAR_{final}	0.09	bw_{final}	0.001
$HMCR_{inital}$	0.99	NHS	200
$HMCR_{final}$	0.9999	HS	200

$2,141.09. Accordingly, the introduced procedure is capable of providing an annual cost saving of $18,755,983.

Fig. 8.2 compares the convergence speed of the WOA and the introduced IHS method for case study 1. The evaluation demonstrates that the implemented IHS method has better convergence characteristics than the WOA.

Table 8.12 gives the optimal solution of the CHPED problem for the second case study by determining the power and heat supply for the thermal plants, cogeneration plants, and boilers for case study 2. The sum of power and heat supplied by all the units satisfies 9400 MW and 5000 MWth, respectively. The operation cost for this case is equal to $236,902.18; however, recent studies in Refs. [18, 31] have obtained operation costs of $239,139.5018 and $239,896.40826, respectively. The comparison of the results shows a decreased operation cost of the test system with respect to Refs. [18, 31].

Table 8.10 Simulation results for case study 1.

Output	WOA [17]	Introduced IHS method	Output	WOA [17]	Introduced IHS method	Output	WOA [17]	Introduced IHS method
Thermal 1	110.8794	112.4460	Thermal 37	89.7967	107.2993	H CHP 49	130.6574	113.9083
Thermal 2	112.2931	114.0000	Thermal 38	109.9979	109.8760	H CHP 50	97.0303	78.2279
Thermal 3	98.1159	98.6128	Thermal 39	109.9994	110.0000	H CHP 51	44.5495	40.1564
Thermal 4	129.8682	179.6790	Thermal 40	516.5065	512.8476	H CHP 52	29.7277	20.0350
Thermal 5	88.6586	87.7731	P CHP 41	112.3421	124.6826	H CHP 53	154.1830	128.3412
Thermal 6	139.9998	140.0000	P CHP 42	50.4459	44.3213	H CHP 54	75.3759	92.2254
Thermal 7	196.1145	265.8195	P CHP 43	131.6591	126.5941	H CHP 55	116.1224	131.1330
Thermal 8	295.0226	287.7383	P CHP 44	57.3384	43.4304	H CHP 56	88.6239	89.9072
Thermal 9	284.6146	297.3969	P CHP 45	10.0991	17.2956	H CHP 57	39.7995	43.1725
Thermal 10	279.6016	131.0594	P CHP 46	44.2424	41.8988	H CHP 58	33.3716	25.2276
Thermal 11	318.4002	318.2740	P CHP 47	103.2198	104.5039	H CHP 59	144.7767	120.6046
Thermal 12	318.4004	169.6091	P CHP 48	40.2287	52.7798	H CHP 60	88.9117	104.1515
Thermal 13	394.2803	393.6517	P CHP 49	127.0797	100.5562	H CHP 61	128.4364	120.6427

Continued

Table 8.10 Simulation results for case study 1—cont'd

Output	WOA	Introduced IHS method	Output	WOA	Introduced IHS method	Output	WOA	Introduced IHS method
Thermal 14	484.2888	394.3573	P CHP 50	65.5205	43.9310	H CHP 62	81.0055	93.6572
Thermal 15	484.0382	483.8590	P CHP 51	20.6157	10.5407	H CHP 63	40.8289	44.1388
Thermal 16	304.6382	393.8653	P CHP 52	56.4012	35.7095	H CHP 64	30.0546	24.4931
Thermal 17	489.6014	489.0722	P CHP 53	168.9969	124.9562	Boiler 65	383.1144	377.3279
Thermal 18	489.2782	489.9174	P CHP 54	40.4449	60.0061	Boiler 66	59.9997	60.0000
Thermal 19	511.9256	511.3098	P CHP 55	101.1777	129.5554	Boiler 67	60.0000	59.8920
Thermal 20	511.3778	511.0001	P CHP 56	55.7835	57.7166	Boiler 68	119.9855	120.0000
Thermal 21	433.5245	523.7579	P CHP 57	10.0560	17.4558	Boiler 69	119.9990	120.0000
Thermal 22	433.5316	524.2985	P CHP 58	64.4182	49.0977	Boiler 70	402.6530	408.3564
Thermal 23	523.2806	525.2782	P CHP 59	152.2648	110.1608	Boiler 71	59.8978	59.1514
Thermal 24	523.2888	527.0732	P CHP 60	56.1164	73.8634	Boiler 72	59.9998	60.0000
Thermal 25	523.2889	524.0086	P CHP 61	123.1718	110.0874	Boiler 73	119.9996	120.0000
Thermal 26	524.0590	524.3409	P CHP 62	46.9576	61.7323	Boiler 74	119.9999	117.7786

Thermal 27	10.0000	10.0000	P CHP 63	11.9343	19.9982	Boiler 75	385.5340	380.5771
Thermal 28	10.0000	10.7004	P CHP 64	57.1203	45.8119	Boiler 76	58.9464	59.9714
Thermal 29	11.2485	10.0000	H CHP 41	122.3890	128.5106	Boiler 77	59.9995	59.4874
Thermal 30	91.0437	92.3385	H CHP 42	84.0170	78.5967	Boiler 78	119.7982	120.0000
Thermal 31	189.9732	189.6967	H CHP 43	133.2295	130.1290	Boiler 79	119.9988	120.0000
Thermal 32	189.9997	190.0000	H CHP 44	89.9664	77.9599	Boiler 80	381.8642	455.8285
Thermal 33	163.6655	190.0000	H CHP 45	40.0414	43.0478	Boiler 81	59.9999	60.0000
Thermal 34	165.1037	200.0000	H CHP 46	24.2001	22.7419	Boiler 82	59.9991	59.9404
Thermal 35	166.7651	171.6001	H CHP 47	115.7242	116.5395	Boiler 83	119.9980	110.1407
Thermal 36	165.8941	171.2335	H CHP 48	75.1974	85.8482	Boiler 84	119.9982	118.1616

Table 8.11 Evaluation of the generation cost for case study 1.

Output	WOA [17]	Traditional HS procedure	The introduced IHS procedure
Time (s)	158.18	96.8536	59.955205
Min cost ($)	290,123.97424	296,696.9518	287,982.8803

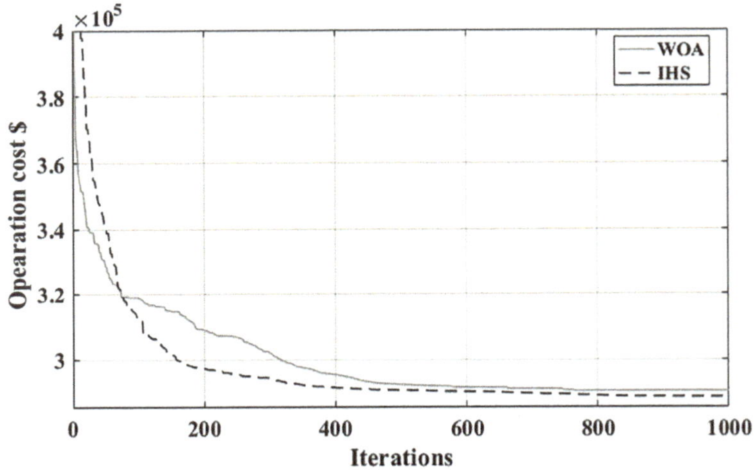

Fig. 8.2 Evaluation of convergence speed for case study 1.

8.5 Conclusion

In this chapter, a new improved HS (IHS) procedure is applied to solve the CHPED problem to obtain a better solution in optimal scheduling of the units. The traditional HS procedure selects *NH* randomly from *HM*. On the other hand, the chance of selection of memories with lower fitness in the presented IHS method is higher with respect to the conventional HS method. The IHS concept is applied in two large-scale case studies for evaluating the performance of the proposed IHS procedure in handling the studied optimization problem with minimum generation cost. The results obtained show an annual cost savings of $18,755,983 for the studied test system 1 with respect to the best solution mentioned by WOA in recent publications. Accordingly, the implemented IHS procedure shows great strength in solving the optimization problems of power networks.

Table 8.12 Simulation results for case study 2.

Output	Introduced method	Output	Introduced method	Output	Introduced method
Thermal 1	629.29	P CHP 53	179.09	H CHP 57	41.39628
Thermal 2	298.80	P CHP 54	94.21	H CHP 58	17.15724
Thermal 3	226.78	P CHP 55	225.22	H CHP 59	116.1711
Thermal 4	160.94	P CHP 56	125.8	H CHP 60	92.61052
Thermal 5	172.52	P CHP 57	10.03	H CHP 61	167.3967
Thermal 6	112.52	P CHP 58	104.98	H CHP 62	86.17834
Thermal 7	159.91	P CHP 59	159.43	H CHP 63	43.17458
Thermal 8	176.3920704	P CHP 60	122.18	H CHP 64	11.36008
Thermal 9	160.93	P CHP 61	210.86	H CHP 65	87.57
Thermal 10	77.15	P CHP 62	107.85	H CHP 66	37.96
Thermal 11	119.94	P CHP 63	47.67	H CHP 67	175.37
Thermal 12	119.17	P CHP 64	78.76	H CHP 68	73.7
Thermal 13	55.61	P CHP 65	81	H CHP 69	55
Thermal 14	458.7843024	P CHP 66	40	H CHP 70	32.84
Thermal 15	159.79	P CHP 67	81.04	H CHP 71	104.84
Thermal 16	317.90	P CHP 68	87.67	H CHP 72	135.54
Thermal 17	116.86	P CHP 69	60	H CHP 73	134.86
Thermal 18	108.77	P CHP 70	105	H CHP 74	128.917
Thermal 19	113.77	P CHP 71	116.43	H CHP 75	49.89
Thermal 20	114.98	P CHP 72	104.25	H CHP 76	40.40
Thermal 21	165.33	P CHP 73	207.83	Boiler 77	952.06
Thermal 22	171.64	P CHP 74	106.31	Boiler 78	0
Thermal 23	91.42	P CHP 75	50.54	Boiler 79	0
Thermal 24	75.25	P CHP 76	101.68	Boiler 80	78.353

Continued

Table 8.12 Simulation results for case study 2—cont'd

Output	Introduced method	Output	Introduced method	Output	Introduced method
Thermal 25	74.64	H CHP 53	161.53	Boiler 81	63.73
Thermal 26	109.65	H CHP 54	112.31	Boiler 82	55.07
Thermal 27	96.60310893	H CHP 55	164.1356	Boiler 83	11.29
Thermal 28	0.135283837	H CHP 56	122.0462	Boiler 84	0
Thermal 29	11.90389811	Thermal 41	280.21	Boiler 85	92.03
Thermal 30	60.00044876	Thermal 42	260.36	Boiler 86	41.51
Thermal 31	84.75420318	Thermal 43	175.12	Boiler 87	320.11
Thermal 32	60.0832715	Thermal 44	102.90	Boiler 88	40.79
Thermal 33	60	Thermal 45	84.89	Boiler 89	31.07
Thermal 34	60	Thermal 46	96.70	Boiler 90	102.44
Thermal 35	60	Thermal 47	108.16	Boiler 91	90.018
Thermal 36	40	Thermal 48	156.34	Boiler 92	196.36
Thermal 37	83.09	Thermal 49	120	Boiler 93	59.63
Thermal 38	55	Thermal 50	118.44	Boiler 94	59.98
Thermal 39	82.93	Thermal 51	106.73	Boiler 95	94.34
Thermal 40	179.06	Thermal 52	106.24	Boiler 96	113.44

References

[1] Nazari-Heris M, Abapour S, Mohammadi-Ivatloo B. Optimal economic dispatch of FC-CHP based heat and power micro-grids. Appl Therm Eng 2017;114:756–69.
[2] Narang N, Sharma E, Dhillon J. Combined heat and power economic dispatch using integrated civilized swarm optimization and Powell's pattern search method. Appl Soft Comput 2017;52:190–202.
[3] Basu M. Group search optimization for combined heat and power economic dispatch. Int J Electr Power Energy Syst 2016;78:138–47.
[4] Nazari-Heris M, Mohammadi-Ivatloo B, Gharehpetian G. Short-term scheduling of hydro-based power plants considering application of heuristic algorithms: a comprehensive review. Renew Sustain Energy Rev 2017;74:116–29.
[5] Nazari-Heris M, Mohammadi-Ivatloo B, Gharehpetian GB. Optimal short-term scheduling of combined heat and power systems considering electrical energy storage, In: 3rd International conference of IEA; 2017.
[6] Gopalakrishnan H, Kosanovic D. Operational planning of combined heat and power plants through genetic algorithms for mixed 0–1 nonlinear programming. Comput Oper Res 2015;56:51–67.
[7] Sudhakaran M, Slochanal SMR. Integrating genetic algorithms and tabu search for combined heat and power economic dispatch, In: TENCON 2003. Conference on convergent technologies for the Asia-Pacific regionvol. 1. IEEE; 2003. p. 67–71.
[8] Sashirekha A, Pasupuleti J, Moin N, Tan C. Combined heat and power (CHP) economic dispatch solved using Lagrangian relaxation with surrogate subgradient multiplier updates. Int J Electr Power Energy Syst 2013;44(1):421–30.
[9] Abdolmohammadi HR, Kazemi A. A benders decomposition approach for a combined heat and power economic dispatch. Energy Convers Manage 2013;71:21–31.
[10] Moradi-Dalvand M, Nazari-Heris M, Mohammadi-Ivatloo B, Galavani S, Rabiee A. A two-stage mathematical programming approach for the solution of combined heat and power economic dispatch. IEEE Syst J 2019;.
[11] Lin C, Wu W, Zhang B, Sun Y. Decentralized solution for combined heat and power dispatch through benders decomposition. IEEE Trans Sustainable Energy 2017;8 (4):1361–72.
[12] Nazari-Heris M, Fakhim-Babaei A, Mohammadi-Ivatloo B. A novel hybrid harmony search and particle swarm optimization method for solving combined heat and power economic dispatch, In: Smart grid conference (SGC), 2017 IEEE; 2017. p. 1–9.
[13] Pattanaik JK, Basu M, Dash DP. Modified teaching-learning-based optimization for combined heat and power economic dispatch. Int J Emerg Electr Power Syst 2017;18(5).
[14] Wu Y, Wang X, Fu Y, Xu Y. Difference brain storm optimization for combined heat and power economic dispatch, In: International conference in swarm intelligenceSpringer; 2017. p. 519–27.
[15] Mehdinejad M, Mohammadi-Ivatloo B, Dadashzadeh-Bonab R. Energy production cost minimization in a combined heat and power generation systems using cuckoo optimization algorithm. Energ Effic 2017;10(1):81–96.
[16] Feng Z-Y, Guo H, Liu Z-T, Xu L, She J. Hybridization of harmony search with Nelder-Mead algorithm for combined heat and power economic dispatch problem, In: 36th Chinese control conference (CCC), 2017 IEEE; 2017. p. 2790–5.
[17] Sun J, Li Y. Social cognitive optimization with tent map for combined heat and power economic dispatch. Int Trans Electr Energy Syst 2018;e2660.
[18] Haghrah A, Nazari-Heris M, Mohammadi-Ivatloo B. Solving combined heat and power economic dispatch problem using real coded genetic algorithm with improved Mühlenbein mutation. Appl Therm Eng 2016;99:465–75.

[19] Nazari-Heris M, Mehdinejad M, Mohammadi-Ivatloo B, Babamalek-Gharehpetian G. Combined heat and power economic dispatch problem solution by implementation of whale optimization method. Neural Comput Applic 2017;1–16.
[20] Meng A, Mei P, Yin H, Peng X, Guo Z. Crisscross optimization algorithm for solving combined heat and power economic dispatch problem. Energy Convers Manage 2015;105:1303–17.
[21] Mellal MA, Williams EJ. Cuckoo optimization algorithm with penalty function for combined heat and power economic dispatch problem. Energy 2015;93:1711–8.
[22] Shi B, Yan L-X, Wu W. Multi-objective optimization for combined heat and power economic dispatch with power transmission loss and emission reduction. Energy 2013;56:135–43.
[23] Niknam T, Azizipanah-Abarghooee R, Roosta A, Amiri B. A new multi-objective reserve constrained combined heat and power dynamic economic emission dispatch. Energy 2012;42(1):530–45.
[24] Geem ZW, Kim JH, Loganathan GV. A new heuristic optimization algorithm: harmony search. Simulation 2001;76(2):60–8.
[25] Nazari-Heris M, Babaei AF, Mohammadi-Ivatloo B, Asadi S. Improved harmony search algorithm for the solution of non-linear non-convex short-term hydrothermal scheduling. Energy 2018;151:226–37.
[26] Nazari-Heris M, Mohammadi-Ivatloo B. Optimal placement of phasor measurement units to attain power system observability utilizing an upgraded binary harmony search algorithm. Energy Syst 2015;6(2):201–20.
[27] Maleki A, Pourfayaz F. Sizing of stand-alone photovoltaic/wind/diesel system with battery and fuel cell storage devices by harmony search algorithm. J Energy Storage 2015;2:30–42.
[28] Jeddi B, Vahidinasab V. A modified harmony search method for environmental/economic load dispatch of real-world power systems. Energy Convers Manage 2014;78:661–75.
[29] Elattar EE. Modified harmony search algorithm for combined economic emission dispatch of microgrid incorporating renewable sources. Energy 2018;.
[30] Nazari-Heris M, Mohammadi-Ivatloo B, Gharehpetian G. A comprehensive review of heuristic optimization algorithms for optimal combined heat and power dispatch from economic and environmental perspectives. Renew Sustain Energy Rev 2017;.
[31] Mohammadi-Ivatloo B, Moradi-Dalvand M, Rabiee A. Combined heat and power economic dispatch problem solution using particle swarm optimization with time varying acceleration coefficients. Electr Power Syst Res 2013;95:9–18.

CHAPTER 9

Hybrid solar-heated anaerobic digestion system for electricity and heat cogeneration

Mohammed El Ibrahimi, Ismail Khay, Anas El Maakoul, Asmae Berrada, and Mohamed Bakhouya
College of Engineering, LERMA Lab, International University of Rabat, Sala El Jadida, Morocco

9.1 Introduction

Renewable energy sources (RES) mainly consist of solar, wind, biomass, hydroelectric, geothermal, and tidal energies. These energies are said to be renewable because they are natural, clean, and inexhaustible [1]. In the last decades, the worldwide focus has been shifting toward RES as means of energy generation because of the rapidly dwindling fossil fuel reserves and climate change concerns [2]. However, due to the intermittent nature of the natural resources (e.g., sun and wind), the low efficiency (relative to fossil fuels), and the expensive deployment costs of the renewable energy technologies (RET), the transition to renewable energy sources is not as seamless as it needs to be [3].

Biomass is currently deployed in an attempt to overcome these restrictions since it is less costly and more efficient than conventional RET, and independent from natural resources [4]. There are two main methods through which this renewable energy can be harnessed, namely combustion and anaerobic digestion (AD). Combustion is carried out by burning biomass (organic waste) and recovering energy in the form of heat, which could be used directly for heating or be further converted into electricity. As for AD, it involves the biodegradation of organic matter (farm produce, paper waste, etc.) by bacteria (made available by the addition of animal manures or municipal wastewater) in the absence of oxygen. This biological process allows recovering energy in the form of biogas (a mixture of methane and carbon dioxide). Compared to combustion, AD offers superior potential energy for herbaceous biomass as reported in [5]. The AD process has proven to be a reliable and sustainable method of producing energy while,

Hybrid Energy System Models
https://doi.org/10.1016/B978-0-12-821403-9.00009-3

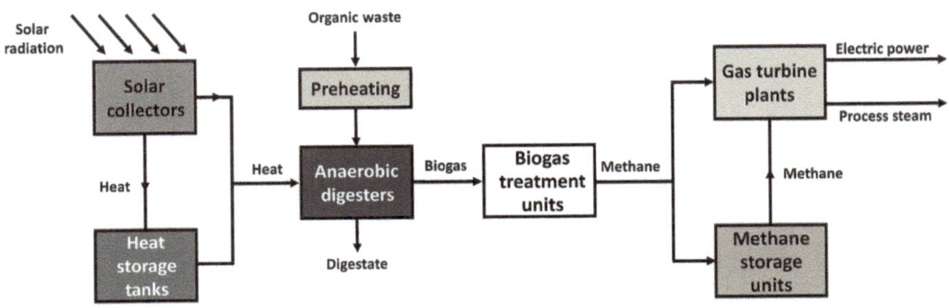

Fig. 9.1 Global diagram of a hybrid renewable energy system for power production, containing mainly: anaerobic digesters, thermal solar systems, and gas-turbine plants.

at the same time, reducing bad odors, eliminating pathogens, and generating good-quality soil fertilizer [6].

A key parameter that greatly affects the overall rate and yield of the AD process is heating [7]. As a consequence, in order to render anaerobic digesters more efficient, heating can be carried out through hybrid renewable energy systems (HRES). In fact, these hybrid systems operate by combining two or more renewable energy technologies (RETs) to achieve higher efficiency than that of a single RET [8]. Furthermore, well-sized hybrid systems may allow total independence from other nonrenewable energy sources [9]. When coupled with solar thermal collectors, the heating needs of anaerobic digesters can be met without resorting to electricity nor to conventional fuels (Fig. 9.1). These solar/anaerobic digestion hybrid systems are truly renewable and can be completely independent of the electric grid, which is particularly useful in remote and isolated locations. In order to achieve total autonomy, energy storage solutions must be integrated.

In this chapter, the basic principles and parameters of the AD process are detailed, and the commonly used and most practical AD models are presented and described. Additionally, fundamental thermal equations and correlations are presented and applied to anaerobic digesters and solar thermal energy systems (thermal collectors and heat storage tanks) for energetic modeling purposes. Thereafter, thermodynamic cycles for power generation using methane (produced through AD) as a fuel are presented and discussed. Finally, a case study investigating heat and electricity production using a hybrid solar-anaerobic digestion system is carried out.

9.2 Anaerobic digestion process

The anaerobic digestion (AD) process takes place following four main chronological stages, namely the (1) hydrolysis, (2) acidogenesis, (3) acetogenesis, and (4) methanogenesis, as illustrated in Fig. 9.2. Hydrolysis starts taking place as soon as the organic matter is introduced inside the reactor and comes into contact with water. In this first stage, the larger organic polymers (made of proteins, lipids, and polysaccharides) are broken down into smaller and more accessible parts as an energy source (e.g., amino acids, fatty acids, and simple sugars) for bacteria [10]. Once this process is completed, the acidogenesis stage begins, where the simpler monomers produced during hydrolysis are turned into alcohols, acetate, and volatile fatty acids (VFAs) by the acidogenic bacteria [11]. The acetogenic bacteria then take over to convert the previously produced components into formate (salt of the formic acid), methanol, H_2, CO_2, and acetate in the acetogenesis stage [12]. The last stage in AD is the methanogenesis, in which the methanogenic bacteria use the byproducts of the previous steps to produce CH_4, CO_2, and H_2O [13].

9.2.1 AD parameters

Several parameters influence the performance and stability of anaerobic digesters, namely the temperature, pH, VFA concentration, nutrients, solid content, C/N content, and mixing.

- Temperature

Similar to other biological processes, the temperature has a significant impact on the microbial activities in the AD process [14]. In fact, temperature affects the global methane yield by directly influencing the microbial community's diversity and growth rate, the process' kinetics and stability, and the equilibrium of the reactions taking place during AD [15]. The AD process is usually carried out in one of two temperature ranges, namely mesophilic (between 25°C and 40°C) and thermophilic (over 45°C), each offering several benefits and drawbacks.

- pH

The pH is a crucial parameter in the AD process. Actually, both the hydrolysis and acidogenesis rates are dependent on the value of pH. At an optimal pH (of about 7), more of the chemical oxygen demand (COD) and total organic carbon (TOC) are soluble and higher volatile fatty acids (VFA) concentrations are attained in shorter times. Furthermore, at optimal pH, lower concentrations of lactic acid are present in the acidified products, which translates to better biogas production [16].

- Nutrients

The anaerobic bacteria present in digesters necessitate nutrients, mainly C, H_2, O_2, N_2, P, and S. For the municipal organic waste, it does not have

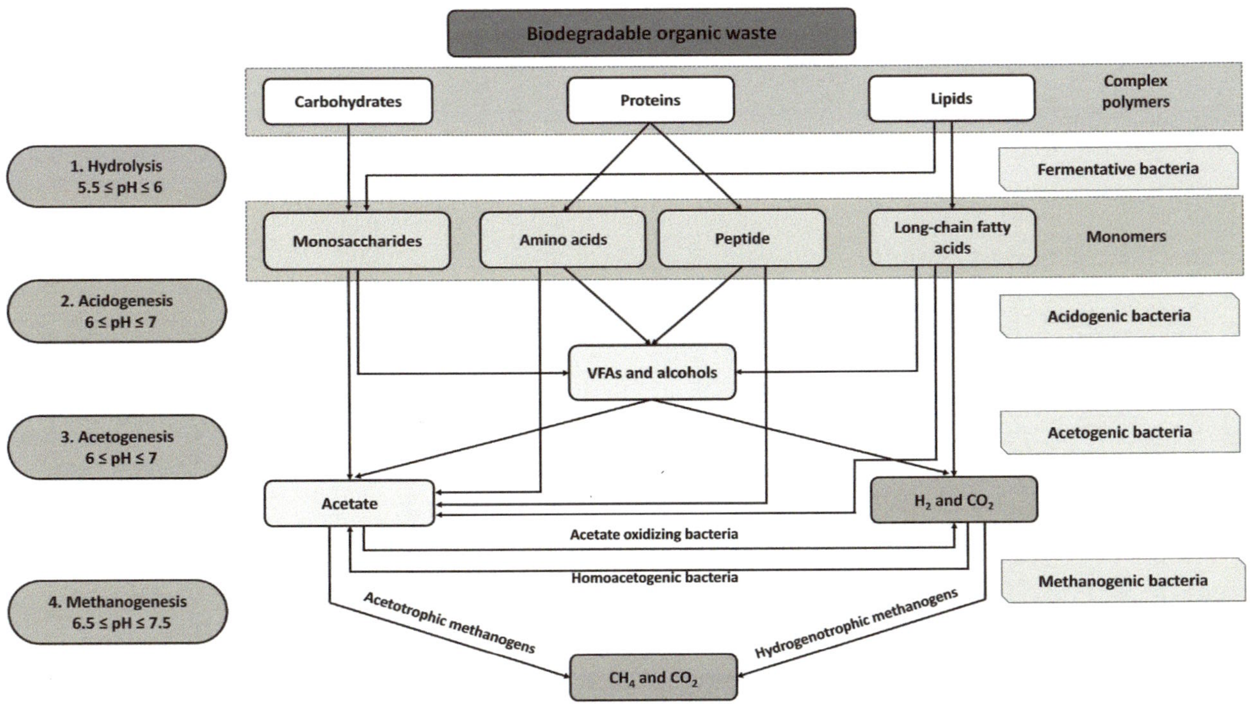

Fig. 9.2 The chronological stages of the anaerobic digestion process. *(Adapted from Wang P, Wang H, Qiu Y, Ren L, Jiang B. Microbial characteristics in anaerobic digestion process of food waste for methane production—a review. Bioresour Technol 2018;248:29–36. doi:https://doi.org/10.1016/j.biortech.2017.06.152.)*

sufficient quantities of nutrients, namely P and N_2. Therefore, to ensure a balance in the nutrients available for the anaerobes and increase biogas production, nitrogen- and phosphorus-rich substrates such as animal manure and wastewater need to be added [17].

- Total solid content (TS)

The initial TS of substrates influences the anaerobic digestion process. In fact, substrates with lower TS lead to higher COD removal rates and improved methane yields [18]. Additionally, it is easier and less costly to heat digesters loaded with low TS feedstocks [19]. Furthermore, Forster-Carneiro et al. [20] reported that the optimal TS for a mesophilic AD performed on food waste is 20%, since it triggers a shorter start-up time and an enhanced CH_4 production.

- Hydraulic retention time (HRT)

An important parameter that indicates the average length of time that a substrate spends inside an anaerobic digester is named HRT. This parameter is responsible for the stability and performance of continuous anaerobic digesters. If the retention time is too short, bacteria will not be able to digest the feedstock fast enough. However, if the retention time is too long, biogas production will decrease because of the lack of nutrients in the feedstock. Therefore, an optimal HRT must be maintained to provide sufficient time for the bacteria to convert the organic matter into biogas [21].

- Carbon to nitrogen ratio (C/N)

The carbon available in a feedstock represents the main source of energy for the bacteria, while the nitrogen content is responsible for making the anaerobic microbial communities more abundant. In fact, the AD process is affected by the C/N ratio, thus the need for an appropriate ratio. Literature indicates an adequate range for this ratio between 20 and 30, with an optimal value of 25 for maximal microbial growth [22].

- Mixing

To facilitate the access of bacteria to nutrients, proper mixing is employed in anaerobic digesters. It is reported that weaker degrees of mixing enhance the fermentation and, as a consequence, the global biogas production. However, violent mixing leads to adverse results on the AD process [23] as well as the hydraulic performance of anaerobic digesters [24]. Furthermore, mixing helps eliminate the temperature stratification phenomena caused by natural convection in unmixed digesters [25].

9.2.2 Digester sizing

The main design parameter in continuously stirred tank reactors (CSTRs) is their volume. The working volume of a reactor (V) is related to HRT and the flow rate of the feeding (Q). It is expressed as

$$V = \frac{HRT}{Q} \tag{9.1}$$

Furthermore, the volume is dependent on the organic matter concentration (S_0) in the feeding and on the organic loading rate (OLR):

$$OLR = \frac{Q\,S_0}{V} = \frac{S_0}{HRT} \tag{9.2}$$

The required useful volume of a digester is determined based on the selected sizing factor. Two factors are generally used. For the first one, HRT is fixed, and for the second one, OLR is fixed. Actually, HRT (alongside temperature and feedstock composition) has an effect on the methane generation (Fig. 9.3A) and the VS removal efficiency (Fig. 9.3B) [26]. The required useful volume of a digester is determined from Eq. (9.2) with the fixed OLR. Thereafter, HRT is calculated by fixing S_0 or the other way around.

9.2.3 Anaerobic digestion models

Since it was not our aim in this section to delve into details about all the developed AD models, only brief summaries concerning the kinetics of basic processes and modeling in anaerobic digestion are discussed in the next

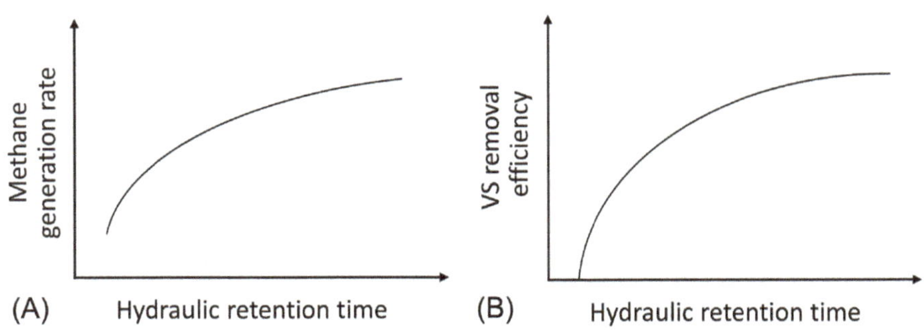

Fig. 9.3 Typical evolution of (A) methane production as a function of HRT at 35°C and (B) total volatile solids removal efficiency as a function of HRT at 35°C.

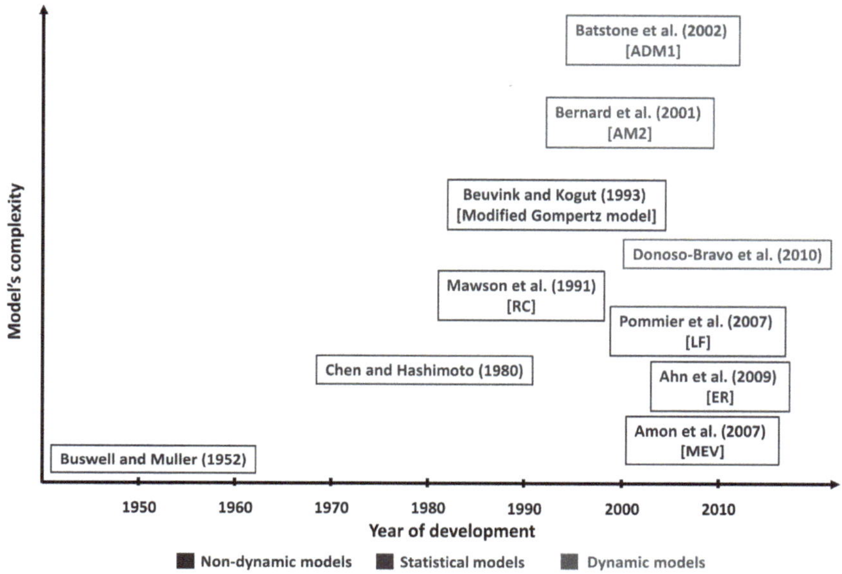

Fig. 9.4 Complexity of the anaerobic digestion models as a function of the year of their development.

paragraphs. The most widespread AD models over the last decades (based on their complexity) are illustrated in Fig. 9.4.

9.2.3.1 Nondynamic models

- Theoretical biogas yield

In nondynamic models, the substrates are linked to the products based on stoichiometry. It is usually possible to calculate the biogas composition and its yield through the AD process from the feedstock's chemical composition. Actually, Buswell and Mueller [27] developed a straightforward method for calculating biogas (CH_4 and CO_2) yield with a 5% uncertainty (Eq. 9.3):

$$C_aH_bO_c + \left(a - \frac{b}{4} - \frac{c}{2}\right)H_2O \rightarrow \left(\frac{a}{2} + \frac{b}{8} - \frac{c}{4}\right)CH_4 + \left(\frac{a}{2} - \frac{b}{8} + \frac{c}{4}\right)CO_2 \tag{9.3}$$

To consider the presence of H_2S (hydrogen sulfide) and NH_3 (ammonia) in the produced biogas, S (sulfur) and N (nitrogen) are taken into account in the formula [28] as shown in Eq. (9.4):

Table 9.1 Methane potential determined theoretically of organic substrates [29].

Organic component	Formula	CH_4 yield STP L/kg_{VS}	CH_4 %
Carbohydrate	$(C_6H_{10}O_5)_n$	415	50
Protein[a]	$C_5H_7NO_2$	496	50
Lipids	$C_{57}H_{104}O_6$	1014	70
Acetate	$C_2H_4O_2$	373	50

[a]N is converted into NH_3.
Notes: STP is the standard temperature and pressure (0°C and 1 atm).

$$C_aH_bO_cN_dS_e + \left(a - \frac{b}{4} - \frac{c}{2} + \frac{3d}{4} + \frac{e}{2}\right)H_2O \rightarrow \left(\frac{a}{2} + \frac{b}{8} - \frac{c}{4} - \frac{3d}{8} - \frac{e}{4}\right)CH_4 + \left(\frac{a}{2} - \frac{b}{8} + \frac{c}{4} + \frac{3d}{8} + \frac{e}{4}\right)CO_2 + d\,NH_3 + e\,H_2S \tag{9.4}$$

By applying Eq. (9.4), the CH_4 potential for each organic component is calculated and shown in Table 9.1.

- Methane energy value model (*MEV*)

The *MEV* model was built based on full regression models analysis that approximates CH_4 yields based on the composition of nutrients in energy crops being fermented [30]. This model examines the influence of the protein (*XP*), lipids (*XL*), fiber (*XF*), and N-free extracts (*XX*) contents on CH_4 production through Eq. (9.5):

$$MEV = x_1\,XP + x_2\,XL + x_3\,XF + x_4\,XX \tag{9.5}$$

where *MEV* is the methane energy value in NL CH_4/kg_{VS}, *XP* is the protein content in percentage of dry matter, *XL* is the fat content in percentage of dry matter, *XF* is the fiber content in percentage of dry matter, and *XX* is the N-free extract content in percentage of dry matter. x_1, x_2, x_3, and x_4 are the regression coefficients determined through batch assays. It should be noted that the presence of nonbiodegradable matter complicates this stoichiometric method, making it less trustworthy and thus not frequently used.

9.2.3.2 Kinetics of basic processes

All of the anaerobic digestion phases (except the hydrolysis phase) are considered as metabolic stages. Biological kinetics are based on two principles: bacterial growth and consumption of the substrate.

- Bacterial growth rate

The bacterial growth is represented by the following first-order equation (Eq. 9.6) [31]:

$$\frac{dX}{dt} = (\mu - k_d)X \tag{9.6}$$

In the case of a continuous AD process, Eq. (9.7) is used instead:

$$\frac{dX}{dt} = \mu X - DX \tag{9.7}$$

where X is the concentration of biomass (kg/m^3), μ is the specific bacterial growth rate (d^{-1}), D is the dilution rate (flow per reactor volume) (d^{-1}), and k_d is the death rate (d^{-1}).

- Substrate utilization rate

The consumption of the substrate is described according to Eq. (9.8):

$$\left(\frac{dX}{dt}\right)_{growth} = -Y\left(\frac{dS}{dt}\right)_{utilization} \tag{9.8}$$

where S is the concentration of the substrate (acetate) (g/L), and Y is the fraction of the substrate converted into biomass.

Furthermore, the substrate utilization rate is proportional to the biomass concentration as shown in Eq. (9.9):

$$\left(\frac{dS}{dt}\right)_{utilization} = -KX \tag{9.9}$$

where K is the specific substrate utilization rate. Therefore, based on the relation between K and μ, the substrate consumption may be determined for different values of μ (Eq. 9.10).

$$Y = \frac{\mu}{K} \tag{9.10}$$

It is worth mentioning that when the substrate is in excess, K remains constant.

- Models for the bacterial growth

In the methods described in the preceding sections, only feedstocks constituted of a single substrate were considered. However, the AD of complex substrates (containing two or more substrates) involves the action of diversified microbial communities which prompts numerous transformations happening at the same time. In the reactions where the bacterial growth intervenes, the kinetics are generally expressed as a function of the substrate utilization and the rate of biomass growth.

For kinetic models without inhibition, Monod proposed that the specific growth rate coefficient (μ) is dependent on the substrate concentration (S). They developed (Eq. 9.11) as given below:

$$\mu = \mu_{\max} \frac{S}{K_s + S} \tag{9.11}$$

where $\mu_{\max}$ is the maximum specific bacterial growth rate (d^{-1}). The Monod constant (K_s) is the concentration of the substrate at half the maximum specific growth rate ($\mu_{\max}/2$).

As for Contois [32], μ is a function of the biomass concentration (X) as shown in Eq. (9.12):

$$\mu = \mu_{\max} \frac{S}{K X + S} \tag{9.12}$$

On the other hand, Chen and Hashimoto [33] have modified the Contois kinetic equation (Eq. 9.12) since the biomass concentration (X) is difficult to determine for a biological treatment system. The modified specific bacterial growth rate coefficient is expressed as Eq. (9.13) where S_0 is the initial substrate concentration:

$$\mu = \mu_{\max} \frac{S}{K S_0 + (1 - K) S} \tag{9.13}$$

Additionally, Grau et al. [34] based their kinetic model on the linear removal concept. This concept represents a distinct case of the broader Monod equation (Eq. 9.11). The specific growth rate coefficient is expressed as

$$\mu = \mu_{\max} \frac{S}{S_0} \tag{9.14}$$

As for Moser [35], the specific growth rate coefficient is expressed as (Eq. 9.15):

$$\mu = \mu_{\max} \frac{S^n}{S^n + K_s} \tag{9.15}$$

where n is the degree of inhibition.

The Tessier kinetic model [36] takes into consideration both the microorganism decay coefficient and the maintenance energy requirement (Eq. 9.16). The specific microbial growth rate coefficient is expressed as

Table 9.2 Equations describing the specific growth rates for the anaerobic bacteria according to different models.

Model	Specific growth rate (μ)
Monod	$\mu_{max}\,\frac{S}{K_s+S}$
Contois [32]	$\mu_{max}\,\frac{S}{K\,X+S}$
Chen and Hashimoto [33]	$\mu_{max}\,\frac{S}{K\,S_0+(1-K)S}$
Grau et al. [34]	$\mu_{max}\,\frac{S}{S_0}$
Moser [35]	$\mu_{max}\,\frac{S^n}{S^n+K_s}$
Tessier [36]	$\mu_{max}\left(1-\exp\left[-\frac{S}{K_s}\right]\right)$

$$\mu = \mu_{max}\left(1-\exp\left[-\frac{S}{K_s}\right]\right) \tag{9.16}$$

The expressions of specific bacterial growth rates according to the different kinetic models are summarized in Table 9.2.

On the other hand, for kinetic models with inhibition, not only the microbial population and substrate utilization are considered for determining the bacterial growth rate. It is as well dependent on other aspects, mainly inhibition. Inhibition might be induced either by components already present in the initial feedstocks or by-products created during AD. Concerning the latter case of inhibition, two types exist, being an enzymatic (or substrate) inhibition and product inhibition. A number of methods consider the influence of substrate inhibition on the bacterial growth rate [37]. Furthermore, since the consequences of both product and substrate inhibitions are alike, certain models (e.g., Refs. [38,39]) can be used for both types of inhibition.

9.2.3.3 Process kinetics in anaerobic digestion

Anaerobic digestion is considered a highly nonlinear and intricate process due to the complexity in obtaining significant experimental data or model development. A multitude of mathematical dynamic models of the AD process have been developed in the last twenty years, and several approaches were used to validate these models and estimate their parameters. These models are classified into single-, two-, three-, or four-stage models (Fig. 9.5).

- Single-stage models

This type of model is based on the establishment of the rate-limiting phase in AD, which can either be the hydrolysis phase, the VFA to biogas conversion,

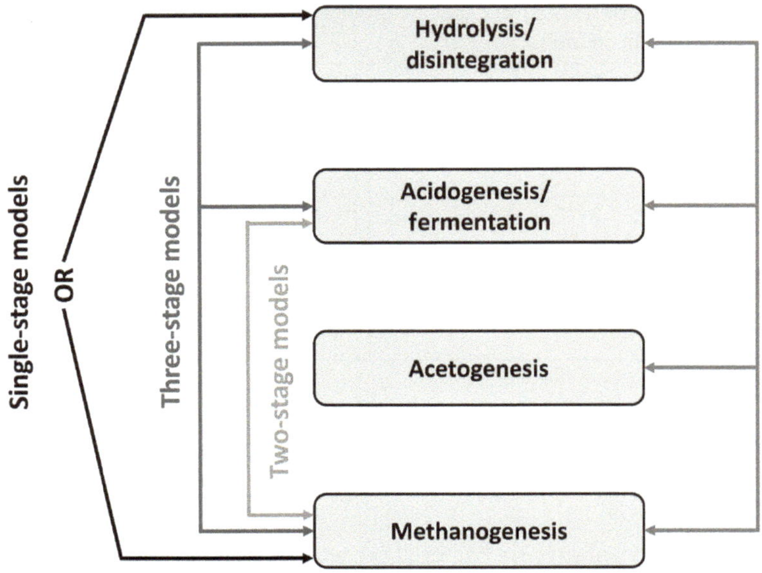

Fig. 9.5 Diagram of the two-, three-, and four-stage models of the anaerobic digestion process.

or the methanogenic phase [40]. Since hydrolysis is generally presumed as the limiting phase, and assuming 1st-order kinetics, the cumulated CH_4 production is expressed as:

$$G(t) = G_0\,(1 - \exp\,(-Kt)) \tag{9.17}$$

where $G(t)$ is the cumulated CH_4 yield ($\text{L}\,\text{kg}_{\text{VS}}{}^{-1}$), t is the duration of AD (d), G_0 is the CH_4 potential of the feedstock ($\text{L}\,\text{kg}_{\text{VS}}{}^{-1}$), and K is the first-order constant of disintegration rate (d^{-1}). It is worth mentioning that K is calculated as the reciprocal of time from the start of the BMP (biochemical methane production) experiment until when the cumulative methane yield is equal to $0.632G_0$ [41].

- Two-stage models

These models consider the acidogenic and methanogenic stages separately. This approach allows for predictions of the behavior of each phase in a distinct way. These two-stage models are closer to experimentation since they take into consideration the conversion of the complex substrate (organic matter) into VFAs, which will be in the most part converted into biogas. A widely used two-stage model is the Anaerobic Model no. 2 (AM2). This model has been proposed by Bernard et al. [42] for the AD of wastewater. This model offers a good reality-simplicity compromise since it only involves two reactions. These reactions represent the first step of AD, which is the conversion of the organic matter to CO_2 and VFAs by the acidogenic bacteria, and the second step, which manifests in the conversion of VFAs to CH_4 and CO_2 by the methanogenic bacteria. Because of its relative simplicity and accuracy, this model is commonly employed.

- Multistage models

The three-stage AD model takes into consideration the hydrolysis, acidogenesis, and methanogenesis phases [43,44]. As for the four-stage models, they consider the kinetics of hydrolysis and acetogenesis in addition to the other anaerobic digestion stages. Several studies reported in literature led to the development of these models [45]. Actually, the multistage models consider various substrates, take into consideration further processes and species, and involve kinetics with inhibition. A popular multiphase model, which was developed by the IWA Task Group for Mathematical Modeling of Anaerobic Digestion Processes as a response to the need for a generic model, is the Anaerobic Digestion Model no. 1 (ADM1).

ADM1 is a well-defined generic model having numerous steps that represent the physical, chemical, and biological processes of the anaerobic digestion process of complex organic substrates [46]. Since its development in 2002, the ADM1 model was considered as the most complete anaerobic digestion model. It has been used in many applications and has been validated for several industrial processes [47].

In fact, the substrates are characterized by their organic and inorganic compositions in the ADM1. This model takes into consideration carbs, proteins, fats, amino acids, and LCFAs and VFAs (Long-Chain and Volatile Fatty Acids), in addition to inert soluble and particulates for the organic fraction. As for the inorganic fraction, anions and cations, such as HCO_3^- (bicarbonate), PO_4^{3-} (phosphate), NO_3^- (nitrate), SO_4^{2-} (sulfate), K^+ (potassium), Ca^{2+} (calcium), and Mg^{2+} (magnesium), are considered. Concerning the dead biomass, it is treated as complex particulate material [46]. The biological and chemical conversion processes according to the ADM1 are given in detail by Ref. [48].

- Statistical models

The main advantage of ADM1 is its ability of predicting the AD of various substrates. However, its main drawback manifests in its complexity since it requires numerous parameters as inputs, which makes this model time-consuming and costly from an experimental standpoint. In order to eliminate such inconveniences, regression statistical models, such as Logistic function, modified Gompertz model, reaction curve type model, and transference function, were established in order to predict the AD performances. For instance, the modified Gompertz model has a large scope of applications for predicting CH_4 yield [49,50] and is expressed as Eq. (9.18):

$$G(t) = G_0 \exp\left\{-\exp\left[\frac{R_{\max}\, e}{G_0}(\lambda - t) + 1\right]\right\} \tag{9.18}$$

where $R_{\max}$ is the CH_4 maximal yield ($\mathrm{L\,kg_{VS}}^{-1}\,\mathrm{d}^{-1}$), e is the exponential constant (2.718), and λ is the lag phase (d).

Furthermore, the logistic function equation (*LF*) is used to predict the CH_4 yield from landfill leachate biodegradation (Eq. 9.19) [51]. Additionally, the reaction curve-type model (*RC*) is usually employed for control processes of AD (Eq. 9.20) [40]. Moreover, the exponential rise to maximum (*ER*) is established from the 1st-order equation (FO) to predict the maximal biogas production while considering that the hydrolysis rate coefficient (k) is a limiting factor for AD (Eq. 9.21) [52]:

$$y_{LF} = \frac{A}{1 + \exp\left(4A_m \dfrac{\lambda - t}{A} + 2\right)} \tag{9.19}$$

$$y_{RC} = A\left[1 - \exp\left(-\frac{A_m(t - \lambda)}{A}\right)\right] \tag{9.20}$$

$$y_{ER} = A[1 - \exp(-kt)] \tag{9.21}$$

where y is the cumulated biogas yield ($\mathrm{mL\,g^{-1}}$), A is the production potential ($\mathrm{mL\,g^{-1}}$), A_m is the biogas maximal production rate ($\mathrm{mL\,g^{-1}\,d^{-1}}$), and k is the hydrolysis rate constant ($\mathrm{d^{-1}}$).

- Chen and Hashimoto model

Based on the Contois model as well as the fundamental biochemical principles, the Chen and Hashimoto model was used for the batch and continuous AD processes with acceptable results [33,49,53]. The parameters of this model, in particular the critical retention time ($HRT_{critical}$), have a considerable importance for assessing the AD reactions. This model's equation is expressed as (Eq. 9.22):

$$G(t) = G_0\left(1 - \frac{K_{CH}}{HRT\,\mu_{\max} + K_{CH} - 1}\right) \tag{9.22}$$

where K_{CH} is the kinetic constant of Chen and Hashimoto (–). Eq. (9.22) can be converted to Eq. (9.23):

$$HRT = \frac{1}{\mu_{\max}} + \frac{K_{CH}}{\mu_{\max}}\frac{G(t)}{G_0 - G(t)} \tag{9.23}$$

$HRT_{critical}$ is obtained by considering $\frac{G(t)}{G_0 - G(t)}$ as the variable, $\frac{K_{CH}}{\mu_{max}}$ as the slope, and $\frac{1}{\mu_{max}}$ as the intercept. It is reported that anaerobic digesters function better at shorter $HRT_{critical}$ values [41]. $HRT_{critical}$ is expressed as (Eq. 9.24):

$$HRT_{critical} = \frac{1}{\mu_{max}} \tag{9.24}$$

9.3 Heat transfer modeling in anaerobic digesters

9.3.1 Heat balance

When developing a thermal model for an anaerobic digester for a specific application, it is necessary to determine the heat exchange (losses and gains) between the many components of the system (Fig. 9.6). Since anaerobic digesters have distinct designs and operate under various climatic conditions, a universal model that can be applied to accurately predict the heat transfer in any given digester does not exist to date. Therefore, an adequate model should be developed for each digester while considering numerous factors specific to its modes of operation, such as its placement (aboveground, semiburied, or buried), feeding mode (continuous, semicontinuous, or batch), temperature range (mesophilic or thermophilic), climatic conditions, etc.

The global energy balance of the studied anaerobic digester needs to be established. This balance is the difference between the energy input (i.e., heating/preheating) and output (i.e., heat loss) for each component of the digester (e.g., walls, cover, biogas, slurry feed). It is generally expressed as (Eq. 9.25):

$$Q_{heating} + Q_{mass\ flow} + Q_{sol} + Q_{chem} + Q_{loss} = 0 \tag{9.25}$$

where $Q_{heating}$ is the heat required for maintaining the digester at the operating temperature (W), $Q_{mass\ flow}$ is the heat exchange rate due to feedstock

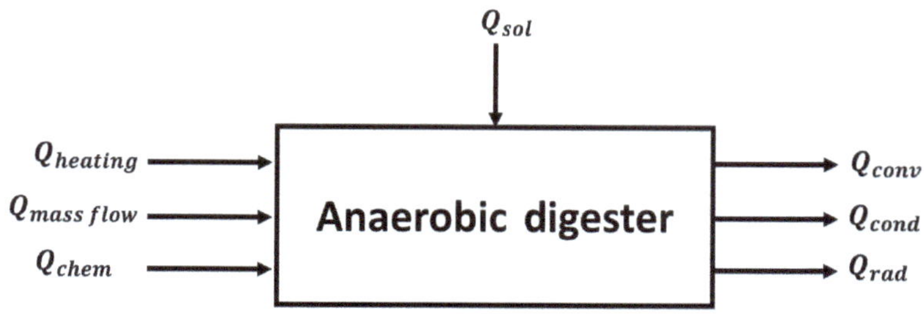

Fig. 9.6 Diagram of the heat balance of an anaerobic digester.

feeding and evacuation (W), Q_{sol} is the absorbed solar radiation heat rate (W), Q_{chem} is the heat exchange due to the chemical reactions taking place throughout the AD process, and Q_{loss} is the heat loss due to convection, conduction, and radiation (W). It is worth mentioning that for anaerobic digesters operating under batch conditions, $Q_{mass\ flow}=0$, and for indoor digesters (not exposed to sunlight), $Q_{sol}=0$.

In fact, heat is exchanged between the different components of the digesters through different transfer modes, i.e., convection, conduction, and radiation. The heat transfer rates are determined from the well-known heat transfer equations. In anaerobic digesters, convection generally occurs between the slurry and the internal walls and between the ambient air and the external walls. As for radiation, it constantly occurs between the different components of anaerobic digesters and the ambient air. Finally, the conductive heat transfer occurs through the digester's walls. It is worth mentioning that in order to decrease these conductive heat losses, the external walls of anaerobic digesters are usually encased in thermally insulating materials (e.g., fiberglass, foamed plastic). This thermal insulation effectively decreases the heat loss through conduction by reducing the conductive heat transfer coefficient through the digesters' walls.

9.3.2 Heat loss

The convective, conductive, and radiative thermal resistances (respectively R_{conv}, R_{cond}, and R_{rad}) are illustrated in Fig. 9.7 and are given by Eqs. (9.26), (9.27), and (9.28), respectively:

$$R_{conv}=\frac{1}{h_c A_{conv}} \tag{9.26}$$

$$R_{cond}=\frac{e}{k A_{cond}} \tag{9.27}$$

$$R_{rad}=\frac{1}{\sigma \varepsilon A_{rad}\left(T_s^2+T_{surr}^2\right)\left(T_s+T_{surr}\right)} \tag{9.28}$$

where h_c is the convective heat transfer coefficient (W/m^2 K), k is the thermal conductivity (W/mK), e is the thickness of the material through which conduction occurs (m), A_{conv}, A_{cond}, and A_{rad} are the areas through which the convection, conduction, and radiation occur (m^2), T_s is the temperature of the emitting surface (K), T_{surr} is the surrounding temperature (K), σ is the Stefan-Boltzmann constant (W/m^2 K^4), and ε is the surface's emissivity (–).

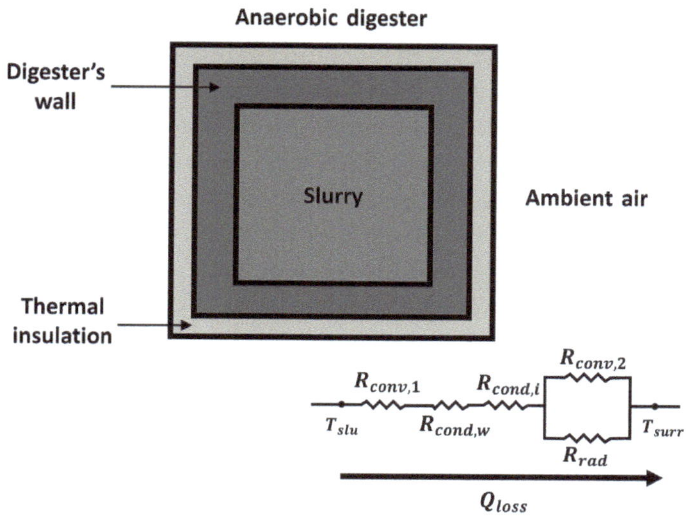

Fig. 9.7 Thermal resistance network diagram for an insulated anaerobic digester exposed to ambient air.

It is worth mentioning that the conductive resistance for cylindrical geometries is expressed as (Eq. 9.29):

$$R_{cond} = \frac{\ln(r_2/r_1)}{2\pi L k} \tag{9.29}$$

where r_1 and r_2 are, respectively, the internal and external diameters of the cylinder, and L is its height.

Therefore, the global heat loss for an anaerobic digester can be determined using Eqs. (9.30), (9.31), where T_{slu} is the slurry's temperature, $R_{cond,w}$ and $R_{cond,i}$ are the thermal resistances of conduction through the walls of the digester and the insulation, and R_{glob} is the global thermal resistance.

$$Q_{loss} = \frac{T_{slu} - T_{surr}}{R_{glob}} \tag{9.30}$$

$$R_{glob} = R_{conv} + R_{cond,w} + R_{cond,i} + \frac{R_{conv}\, R_{rad}}{R_{conv} + R_{rad}} \tag{9.31}$$

To determine h_c, Eq. (9.32) is used:

$$h_c = Nu\frac{k}{L} \tag{9.32}$$

where Nu is the Nusselt number (–), and L is the characteristic length (m).

Many empirical correlations have been developed to calculate the heat transfer coefficients. These correlations are specific to certain flow conditions (e.g., laminar, turbulent) and geometries (e.g., planes, cylinders, spheres). For instance, if we consider the digester's cover, it is simultaneously in contact with the trapped biogas (internally) and ambient air (externally). The heat transfer between the cover and the trapped biogas occurs through natural convection, whereas between the cover and air, the exchange happens through forced convection (presence of wind). To calculate the convective heat transfer through the cover, it is usually considered as a horizontal surface [54], which allows using the horizontal plate correlations for determining the Nusselt number [55]. In this case, for the heat transfer through natural convection, h_c is determined based on the Rayleigh (Ra) and Nusselt numbers (Eqs. 9.33, 9.34, respectively):

$$Ra = \frac{g\beta(T_{wall} - T_a)L^3}{\alpha\nu} \tag{9.33}$$

$$Nu = \begin{cases} 0.54\,Ra^{1/4} & \text{for } 10^4 \le Ra \le 10^7 \text{ and } Pr \ge 0.7 \\ 0.15\,Ra^{1/3} & \text{for } 10^7 \le Ra \le 10^{11} \text{ and all } Pr \end{cases} \tag{9.34}$$

where Ra is the Rayleigh number (–), g is the gravitational acceleration (m/s^2), β is the volumetric thermal expansion coefficient (K^{-1}), ($T_{wall} - T_a$) is the temperature difference between the wall and the ambient fluid, L is the characteristic length (m), α is the thermal diffusivity (m^2/s), ν is the kinematic viscosity (m^2/s), and Pr is the Prandtl number (–). The Prandtl number is calculated using Eq. (9.35), where c_p is the specific heat capacity (J/kg K) and μ is the dynamic viscosity (Pa s):

$$Pr = c_p\frac{\mu}{k} \tag{9.35}$$

As for the heat transfer through forced convection, h_c is determined based on the Reynolds (Re) and Nusselt numbers (Eqs. 9.36, 9.37, respectively):

$$Re = U\frac{L}{\nu} = \rho\,U\frac{L}{\mu k} \tag{9.36}$$

$$Nu = \begin{cases} 0.628\,Re^{1/2}\,Pr^{1/3} & \text{for } Re \le 5\times10^5 \text{ and } 0.5 \le Pr \le 10 \\ 0.035\,Re^{4/5}\,Pr^{1/3} & \text{for } Re > 5\times10^5 \text{ and } Pr \ge 0.5 \end{cases} \tag{9.37}$$

where Re is the Reynolds number (–), U is the fluid's velocity (m/s), and ρ is the fluid's density (kg/m^3).

Table 9.3 Values of "C" and "n" for different Reynolds number ranges.

Re	*C*	*n*
0.4–4	0.989	0.330
4–40	0.911	0.385
40–4000	0.683	0.466
4000–40,000	0.193	0.618
40,000–250,000	0.026	0.805

It is worth mentioning that for a mechanically stirred (mixed) anaerobic digester, Re is calculated using Eq. (9.38) [56]:

$$Re = N\, D_{imp}{}^{2}\frac{\rho}{\mu} \tag{9.38}$$

where N and D_{imp} are the impeller's rotational speed and diameter, respectively.

Furthermore, the heat transfer rate due to forced convection varies with different geometries. For instance, in order to determine the convective heat transfer coefficient for a cylindrical digestion tank, Re number is first calculated using the appropriate equation (Eq. 9.36 or Eq. 9.38). According to this number's value, the coefficients C and n are determined from Table 9.3 and inserted in Eq. (9.39):

$$Nu = C\, Re^{n}\, Pr^{1/3} \mu k \tag{9.39}$$

As for digesters with vertical walls, the following correlations are used (Eq. 9.40):

$$Nu = \begin{cases} 0.68 + \dfrac{0.67\, Ra^{1/4}}{\left[1 + (0.492/Pr)^{9/16}\right]^{4/9}} & \text{for } Ra \leq 10^{9} \\ \left\{ 0.825 + \dfrac{0.387\, Ra^{1/6}}{\left[1 + (0.492/Pr)^{9/16}\right]^{8/27}} \right\}^{2} & \text{for } Ra > 10^{9} \end{cases} \tag{9.40}$$

9.3.3 Solar radiation heat transfer

For outdoor anaerobic digesters (i.e., exposed to sunlight), a term corresponding to the absorbed heat rate from the solar radiation (Q_{sol}) (Eq. 9.41) should be included in the energy balance equation:

$$Q_{sol} = \acute{\alpha}\, A_{rad,s}\, F\, I, \tag{9.41}$$

where á is the absorbance (–), I is the solar irradiation (W/m^2), and F is a shading reduction factor (–). If the digester is fully exposed to sunlight (i.e., no shading), $F=1$.

9.3.4 Microbial heat transfer

Some of the chemical reactions that occur during the AD process generate heat (exothermic) and some absorb heat (endothermic). In literature, the heat exchange due to these reactions is generally neglected because of its insignificance compared to the other modes of heat transfer [54,57–59]. However, authors in Ref. [60] calculated the microbial heat exchange by determining the enthalpies of the reactions based on stoichiometric degradation. For that purpose, glucose, alanine, and palmitic acid were considered as reference substances for carbohydrates, proteins, and fats, respectively. These equations for a certain substrate are given below (Eqs. 9.42–9.44).

$$C_6H_{12}O_6 \rightarrow 3\,CO_2 + 3\,CH_4 \tag{9.42}$$

$$2\,C_3H_7NO_2 + 2\,H_2O \rightarrow 3\,CO_2 + 3\,CH_4 + 2\,NH_3 \tag{9.43}$$

$$2\,C_{16}H_{32}O_2 + 14\,H_2O \rightarrow 9\,CO_2 + 23\,CH_4 \tag{9.44}$$

9.3.5 Feedstock preheating

Continuous anaerobic digestion is characterized by a constant flow of fresh material (in the form of a slurry) entering the digester, and an equal mass of digested material leaving it. This type of digesters loses (or gains) heat through feedstock addition (feeding) and/or digestate removal (purging) [61]. The heat transfer rate due to mass flow is calculated using Eq. (9.45):

$$Q_{mass\,flow} = \dot{m}_{feed} c_p \left(T_{feed} - T_0\right) \tag{9.45}$$

where $\dot{m}_{feed}$ is the feedstock's mass flow rate (kg/s), and T_0 and T_{feed} are the digester's operating temperature and the feedstock's temperature, respectively (K).

It is evident, from Eq. (9.45), that when the feedstock's temperature is lower than the operating temperature of the digester, the heat exchange is negative (i.e., the digester loses heat). This leads to a decrease in the operating temperature of the digester at every feeding, causing detrimental effects on the digester's stability and biogas yield. In order to overcome this issue, the feedstock can be preheated using a heat exchanger before being introduced into the digester. The total heat transfer rate between the feedstock and the hot fluid (in the exchanger) is determined from Eqs. (9.46), (9.47):

$$Q_{exch} = U\,A\,\Delta T_{lm} \tag{9.46}$$

$$\Delta T_{lm} = \frac{\left(T_{fluid,in} - T_{feed,out}\right) - \left(T_{fluid,out} - T_{feed,in}\right)}{\ln\left[\left(T_{fluid,in} - T_{feed,out}\right)/\left(T_{fluid,out} - T_{feed,in}\right)\right]} \tag{9.47}$$

where Q_{exch} is the heat transfer rate in the heat exchanger (W), U is the overall heat transfer coefficient (W/m^2 K), ΔT_{lm} is the log mean temperature difference (K), $T_{fluid,\,in}$, $T_{fluid,\,out}$ are, respectively, the fluid's temperatures at the inlet and outlet of the exchanger, and $T_{feed,in}$, $T_{feed,out}$ are, respectively, the feedstock's temperature at the inlet and outlet of the exchanger.

After determining the various terms of the energy balance equation, the required energy rate for heating an anaerobic digester is expressed as (Eq. 9.48):

$$Q_{heating} = \frac{T_{slu} - T_{surr}}{R_{glob}} - \acute{\alpha}\,A_{rad,s}\,F\,I - \dot{m}_{feed}\,c_p\left(T_{feed} - T_0\right) - Q_{chem} \tag{9.48}$$

9.4 Solar thermal energy systems

Solar collectors work by converting the energy of the solar radiation into thermal energy. This energy is absorbed by a fluid (usually oil or water) circulating through the solar collector. The generated heat can be used directly or stored for later use. This renewable energy technology is extensively used in many applications, including heating anaerobic digesters. As detailed before, digesters require permanent heating to increase their stability and improve their energy production. When coupled with solar thermal collectors, their heating needs are met without resorting to electricity nor to conventional fuels.

Solar/anaerobic digestion hybrid systems are truly renewable and can be completely independent from the electric grid, which is particularly useful in remote and isolated locations. In order to achieve total energy autonomy, energy storage solutions are integrated. During cloudy days and after sunset, the solar resource can no longer be used for heat production, which is why the heated fluid (during the sunny conditions) is stored in thermal storage units for later use. A typical solar thermal energy system containing a flat-plate solar collector coupled with a heat storage tank is illustrated in Fig. 9.8.

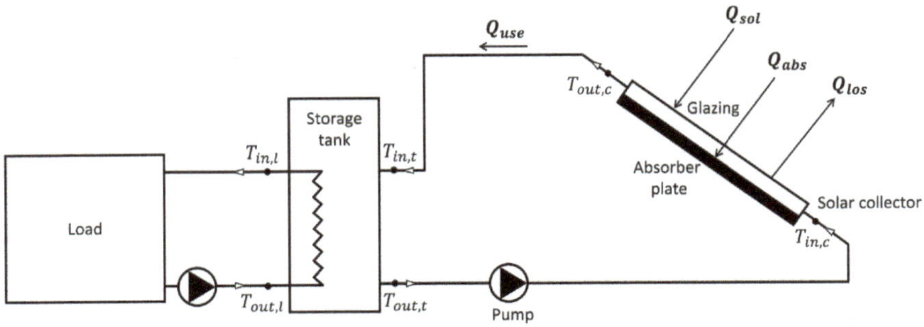

Fig. 9.8 Illustration of a solar thermal energy system composed of a glazed flat-plate solar collector and a storage tank connected to a load.

9.4.1 Flat-plate thermal collectors

Flat-plate collectors are equipped with an absorber plate characterized by a high absorptivity (achieved through black paint or special coatings) which is exposed to solar radiation. This plate absorbs a large portion of the incoming energy and transfers it to the circulating fluid going through internal passageways. These collectors are usually thermally insulated to reduce heat loss. This type of solar collectors is not expensive to procure and install, does not require sun tracking, and allows capturing both the diffuse and beam radiation.

The amount of solar radiation reaching the solar collector is expressed as (Eq. 9.49):

$$Q_{sol} = I\,A \tag{9.49}$$

where Q_{sol} is the solar radiation reaching the surface of the solar collector (W), I is the solar radiation intensity ($\mathrm{W\,m^{-2}}$), and A is the solar collector's aperture surface area ($\mathrm{m^2}$). However, only a portion of this radiation reaches the absorber plate, since a fraction gets reflected and another gets absorber by the glazing. The radiation that reaches the absorber plate (after passing through the glazing) is expressed as (Eq. 9.50):

$$Q_{abs} = I\,A\,\tau\,\acute{\alpha} \tag{9.50}$$

where Q_{abs} is the solar radiation reaching the absorber plate (W), á is the absorbance (–), and τ is the transmissivity (–). After absorbing heat from the solar radiation, the flat-plate collector's temperature becomes higher than the ambient temperature. This difference in temperature results in heat loss through convection and radiation. This heat loss is expressed as (Eq. 9.51):

$$Q_{los} = A\,U_c\left(T_{col} - T_{amb}\right) \tag{9.51}$$

where Q_{los} is the heat loss through the top of the solar collector (W), T_{col} is the collector's temperature (K), T_{amb} is the ambient temperature (K), and U_c is the overall heat transfer coefficient of the collector (W/m^2 K). It is worth mentioning that the heat loss through the bottom and sides of the collector is always present; however, due to their insignificant contribution compared to the collector's top, they are usually neglected [62]. As a result, the available energy rate to the collector (Q_{ava}) is expressed as (Eq. 9.52):

$$Q_{ava} = Q_{abs} - Q_{los} = A\left[I\,\tau\,\acute{\alpha} - U_c(T_{col} - T_{amb})\right] \tag{9.52}$$

Additionally, the useful rate of heat absorbed by the circulating fluid is expressed as (Eq. 9.53):

$$Q_{use} = \dot{m}_{flu1}\,c_{p,flu1}\,(T_{out,c} - T_{in,c}) \tag{9.53}$$

where Q_{use} is the heat absorption rate by the fluid (W), $\dot{m}_{flu1}$ is the circulating fluid's mass flow rate (kg/s), $c_{p,\ flu1}$ is the circulating fluid's specific heat capacity (J/kgK), and $T_{in,\ c}$ and $T_{out,\ c}$ are, respectively, the temperatures of the fluid at the inlet and outlet of the collector (K). For simplification, the average temperature of the collector (T_{col}) is considered equal to the fluid's temperature at the inlet ($T_{in,\ c}$). A ratio named the collector heat removal factor (F_r) is defined. This quantity is expressed as the ratio between the useful heat rate (Q_{use}) and the available energy rate (Q_{ava}):

$$F_r = \frac{Q_{use}}{Q_{ava}} = \frac{\dot{m}_{flu1}\,c_{p,flu1}\,(T_{out,c} - T_{in,c})}{A\left[I\,\tau\,\acute{\alpha} - U_c(T_{in,c} - T_{amb})\right]} \tag{9.54}$$

By substituting the available energy rate to the collector (Q_{ava}) in Eq. (9.54) with its expression in Eq. (9.52), the useful heat rate can be expressed equally as

$$Q_{use} = A\,F_r\left[I\,\tau\,\acute{\alpha} - U_c(T_{in,c} - T_{amb})\right] \tag{9.55}$$

This equation (Eq. 9.55) is known as the Hottel-Whillier-Bliss equation. From this equation, the efficiency (Q_{use}/Q_{sol}) of a given flat-plate solar collector can be determined (Eqs. 9.56, 9.57):

$$\eta = F_r\left[\tau\,\acute{\alpha} - \frac{U_c}{I}(T_{in,c} - T_{amb})\right] \tag{9.56}$$

or

$$\eta = \frac{\dot{m}_{flu1}\,c_{p,flu1}\,(T_{out,c} - T_{in,c})}{I\,A} \tag{9.57}$$

From the efficiency equation (Eq. 9.56), it is evident that improving a solar collector's efficiency depends on increasing the absorbance-transmissivity term ($\tau\hat{a}$) and decreasing the overall heat transfer coefficient (U_c). In order to achieve that, high-clarity glass can be used as glazing, and better thermal insulation may be installed to limit heat loss.

9.4.2 Heat storage tanks

For storing the excess thermal energy produced during the day for later use, storage tanks are used. These thermally insulated tanks are filled with a liquid (usually water) that stores usable heat. When favorable heat production conditions are available (i.e., sunny weather), the liquid is pumped from the tanks and circulated through the internal passageways of the solar collector, and thus getting heated. The heated liquid is then pumped back into the tanks and thus raising the overall temperature of the storage unit. The pumping process is maintained until the temperature difference between the circulating fluid and the storage medium is eliminated. At the same time, a different liquid is pumped through a heat exchanger inside the storage tanks to provide heat for the load (Fig. 9.8).

The energy balance for a storage tanks is expressed as (Eq. 9.58):

$$Q_{use} - Q_{load} - Q_{los,t} = 0 \tag{9.58}$$

where Q_{load} is the heat gained by the load (W), and $Q_{los,\ t}$ is the heat loss of the storage tanks (W).

The heat extraction rate of the load can be expressed as (Eq. 9.59):

$$Q_{load} = \dot{m}_{flu2} c_{p,flu2} \left(T_{out,l} - T_{in,l} \right) \tag{9.59}$$

where $\dot{m}_{flu2}$ is the mass flow rate of the fluid circulating in the load (kg/s), $c_{p,flu2}$ is the load fluid's specific heat capacity (J/kgK), and $T_{in,l}$ and $T_{out,l}$ are, respectively, the temperatures of the fluid at the inlet and outlet of the load (K).

Furthermore, the convective and radiative heat losses of the storage tanks are expressed as

$$Q_{los,t} = A_t\, U_t \Delta T_{lm} \tag{9.60}$$

$$\Delta T_{lm} = \frac{\left(T_{in,t} - T_{out,l} \right) - \left(T_{out,t} - T_{in,l} \right)}{\ln\left[\left(T_{in,t} - T_{out,l} \right) / \left(T_{out,t} - T_{in,l} \right) \right]} \tag{9.61}$$

where A_t is the storage tanks' surface area (m^2), U_t is the overall heat transfer coefficient for the tanks (W/m^2 K), and $T_{in,\ t}$ and $T_{out,\ t}$ are, respectively, the temperatures of the fluid at the inlet and outlet of the storage tanks (K).

After determining the different terms of the energy balance equation, it is expressed as (Eq. 9.62):

$$\dot{m}_{flu1}c_{p,flu1}\left(T_{out,c} - T_{in,c}\right) - \dot{m}_{flu2}c_{p,flu2}\left(T_{out,l} - T_{in,l}\right) \\ -A_t\,U_t\frac{\left(T_{in,t} - T_{out,l}\right) - \left(T_{out,t} - T_{in,l}\right)}{\ln\left[\left(T_{in,t} - T_{out,l}\right)/\left(T_{out,t} - T_{in,l}\right)\right]} = 0 \tag{9.62}$$

$Q_{heating}$, which is determined from the heat balance equation of the anaerobic digester (Eq. 9.48), represents the energy rate required to heat the digester while considering the various heat loss modes. During daytime (on a clear day), this energy rate needs to be provided by the solar collector. The heating energy rate in this case is expressed as (Eq. 9.63):

$$Q_{heating} = \dot{m}_{flu1}c_{p,flu1}\left(T_{out,c} - T_{in,c}\right) \tag{9.63}$$

However, during nighttime (or unfavorable weather conditions), the heating energy rate is provided by the storage tanks, expressed as (Eq. 9.64):

$$Q_{heating} = \dot{m}_{flu2}c_{p,flu2}\left(T_{out,l} - T_{in,l}\right) + A_t\,U_t\Delta T_{lm} \tag{9.64}$$

9.5 Thermodynamic cycles

Thermal power plants are power generation systems in which heat, obtained through the combustion of carbon-based fuels such as natural gas, is converted to electric power. Around two-thirds of the electricity produced worldwide is generated by thermal plants. These plants operate on thermodynamic cycles (also called power cycles). In this section, we will present and discuss different thermodynamic cycles for power generation using the methane produced through anaerobic digestion as a carbon-based fuel. The heat released through methane combustion is expressed as (Eq. 9.65):

$$Q_{in} = \eta_{combustion}\left(\dot{m}_{fuel} \times LHV\right) \tag{9.65}$$

where $\eta_{combustion}$ represents the combustion efficiency, $\dot{m}_{fuel}$ is the methane mass flow rate (kg s^{-1}), and LHV is the lower heating value, which represents the amount of heat released when methane is completely burned per unit mass (kJ/kg).

The two major thermal power cycles are considered: vapor and gas-turbine power plants. The modeling of each cycle is discussed in the following subsections.

9.5.1 Vapor power plant (Rankine cycle)

Vapor power plants are based on Rankine cycle where the working fluid is vaporized and condensed alternatively to produce power. Most plants use steam as the working fluid because of its various desired attributes (abundance, low cost, and high enthalpy of vaporization). The basic Rankine cycle operating with steam as the working fluid consists of these following components (Fig. 9.9):

- Steam generator (or boiler)
- Condenser
- Steam turbine
- Liquid water pump

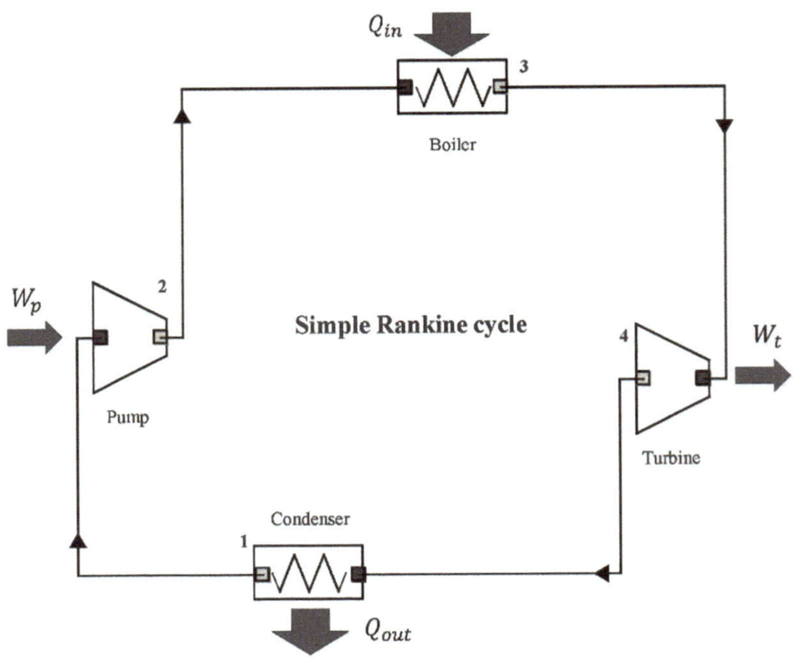

Fig. 9.9 Simple Rankine cycle.

In its ideal form, the simple Rankine cycle involves these 4 steady-flow processes:

- 1-2 Isentropic (reversible and adiabatic) compression in a pump
- 2-3 steam generation in a boiler
- 3-4 Isentropic (reversible and adiabatic) expansion in a steam turbine
- 4-1 condensation in a condenser

The main work and heat transfers of Rankine cycle are portrayed in Fig. 9.9. Based on the 1st law of thermodynamics and the conservation of mass, the energy balance is applied to each component of the ideal Rankine cycle. The following expressions are obtained (in watts):

Input work to the ideal pump (1−2)

$$W_{v,ideal\,p} = \dot{m}_v(h_2 - h_1) \tag{9.66}$$

Added heat to the boiler (2−3)

$$Q_{v,in} = \dot{m}_v(h_3 - h_2) \tag{9.67}$$

Output work of the ideal turbine (3−4)

$$W_{v,ideal\,t} = \dot{m}_v(h_3 - h_4) \tag{9.68}$$

Rejected heat in the condenser (4−1)

$$Q_{v,out} = \dot{m}_v(h_4 - h_1) \tag{9.69}$$

Subscript v refers to vapor cycle, $\dot{m}_v$ is the steam mass flow rate ($kg\,s^{-1}$), and h is the fluid enthalpy ($kJ\,kg^{-1}$) which can be obtained from the fluid properties tables. $Q_{v,\,in}$ is the transferred energy by heat from the energy source into the working fluid which represents the fuel consumption (in this case methane).

Naturally, the work product of a real Rankine cycle is reduced due to irreversibilities and loss. A real pump needs a higher input work, and a real turbine generates a reduced output work due to heat loss and fluid friction. This loss is accounted for in the energy balance using the isentropic efficiencies of the pump (η_p) and the turbine (η_t), therefore:

Real work input to the pump (1−2)

$$W_{v,p} = \frac{W_{v,ideal\,p}}{\eta_p} \tag{9.70}$$

Real work output of the turbine (3−4)

$$W_{v,t} = \eta_t W_{v,ideal\,t} \tag{9.71}$$

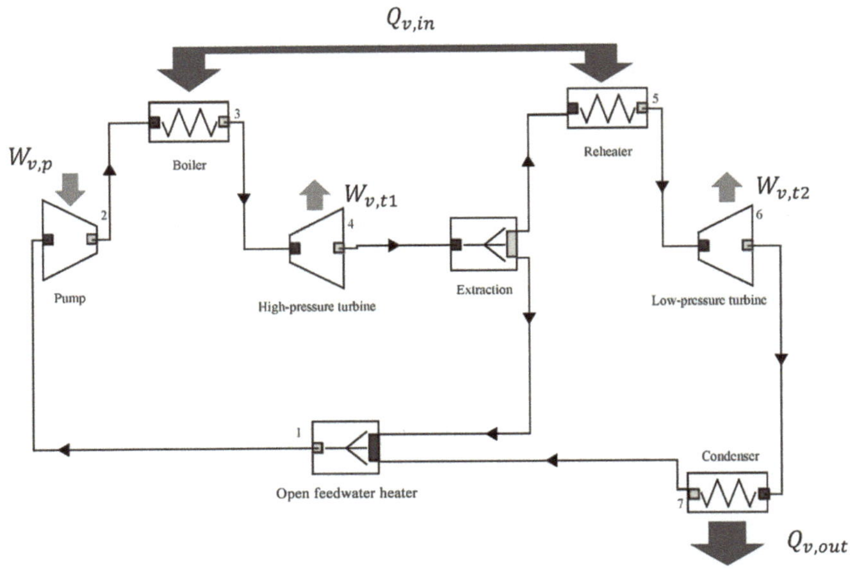

Fig. 9.10 Rankine cycle with a single reheat and one-stage regeneration.

For instance, well-designed turbines for large-scale plants are characterized by isentropic efficiencies η_t over 90%, while for small plants, $\eta_{is,\ t}$ might fall beneath 70%.

The simple Rankine cycle shown in Fig. 9.9 can be modified to increase its thermal efficiency. Real plants built based on Rankine cycle may contain multiple turbines, reheaters, and regeneration stages as illustrated in Fig. 9.10. Part of the steam expanding in the high-pressure turbine is sent back to the reheater, and then the reheated steam is led to the lower-pressure turbine. The ideal reheating pressure is ≈1/4 of the greatest cycle pressure, as well as the total output turbine work for a reheat cycle is obtained by the sum:

$$W_{v,t} = W_{v,t1} + W_{v,t2} \text{ and } Q_{v,in} = \dot{m}_v[(h_3 - h_2) + (h_5 - h_4)]$$

.

In the regeneration process, part of steam is extracted to heat-up feed-water to the boiler. Most of the large thermal plants have a single reheat with 4–9 regeneration stages. However, for industrial power plants, a nonreheat cycle is often adopted for economic reasons [63].

The performance parameters for any Rankine cycle are expressed as

Cycle's net output power

$$W_{v,net} = \sum W_{v,t} - \sum W_{v,p} \tag{9.72}$$

Thermal efficiency

$$\eta_{th,v} = \frac{W_{v,net}}{Q_{v,in}} \tag{9.73}$$

In these performance parameters, one might also include the station service power consumption and electrical loss, additionally known as auxiliary consumption W_{aux}, and additional/supplementary firing fuel consumption Q_{sf}. The thermal efficient becomes:

Net thermal efficiency

$$\eta_{th,v} = \frac{W_{v,net} - W_{aux}}{Q_{v,in} + Q_{sf}} \tag{9.74}$$

Currently, the most advanced Rankine cycle is the supercritical Rankine cycle, which is characterized by a principal steam pressure of ≈290 bar, temperature of 600°C, and reheat temperature of 620°C. Considering an isentropic pump ($\eta_p = 1$) and an isentropic efficiency of $\eta_t = 0.85$ at each turbine stage, the thermal efficiency of the supercritical Rankine cycle is ≈40%.

9.5.2 Simple gas-turbine power plant (Brayton cycle)

The simple gas-turbine cycle is based on the Brayton cycle. In this cycle, the working fluid stays in its gaseous form during the whole cycle as opposed to a vapor cycle. In addition, gas turbines tend to be less heavy and occupy less space than vapor power plants. The operating temperatures of gas-turbine cycles are usually much higher than those of vapor cycles, as the maximal temperature of the fluid at the inlet of the turbine is 1425°C for gas-turbine power plants compared to 620°C for current steam power plants. Actually, a modern gas turbine is characterized by three important modules, namely a turbine, a compressor, and a combustion chamber as portrayed in Fig. 9.11. The power production process starts by compressing air using a compressor, then feeding it to a combustion chamber where fuel injection happens constantly, thus releasing heat and raising air's pressure and temperature. Thereafter, the issue stream of high pressure and temperature air is injected into the turbine for power production.

Based on the first law of thermodynamics and the conservation of mass, the energy balance is applied to each component of the Brayton cycle. The following expressions are obtained (in watts):

Work input to the compressor (1–2)

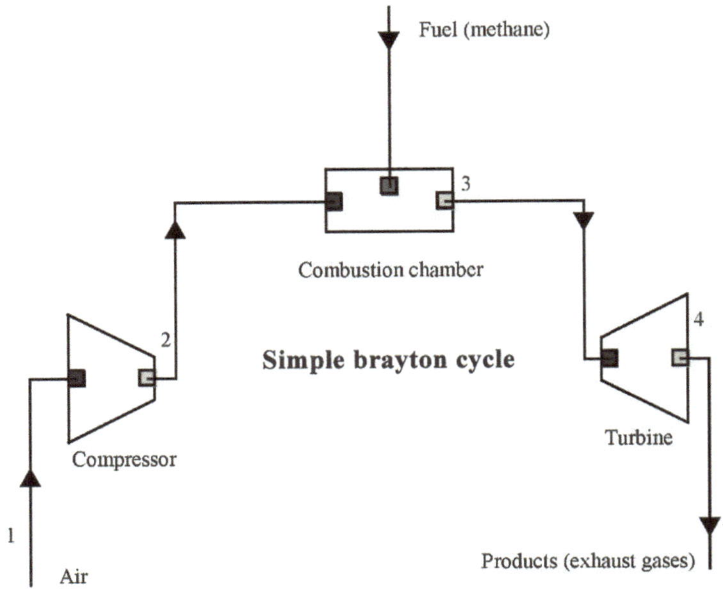

Fig. 9.11 Simple open cycle gas turbine.

$$W_{g,c} = \frac{\dot{m}_g(h_2 - h_1)}{\eta_c} \tag{9.75}$$

Heat supplied to the cycle (2–3)

$$Q_{g,in} = \dot{m}_g(h_3 - h_2) \tag{9.76}$$

Work output from turbine (3–4)

$$W_{g,t} = \dot{m}_g(h_3 - h_4)\eta_t \tag{9.77}$$

Subscript g refers to gas cycle, $\dot{m}_g$ is the gas mass flow rate ($\mathrm{kg\,s^{-1}}$), h is the enthalpy of air ($\mathrm{kJ\,kg^{-1}}$). η_c and η_t are the isentropic efficiencies of the compressor and turbine, respectively. The performance parameters are expressed as:

Cycle's net output power

$$W_{g,net} = W_{g,t} - W_{g,c} \tag{9.78}$$

Thermal efficiency

$$\eta_{th,g} = \frac{W_{g,net}}{Q_{g,in}} \tag{9.79}$$

The simple Brayton cycle can be modified to improve its efficiency. The temperature of exhaust gases in a simple gas turbine is usually much higher than ambient temperature (superior to 500°C), and thus the possibility to use these exhaust gases as heat source for another vapor cycle. The resulting cycle is called a gas-turbine combined cycle, where the cycle with the higher operating temperature is named the topping cycle. The waste heat it produces is transferred, using a heat exchanger, to a second cycle named the bottoming cycle since its operating temperature is lower (Fig. 9.12). Currently, the used arrangement for power generation at the industrial scale is that of a gas topping cycle with a vapor bottoming cycle.

The performance parameters for a gas-turbine combined cycle are expressed as

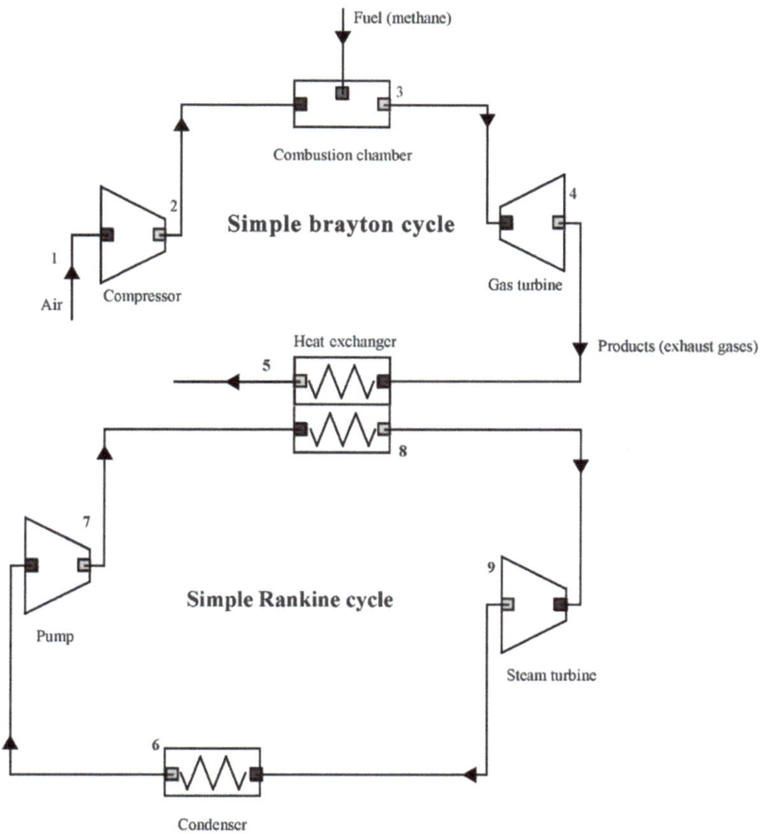

Fig. 9.12 Simple gas-turbine combined cycle.

Cycle's net output power

$$W_{net} = W_{v,t} + W_{g,t} - W_{g,c} - W_{v,p} \tag{9.80}$$

Combined cycle thermal efficiency

$$\eta_{th,g} = \frac{W_{net}}{Q_{g,in}} \tag{9.81}$$

Currently, sophisticated combined-cycle plants are able to achieve a net efficiency of 63%, which is higher than the efficiency of vapor cycles.

9.5.3 Cogeneration/trigeneration situations

In various industrial applications (e.g., textile industries, food processing, steel making and forming, oil production and refining, pulp and paper), many processes necessitate heat input. The industries that require substantial quantities of process heat also need a high quantity of electric power. It is in fact thermodynamically (and usually economically) better to produce both products (heat and electricity) in a single plant. Cogeneration plants simultaneously generate electricity and heat to satisfy the process-heat needs of a variety of industrial processes. Similarly, trigeneration is the production of electricity in addition to cooling and heating.

For a cogeneration cycle, the heating requirements are usually in the form of supplying steam to facilities (generally at 5–7 bars and 150–200°C). There are two types of cogeneration system. In the first, steam is generated using the exhaust gases of a simple gas-turbine plant (Fig. 9.13A). This is appropriate when a large quantity of process steam is required at high pressure and temperature. In the second, steam is extracted from the steam turbine of a gas-turbine combined cycle (Fig. 9.13B). The steam turbine in this case is called "extraction back pressure turbine" and is designed so that the exhaust pressure matches the process requirement. The performance of a cogeneration cycle is defined not only by the thermal efficiency, but also by the utilization factor (ϵ_u):

$$\epsilon_u = \frac{W_{net} + Q_p}{Q_{in}} \tag{9.82}$$

where W_{net} is the cycle's net power output, Q_p the heat delivered to the facilities, and Q_{in} the total heat input.

In trigeneration plants, cogeneration systems are linked to absorption/adsorption chillers that use waste heat to produced chilled water, which can be used for cooling processes or in air-conditioning systems.

A single-effect absorption chiller requires heat input at a temperature higher than 100°C, has coefficient of performance of 0.8, and produces chilled water at temperatures of about 7–10°C.

9.6 Case study

In the present case study, we investigate the use of methane gas generated through the AD of waste as a fuel for two different thermal power cycles. The first cycle is a gas-turbine cogeneration cycle (Fig. 9.13A) for industrial applications (e.g., a textile factory). It generates 35 MW of electricity and 40 MW of high temperature and pressure steam. As for the second cycle, it represents a gas-turbine combined cycle (Fig. 9.12) that is used to generate 51 MW of electricity. Data for both cycles as well as the required methane flow rate are summarized in Table 9.4.

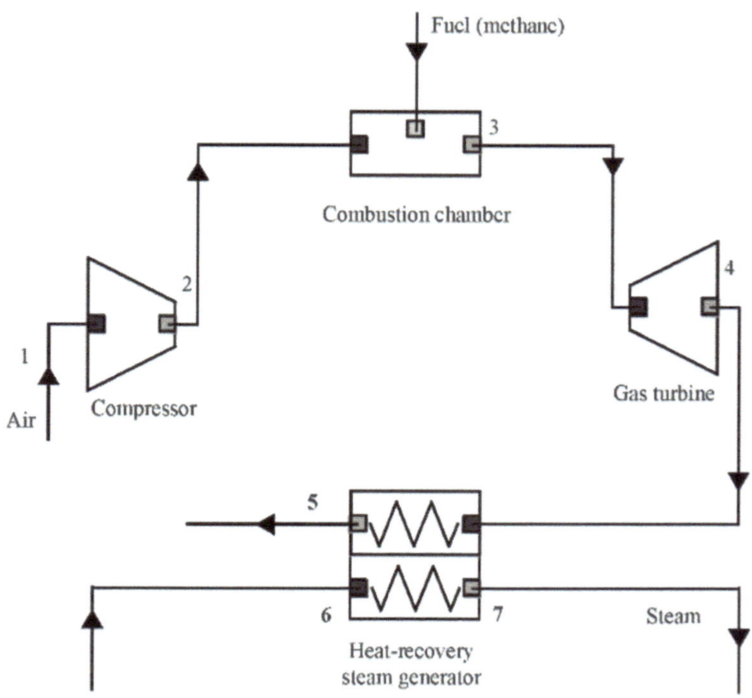

(A)

Fig. 9.13 Cogeneration cycles. (A) A simple gas-turbine plant and

continued

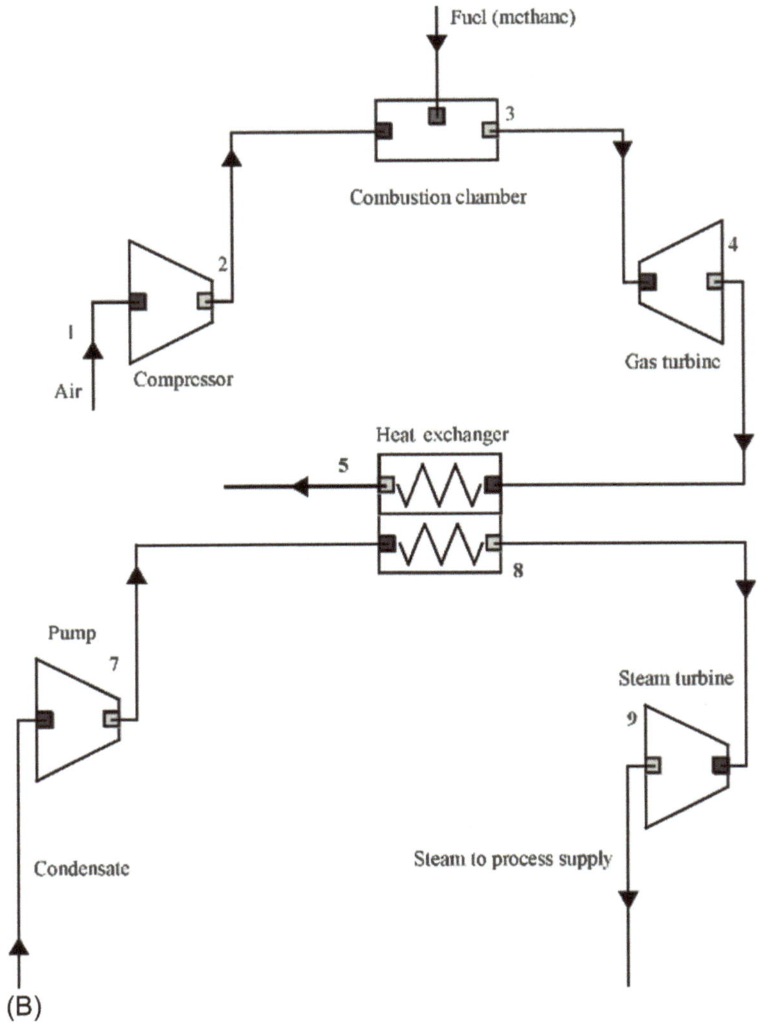

Fig. 9.13—cont'd (B) a gas-turbine combined cycle.

Actually, the methane yield of the anaerobic digestion is greatly dependent on the operating conditions of the digesters, the process parameters, and the waste's composition and quantity. This fact often leads to large discrepancies in the methane production potentials of different anaerobic digesters. A literature survey was conducted to determine the average methane production for different processes (Table 9.5).

Table 9.4 Power generation and fuel (methane) requirement for the two studied processes.

	Cogeneration thermal cycle (electricity + process steam) (Fig. 9.13A)	**Gas-turbine combined cycle (electricity) (Fig. 9.12)**
Produced power (kW_{elec})[a]	34,986.61	50,984.14
Process steam (kW_{th})[a]	39,991.66	–
Steam's temperature and pressure (K; bar)	515; 17.5	–
Thermal efficiency (%)	38.2	55.6
Utilization factor (%)	81.8	–
Total heat input (kW)	91,692.96	91,692.96
Required methane (kg/s)	1.83	1.83
Steam turbine: inlet conditions	–	86 bar/505.4°C
Heat exchanger effectiveness	0.8	0.8
Gas-turbine: inlet temperature (K)	1700	1700
Compression ratio	22.3	22.3

[a] With reference to ISO ambient conditions of 15°C, 1.013 bar, and 60% relative humidity. Turbines and compressors with 0.85 isentropic efficiency.

The calculations regarding digester sizing and methane production rates are based on the experimental tests carried out by Ref. [67]. This specific case was chosen because of the optimal operating conditions and process parameters, which is reflected in the higher reported methane yield. In their experiments, authors in [67] used right cylindrical anaerobic digesters made from stainless steel. The reactors were mechanically stirred (30 rpm) and kept in mesophilic conditions (35°C). The used feedstock was composed of 12 wt % cattle manure (CM), 16 wt% fruit and vegetable waste (FVW), and 72 wt% of water.

In our study, one digester is proposed and examined as a first step, and the number of digesters needed to produce the required methane flow is determined subsequently. The proposed digester for the present case study is supposed indoors (external temperature at 20°C), aboveground, and insulated with 20 cm of fiberglass. Further details about the digester, its operating conditions, and the used feedstock are presented in Table 9.6.

To determine the heating needs of the proposed digester, the terms in the energy balance equation (Eq. 9.25) have to be determined. The heat loss

Table 9.5 Operating conditions, parameters, and average methane yields for different anaerobic digesters (literature survey).

Feedstock	Digester's operating conditions	Process parameters	Model	Average daily methane (kg_{CH4}/kg_{waste} d)	Reference
Cow manure + grass silage	– Continuous lab-scale experiment – Mechanically stirred – Mesophilic temperature (35°C)	– $V = 4$ L – HRT = 18 d – $Q = 0.22\,kg_{waste}/d$ – $OLR = 3\,kg_{VS}/m^3\,d$	–	0.0054	[64]
Food waste from students' restaurants	– Continuous lab-scale experiment – Mechanically stirred – Thermophilic temperature (55°C)	– $V = 10$ L – HRT = 25 d – $Q = 0.45\,kg_{waste}/d$ – $OLR = 9.21\,kg_{VS}/m^3\,d$ – TS = 20% (dry AD)	Modified Gompertz model	0.0438	[65]
Thickened primary sludge + activated sludge	– Continuous lab-scale experiment – Mechanically stirred – Mesophilic temperature (37°C)	– $V = 0.3$ L – SRT = 12 d – $Q = 46.05\,g_{waste}/d$ – $OLR = 2.13\,kg_{VS}/m^3\,d$	Own model (linear regression)	0.0026	[66]
Food and vegetable waste + cattle/swine manure	– Continuous experiment – Lab-scale mechanically stirred – Mesophilic temperature (35°C)	– $V = 2$ L – HRT = 30 d – $Q = 2.72\,g_{waste}/d$ – $OLR = 1.2\,kg_{VS}/m^3\,d$	–	0.1898	[67]

Table 9.6 Operating conditions and characteristics of the proposed digester and feedstock.

	Parameter	Unit	Value
Digester's characteristics and operating conditions	Height	m	20.73
	Diameter (D_{dig})	m	20.73
	Thickness (T_{dig})	cm	4.6
	D_{dig}/T_{dig} ratio [68]	–	450
	Working volume	m^3	7000
	Impeller's diameter (D_{imp})	m	4.14
	D_{dig}/D_{imp} ratio [56]	–	0.2
	Walls' thermal conductivity [69]	$W\,m^{-1}\,K^{-1}$	14.4
	Insulation thickness	m	0.2
	Insulation material thermal conductivity	$W\,m^{-1}\,K^{-1}$	0.04
	Insulation material emissivity	–	0.75
	Operating temperature	°C	35 ± 1
	Mixing speed	rpm	30
	Organic loading rate (OLR)	$kg_{VS}\,m^{-3}\,d^{-1}$	1.2
	Hydraulic retention time (HRT)	d	30
Feedstock's characteristics	Composition (CM:FVW: water)	–	12:16:72
	Flow rate (waste + water)	$kg\,d^{-1}$	33,570
	Total solids (TS) [67]	%	4.3
	Volatile solids (VS) [67]	%	88.3
	Total organic carbon (TOC) [67]	%	14.9
	Total nitrogen (TN) [67]	%	1.8
	Total phosphorus (TP) [67]	%	1.0
	Total potassium (TK) [67]	%	1.5
	Mean viscosity (μ) [70]	Pa s	0.5
	Specific heat (c_p) [71,72]	$J\,kg^{-1}\,K^{-1}$	4180
	Density (ρ) [70]	$kg\,m^{-3}$	1001
	Thermal conductivity (k) [73]	$W\,m^{-1}\,K^{-1}$	0.15

through convection, conduction, and radiation is calculated. Using Eqs. (9.26), (9.32), (9.38), and (9.39), the convective heat transfer between the slurry and the internal walls of the digester is calculated. The convective heat transfer coefficient ($h_{c,\,1}$) and the convective thermal resistance ($R_{conv,\,1}$) were determined ($1.05\,W/m^2K$ and $4.7 \times 10^{-4}\,K/W$, respectively). Furthermore, the conductive heat transfers through the walls of the digester and the insulation are calculated using Eq. (9.29). The conductive thermal resistances through the walls and the insulation ($R_{cond,\ w}$ and $R_{cond,\ i}$, respectively) were found equal to 2.31×10^{-6} and 0.0036 K/W, respectively. As for the convective heat transfer between the surface of the insulation and the ambient air, Eqs. (9.26), (9.32), (9.33), (9.35), and (9.40) are used. The convective heat transfer coefficient ($h_{c,\ 2}$) and the convective thermal resistance ($R_{conv,\ 2}$) were determined (1.66 W/m^{2}K and 2.86×10^{-4} K/W, respectively). Finally, the radiative thermal resistance for the heat transfer between the insulation and the ambient air was determined (0.208 K/W) using Eq. (9.28). Then, using Eqs. (9.30), (9.31), the global thermal resistance (R_{glob}) and the total digester's heat loss (Q_{loss}) were calculated (0.0043 K/W and 3488.37 W, respectively). The heat transfer coefficients and the thermal resistances for the different heat transfer modes are summarized in Table 9.7.

Table 9.7 Thermal parameters for the different heat transfer modes.

Heat transfer mode	Heat transfer coefficient (W/m^{2}K)	Thermal resistance (K/W)	Heat loss (W)	Used equations
Convection (between the slurry and the internal walls)	1.05	4.7×10^{-4}	–	(9.26), (9.32), (9.38), and (9.39)
Conduction (through the digester's walls)	–	2.31×10^{-6}	–	(9.29)
Conduction (through the insulation)	–	0.0036	–	(9.29)
Convection (insulation surface-ambient air)	1.66	2.86×10^{-4}	–	(9.26), (9.32), (9.33), (9.35), and (9.40)
Radiation (insulation surface-surroundings)	–	0.208	–	(9.28)
Digester	–	0.0043	3488.37	(9.30) and (9.31)

Furthermore, the heat transfer due to the slurry feeding is taken into consideration. In fact, it is customary for large anaerobic digesters to use heat exchangers to preheat the slurry before introducing it to the reactor. This preheating brings the temperature of the slurry to the operating temperature of the digester, thus making the heat loss due to feeding nil (by making the temperature difference equal to 0 in Eq. (9.45)). A commonly used heat exchanger in AD processes is the shell-and-tube type [74]. For sizing the heat exchanger adequately, Eq. (9.46) is used to determine the required heat transfer surface area. This area was found to be equal to 0.58 m^2 for preheating ≈35 tons of feedstock (for one digester) from 20°C to 35°C (Table 9.8). The required heat rate ($Q_{mass\ flow}$) was determined (24.36 kW) using Eq. (9.45).

In the present case study, the heat transfer caused by the chemical reactions taking place during the anaerobic digestion process is neglected [54,57–59]. Using the heat balance equation (Eq. 9.25), the global power required to satisfy the heating needs has been calculated. This power is equal to 27.84 kW for one digester. To calculate the total heating needs, the number of required digesters is determined. Actually, based on the experiments carried out by Ref. [67], the methane potential of 1 kg of raw organic waste is 0.1898 kg/d of CH_4 (Table 9.5). Therefore, to meet the needs in methane gas for the thermal power cycles (which is 1.83 kg/s), a mass of about 833 tons of waste must be digested daily. This is the equivalent of 89 anaerobic digesters operating at the same time.

To provide thermal energy and maintain the operating temperature for all the digesters, a solar thermal energy system (composed of solar collectors and storage tanks) is sized using the equations presented in Section 9.4. It is

Table 9.8 Characteristics of the shell-and-tube heat exchanger for preheating the feedstock.

Parameter	Unit	Value
U [74]	W m^{-2} K^{-1}	1803
ΔT_{lm}	°C	23.42
Number of tubes	–	4
Total heat transfer area	m^2	0.58
Tube length	m	1.83
Tube diameter	m	0.025
Tube pattern	–	Square
Tube pass	–	Multiple

worth mentioning that a factor of 0.9 was considered for the efficiency of the heating provided by the storage tanks. The characteristics of the thermal collectors and storage tanks are presented in Table 9.9.

It is worth mentioning that an intermediary tank for methane storage needs to be considered to allow a constant flow of the fuel (CH_4) to the thermal power cycles. The global heating and waste requirements, and the energy production of the 89 digesters are presented in Table 9.10.

Table 9.9 Characteristics of the thermal collectors and storage tanks required for all the digesters.

Thermal collectors	Model	Helio Plan/L20 HTF
	Number	6675 (75 per digester)
	Total area (m^2)	17,422 (196 m^2 per digester)
	Inclination (°)	30
	Annual irradiation (kWh/m^2)	2534
	Conversion factor (%)	80.8
	Heat transfer coefficient ($W/m^2 K$)	3.55
Heat storage tanks	Model	Riello Pufferspeicher-3000
	Number	267 (3 per digester)
	Total capacity (m^3)	801
	Insulation thickness (m)	0.1
	Thermal conductivity (W/m K)	0.065

Table 9.10 Global waste and heating requirements, in addition to the energy generation for the anaerobic digesters.

Parameter		Unit	Value
Number of digesters		–	89
Daily waste feeding		kg_{waste}/d	833,046
Organic loading rate		kg_{VS}/d	747,600
Global heating needs		MW	2.47
Daily methane production		kg/d	158,112
Power production	Cogeneration thermal cycle	MW	34.98 (electric) + 39.99 (steam)
	Gas-turbine combined cycle	MW	50.98 (electric)

9.7 Conclusions

In this chapter, the basic principles of the anaerobic digestion and the main parameters affecting it have been listed and explained. Thereafter, the different types of models used for simulating the biogas formation and waste degradation in the AD process have been categorized and detailed. It is reported that the most used model is the multistage ADM1. However, due to this model's complexity, which is induced by its large number of parameters, simpler statistical models (e.g., Chen and Hashimoto model) are more practical.

Furthermore, a comprehensive approach for modeling anaerobic digesters under different conditions (indoors, outdoors), process parameters (mixed, unmixed), and operating modes (batch, continuous) has been detailed. In fact, well-established equations and correlations for simulating the heat transfer phenomena and determining the heating (and feedstock preheating) needs for anaerobic digesters have been presented and described. Likewise, the fundamental equations for modeling thermal solar collectors and heat storage tanks have been detailed.

Additionally, different thermodynamic cycles for power generation using methane produced through AD as fuel have been presented and discussed. These cycles manifest in vapor power plants (based on Rankine cycle) and simple gas-turbine power plant (based on Brayton cycle). Moreover, cogeneration plants (where heat and electricity are produced) and trigeneration plants (where heat, cold, and electricity are generated) have been illustrated and detailed.

Finally, a case study that investigated a solar/anaerobic digester hybrid system for electricity and heat generation has been carried out. In this study, the required number of digesters, quantity of feedstock (influent waste), and number of thermal collectors and storage tanks have been determined by employing the equations and correlations detailed in the former sections. The processes' parameters, feedstock's composition, and methane potential have been determined based on experimental tests from literature. The size of the anaerobic digesters and the daily quantity of the feedstock have been calculated based on the required methane flow rate to fully power the chosen thermodynamic cycles. A thermal study has been performed on the digesters to determine their heating requirements (as well as the feedstock preheating needs), based on which a solar thermal energy system (composed of thermal collectors and heat storage tanks) has been adequately sized.

References

[1] Bilgen S, Kaygusuz K, Sari A. Renewable energy for a clean and sustainable future. Energy Source 2004;26:1119–29. https://doi.org/10.1080/00908310490441421.
[2] Mohr SH, Wang J, Ellem G, Ward J, Giurco D. Projection of world fossil fuels by country. Fuel 2015;141:120–35. https://doi.org/10.1016/j.fuel.2014.10.030.
[3] Verbruggen A, Fischedick M, Moomaw W, Weir T, Nadaï A, Nilsson LJ, Nyboer J, Sathaye J. Renewable energy costs, potentials, barriers: conceptual issues. Energy Policy 2010;38:850–61. https://doi.org/10.1016/j.enpol.2009.10.036.
[4] Saidur R, Abdelaziz EA, Demirbas A, Hossain MS, Mekhilef S. A review on biomass as a fuel for boilers. Renew Sustain Energy Rev 2011;15:2262–89. https://doi.org/10.1016/j.rser.2011.02.015.
[5] Van Meerbeek K, Appels L, Dewil R, Van Beek J, Bellings L, Liebert K, Muys B, Hermy M. Energy potential for combustion and anaerobic digestion of biomass from low-input high-diversity systems in conservation areas. GCB Bioenergy 2015;7:888–98. https://doi.org/10.1111/gcbb.12208.
[6] Holm-Nielsen JB, Al Seadi T, Oleskowicz-Popiel P. The future of anaerobic digestion and biogas utilization. Bioresour Technol 2009;100:5478–84. https://doi.org/10.1016/j.biortech.2008.12.046.
[7] Merlin G, Kohler F, Bouvier M, Lissolo T, Boileau H. Importance of heat transfer in an anaerobic digestion plant in a continental climate context. Bioresour Technol 2012;124:59–67. https://doi.org/10.1016/j.biortech.2012.08.018.
[8] Khare V, Nema S, Baredar P. Solar–wind hybrid renewable energy system: a review. Renew Sustain Energy Rev 2016;58:23–33. https://doi.org/10.1016/j.rser.2015.12.223.
[9] Boulmrharj S, NaitMalek Y, Elmouatamid A, Bakhouya M, Ouladsine R, Zine-Dine K, et al. Battery Characterization and Dimensioning Approaches for Micro-Grid Systems. Energies 2019;12:1305. https://doi.org/10.3390/en12071305.
[10] Mata-Alvarez J, Dosta J, Romero-Güiza MS, Fonoll X, Peces M, Astals S. A critical review on anaerobic co-digestion achievements between 2010 and 2013. Renew Sustain Energy Rev 2014;36:412–27. https://doi.org/10.1016/j.rser.2014.04.039.
[11] Sarkar O, Butti SK, Mohan SV. Acidogenic biorefinery: food waste valorization to biogas and platform chemicals. In: Waste biorefinery. Elsevier; 2018. p. 203–18. https://doi.org/10.1016/B978-0-444-63992-9.00006-9.
[12] Gavala HN, Angelidaki I, Ahring BK. Kinetics and modeling of anaerobic digestion process. In: Biomethanation I. Berlin, Heidelberg: Springer; 2003. p. 57–93. https://doi.org/10.1007/3-540-45839-5_3.
[13] Ward AJ, Hobbs PJ, Holliman PJ, Jones DL. Optimisation of the anaerobic digestion of agricultural resources. Bioresour Technol 2008;99:7928–40. https://doi.org/10.1016/j.biortech.2008.02.044.
[14] Lin Q, De Vrieze J, Li J, Li X. Temperature affects microbial abundance, activity and interactions in anaerobic digestion. Bioresour Technol 2016;209:228–36. https://doi.org/10.1016/j.biortech.2016.02.132.
[15] Deng L, Yang H, Liu G, Zheng D, Chen Z, Liu Y, et al. Kinetics of temperature effects and its significance to the heating strategy for anaerobic digestion of swine wastewater. Appl Energy 2014;134:349–55. https://doi.org/10.1016/j.apenergy.2014.08.027.
[16] Zhang B, Zhang LL, Zhang SC, Shi HZ, Cai WM. The influence of pH on hydrolysis and acidogenesis of kitchen wastes in two-phase anaerobic digestion. Environ Technol 2005;26:329–40. https://doi.org/10.1080/09593332608618563.
[17] Alburquerque JA, de la Fuente C, Bernal MP. Chemical properties of anaerobic digestates affecting C and N dynamics in amended soils. Agr Ecosyst Environ 2012;160:15–22. https://doi.org/10.1016/j.agee.2011.03.007.

[18] Fernández J, Pérez M, Romero LI. Effect of substrate concentration on dry mesophilic anaerobic digestion of organic fraction of municipal solid waste (OFMSW). Bioresour Technol 2008;99:6075–80. https://doi.org/10.1016/j.biortech.2007.12.048.
[19] El Mashad HE. Solar thermophilic anaerobic reactor (STAR) for renewable energy production. Wageningen University; 2003.
[20] Forster-Carneiro T, Pérez M, Romero L. Influence of total solid and inoculum contents on performance of anaerobic reactors treating food waste. Bioresource technology 2008;99:6994–7002. https://doi.org/10.1016/j.biortech.2008.01.018.
[21] Dahiya A, editor. Bioenergy: biomass to biofuels. Academic Press; 2014.
[22] Khalid A, Arshad M, Anjum M, Mahmood T, Dawson L. The anaerobic digestion of solid organic waste. Waste Manage 2011;31:1737–44. https://doi.org/10.1016/j.wasman.2011.03.021.
[23] Kaparaju P, Buendia I, Ellegaard L, Angelidakia I. Effects of mixing on methane production during thermophilic anaerobic digestion of manure: lab-scale and pilot-scale studies. Bioresour Technol 2008;99:4919–28. https://doi.org/10.1016/j.biortech.2007.09.015.
[24] El Ibrahimi M, Khay I, El Maakoul A, Moussa MO, Barkaoui A, Bakhouya M. Anaerobic co-digestion in a liquid recirculation pilot-scale reactor: thermal and hydraulic study. Energy Rep 2020;6:496–502. https://doi.org/10.1016/j.egyr.2019.09.014.
[25] El Ibrahimi M, Khay I, El Maakoul A, Bakhouya M. Thermal modeling of a pilot-scale liquid recirculation anaerobic digester. In: 2019 7th International renewable and sustainable energy conference (IRSEC); 2019. p. 1–6. https://doi.org/10.1109/IRSEC48032.2019.9078203.
[26] Ai H, Wang Q, Fan X, Xie W, Shinohara R. Effect of hydraulic retention time on the efficiency of vertical tubular anaerobic sludge digester treating waste activated sludge. Environ Technol 2005;26:725–32. https://doi.org/10.1080/09593332608618512.
[27] Buswell AM, Mueller HF. Mechanism of methane fermentation. Ind Eng Chem 1952;44:550–2. https://doi.org/10.1021/ie50507a033.
[28] Boyle WC. Energy recovery from sanitary landfills-a review. In: Microbial energy conversion. Pergamon; 1977. p. 119–38. https://doi.org/10.1016/B978-0-08-021791-8.50019-6.
[29] Angelidaki I, Sanders W. Assessment of the anaerobic biodegradability of macropollutants. Rev Environ Sci Biotechnol 2004;3:117–29. https://doi.org/10.1007/s11157-004-2502-3.
[30] Amon T, Amon B, Kryvoruchko V, Machmüller A, Hopfner-Sixt K, Bodiroza V, Hrbek R, Friedel J, Pötsch E, Wagentristl H, Schreiner M. Methane production through anaerobic digestion of various energy crops grown in sustainable crop rotations. Bioresour Technol 2007;98:3204–12. https://doi.org/10.1016/j.biortech.2006.07.007.
[31] Maier RM, Pepper IL. Bacterial growth. In: Environmental microbiology. Academic Press; 2015. p. 37–56. https://doi.org/10.1016/B978-0-12-394626-3.00003-X.
[32] Contois DE. Kinetics of bacterial growth: relationship between population density and specific growth rate of continuous cultures. Microbiology 1959;21:40–50. https://doi.org/10.1099/00221287-21-1-40.
[33] Chen YR, Hashimoto AG. Substrate utilization kinetic model for biological treatment process. Biotechnol Bioeng 1980;22:2081–95. https://doi.org/10.1002/bit.260221008.
[34] Grau P, Dohanyos M, Chudoba J. Kinetics of multicomponent substrate removal by activated sludge. Water Res 1975;9:637–42. https://doi.org/10.1016/0043-1354(75)90169-4.
[35] Moser H. The dynamics of bacterial populations maintained in the chemostat; 1958.

[36] Nelson MI, Balakrishnan E, Sidhu HS. A fundamental analysis of continuous flow bioreactor and membrane reactor models with Tessier kinetics. Chem Eng Commun 2012;199:417–33. https://doi.org/10.1080/00986445.2010.525155.
[37] Kythreotou N, Florides G, Tassou SA. A review of simple to scientific models for anaerobic digestion. Renew Energy 2014;71:701–14. https://doi.org/10.1016/j.renene.2014.05.055.
[38] Aiba S, Shoda M, Nagatani M. Kinetics of product inhibition in alcohol fermentation. Biotechnol Bioeng 2000;67:671–90. https://doi.org/10.1002/bit.260100610.
[39] Han K, Levenspiel O. Extended Monod kinetics for substrate, product, and cell inhibition. Biotechnol Bioeng 1988;32:430–47. https://doi.org/10.1002/bit.260320404.
[40] Donoso-Bravo A, Pérez-Elvira SI, Fdz-Polanco F. Application of simplified models for anaerobic biodegradability tests. Evaluation of pre-treatment processes. Chem Eng J 2010;160:607–14. https://doi.org/10.1016/j.cej.2010.03.082.
[41] Li P, Li W, Sun M, Xu X, Zhang B, Sun Y. Evaluation of biochemical methane potential and kinetics on the anaerobic digestion of vegetable crop residues. Energies 2019;12:26. https://doi.org/10.3390/en12010026.
[42] Bernard O, Hadj-Sadok Z, Dochain D, Genovesi A, Steyer JP. Dynamical model development and parameter identification for an anaerobic wastewater treatment process. Biotechnol Bioeng 2001;75:424–38. https://doi.org/10.1002/bit.10036.
[43] Haag JE, Wouwer AV, Queinnec I. Macroscopic modelling and identification of an anaerobic waste treatment process. Chem Eng Sci 2003;58:4307–16. https://doi.org/10.1016/S0009-2509(03)00272-0.
[44] Simeonov I, Momchev V, Grancharov D. Dynamic modeling of mesophilic anaerobic digestion of animal waste. Water Res 1996;30:1087–94. https://doi.org/10.1016/0043-1354(95)00270-7.
[45] Donoso-Bravo A, Mailier J, Martin C, Rodríguez J, Aceves-Lara CA, Wouwer AV. Model selection, identification and validation in anaerobic digestion: a review. Water Res 2011;45:5347–64. https://doi.org/10.1016/j.watres.2011.08.059.
[46] Batstone DJ, Keller J, Angelidaki I, Kalyuzhnyi SV, Pavlostathis SG, Rozzi A, et al. The IWA anaerobic digestion model no. 1 (ADM1). Water Sci Technol 2002;45:65–73. https://doi.org/10.2166/wst.2002.0292.
[47] Nordlander E, Thorin E, Yan J. Investigating the possibility of applying an ADM1 based model to a full-scale co-digestion plant. Biochem Eng J 2017;120:73–83. https://doi.org/10.1016/j.bej.2016.12.014.
[48] Boubaker F, Ridha BC. Modelling of the mesophilic anaerobic co-digestion of olive mill wastewater with olive mill solid waste using anaerobic digestion model no. 1 (ADM1). Bioresour Technol 2008;99:6565–77. https://doi.org/10.1016/j.biortech.2007.11.035.
[49] Kafle GK, Chen L. Comparison on batch anaerobic digestion of five different livestock manures and prediction of biochemical methane potential (BMP) using different statistical models. Waste Manage 2016;48:492–502. https://doi.org/10.1016/j.wasman.2015.10.021.
[50] Kafle GK, Kim SH, Sung KI. Ensiling of fish industry waste for biogas production: a lab scale evaluation of biochemical methane potential (BMP) and kinetics. Bioresour Technol 2013;127:326–36. https://doi.org/10.1016/j.biortech.2012.09.032.
[51] Pommier S, Chenu D, Quintard M, Lefebvre X. A logistic model for the prediction of the influence of water on the solid waste methanization in landfills. Biotechnol Bioeng 2007;97:473–82. https://doi.org/10.1002/bit.21241.
[52] Elagroudy S, Radwan AG, Banadda N, Mostafa NG, Owusu PA, Janajreh I. Mathematical models comparison of biogas production from anaerobic digestion of microwave pretreated mixed sludge. Renew Energy 2020;155:1009–20. https://doi.org/10.1016/j.renene.2020.03.166.

[53] Zhao HW, Viraraghavan T. Analysis of the performance of an anaerobic digestion system at the Regina wastewater treatment plant. Bioresour Technol 2004;95:301–7. https://doi.org/10.1016/j.biortech.2004.02.023.
[54] Perrigault T, Weatherford V, Martí-Herrero J, Poggio D. Towards thermal design optimization of tubular digesters in cold climates: a heat transfer model. Bioresour Technol 2012;124:259–68. https://doi.org/10.1016/j.biortech.2012.08.019.
[55] Bergman TL, Incropera FP, DP DW, Lavine AS. Fundamentals of heat and mass transfer. John Wiley & Sons; 2011.
[56] Kresta SM, Etchells III AW, Dickey DS, Atiemo-Obeng VA, editors. Advances in industrial mixing: a companion to the handbook of industrial mixing. John Wiley & Sons; 2015.
[57] Pedersen SV, Martí-Herrero J, Singh AK, Sommer SG, Hafner SD. Management and design of biogas digesters: a non-calibrated heat transfer model. Bioresour Technol 2020;296:122264. https://doi.org/10.1016/j.biortech.2019.122264.
[58] Weatherford VC, Zhai ZJ. Affordable solar-assisted biogas digesters for cold climates: experiment, model, verification and analysis. Appl Energy 2015;146:209–16. https://doi.org/10.1016/j.apenergy.2015.01.111.
[59] Terradas-Ill G, Pham CH, Triolo JM, Martí-Herrero J, Sommer SG. Thermic model to predict biogas production in unheated fixed-dome digesters buried in the ground. Environ Sci Technol 2014;48:3253–62. https://doi.org/10.1021/es403215w.
[60] Lindorfer H, Braun R, Kirchmayr R. Self-heating of anaerobic digesters using energy crops. Water Sci Technol 2006;53:159–66. https://doi.org/10.2166/wst.2006.246.
[61] Gebremedhin KG, Wu B, Gooch C, Wright P, Inglis S. Heat transfer model for plug-flow anaerobic digesters. Trans ASAE 2005;48:777–85. https://doi.org/10.13031/2013.18320.
[62] Struckmann F. Analysis of a flat-plate solar collector. In: Heat and mass transport; 2008 [Project Report 2008MVK160].
[63] Spencer RC, Cotton KC, Cannon CN. A method for predicting the performance of steam turbine-generators … .: 16, 500 kw and larger. J Eng Gas Turbines Power 1963. https://doi.org/10.1115/1.3677341.
[64] Lehtomäki A, Huttunen S, Rintala JA. Laboratory investigations on co-digestion of energy crops and crop residues with cow manure for methane production: effect of crop to manure ratio. Resour Conserv Recycl 2007;51:591–609. https://doi.org/10.1016/j.resconrec.2006.11.004.
[65] Nguyen DD, Chang SW, Jeong SY, Jeung J, Kim S, Guo W, et al. Dry thermophilic semi-continuous anaerobic digestion of food waste: performance evaluation, modified Gompertz model analysis, and energy balance. Energy Convers Manage 2016;128:203–10. https://doi.org/10.1016/j.enconman.2016.09.066.
[66] McLeod JD, Othman MZ, Beale DJ, Joshi D. The use of laboratory scale reactors to predict sensitivity to changes in operating conditions for full-scale anaerobic digestion treating municipal sewage sludge. Bioresour Technol 2015;189:384–90. https://doi.org/10.1016/j.biortech.2015.04.049.
[67] Alvarez R, Liden G. Semi-continuous co-digestion of solid slaughterhouse waste, manure, and fruit and vegetable waste. Renew Energy 2008;33:726–34. https://doi.org/10.1016/j.renene.2007.05.001.
[68] The Steel Construction Institute. Stainless Steel Tanks for Biogas Production. Online article available at: https://www.worldstainless.org/files/issf/non-image-files/PDF/power_generation/Stainless_steel_tanks_for_biogas_production.pdf; 2016. [Accessed 23 May 2020].
[69] Graves RS, Kollie TG, McElroy DL, Gilchrist KE. The thermal conductivity of AISI 304L stainless steel. Int J Thermophys 1991;12:409–15. https://doi.org/10.1007/BF00500761.

[70] Gruber-Brunhumer MR, Montgomery LF, Nussbaumer M, Schoepp T, Zohar E, Muccio M, et al. Effects of partial maize silage substitution with microalgae on viscosity and biogas yields in continuous AD trials. J Biotechnol 2019;295:80–9. https://doi.org/10.1016/j.jbiotec.2019.02.004.
[71] Prapaspongsa T, Poulsen TG, Hansen JA, Christensen P. Energy production, nutrient recovery and greenhouse gas emission potentials from integrated pig manure management systems. Waste Manage Res 2010;28:411–22. https://doi.org/10.1177/0734242X09338728.
[72] Berglund M, Börjesson P. Assessment of energy performance in the life-cycle of biogas production. Biomass Bioenergy 2006;30:254–66. https://doi.org/10.1016/j.biombioe.2005.11.011.
[73] Klejment E, Rosiński M. Testing of thermal properties of compost from municipal waste with a view to using it as a renewable, low temperature heat source. Bioresour Technol 2008;99:8850–5. https://doi.org/10.1016/j.biortech.2008.04.053.
[74] Espinosa-Solares T, Valle-Guadarrama S, Bombardiere J, Domaschko M, Easter M. Effect of heating strategy on power consumption and performance of a pilot plant anaerobic digester. Appl Biochem Biotechnol 2009;156:35–44. https://doi.org/10.1007/s12010-008-8487-6.
[75] Parameswaran P, Rittmann BE. Feasibility of anaerobic co-digestion of pig waste and paper sludge. Bioresour Technol 2012;124:163–8. https://doi.org/10.1016/j.biortech.2012.07.116.

CHAPTER 10

Hydrogen production and derivatives from renewable energy systems for a best valorization of sustainable resources

Rachid El Mrabet[a] and Asmae Berrada[b]
[a]Senior Expert and General Director of the Public Utility Development Company, Morocco
[b]College of Engineering, LERMA Lab, International University of Rabat, Sala El Jadida, Morocco

10.1 Introduction

Hydrogen is currently being significantly consumed. Between 50 and 70 million tons are used annually worldwide. This is expected to increase 10-fold by 2050. Being highly reactive, it serves as a vector gas (designed to transport other active gases) in the manufacturing of electronic components. Associated with nitrogen, it is also involved in the development of ammonia. In metallurgy, it is used in the thermal treatment necessary for the manufacturing of the metal parts. It is also used in textile fibers such as nylon, as well as various plastics. Liquid hydrogen is widely used in the manufacturing and protection of flat glass and flat screens. Other applications of hydrogen are feeding the industry's hopes; among them is the storage of renewable energy.

Not existing in nature in the free state, hydrogen is produced by dissociating it from the atoms with which it is combined. These include oxygen atoms in the case of water, where electrolysis is carried out. Other types include carbon atoms for fossil matter that are dissociated using reforming or oxidation by involving water in the reactions. Apart from water, fossil fuels account for 96% of the world's hydrogen production raw materials.

Usually, produced from hydrocarbons, for industrial uses, hydrogen has begun a new life in the field of energy. Different modes of production, such as electrolysis, allow low-carbon or totally carbon-free hydrogen to be obtained. This is performed using electricity generated either from nuclear

Hybrid Energy System Models
https://doi.org/10.1016/B978-0-12-821403-9.00010-X

power plants or from renewable energies, such as wind, photovoltaic, or hydropower. Multiple uses of this "green" hydrogen are developing. It is already being used for different purposes, whether as a vector for storing or upgrading renewable energy, as a complement to natural gas, or even as a fuel, and for a variety of fuel cell-related applications in stationary or mobile fields. The potential of hydrogen energy is very important. To make full use of it, research and development is current being focused on improving performance and lowering the cost of this technology to produce, store, and distribute it.

Concerning the energy transition, this gas is now added to the panoply of alternatives to carbon energies as it contributes to the energy independence and the economic growth in several countries. Deployed on large scale, green hydrogen, obtained with renewable electricity by electrolysis of water or by reforming biogas from waste treatment or biomass, could account for 18% of the world's final energy consumption by 2050.

This would reduce annual CO_2 emissions by about 6 billion tons, which represents 20% of the effort needed to limit warming to 2°C [1]. Published at COP23, this study confirms the central role of hydrogen in the energy transition and presents a precise roadmap for deploying its applications. Investments in the range of $20 billion to $25 billion per year through 2030 could generate $2500 billion in revenue and create more than 30 million jobs worldwide by 2050. In the transport sector alone, hydrogen could power 10–15 million cars and 500,000 HGVs by 2030, which emit no CO_2 or pollutants.

The 2°C scenario is based on the commitment of the international community to take proactive action on several fronts. Among these components is the massive investment in renewable energy. However, this latter faces a major challenge in the growth of their penetration in today's electricity grids such as intermittency. A number of storage solutions exist with different levels of maturity and application.

One way to store renewable electricity efficiently and flexibly, especially with high energy density, is hydrogen gas (H_2, hydrogen). It is a product derived, among other techniques, from the electrolysis of water. This makes it a prime ally for large photovoltaic and wind power plants, which lack a large-scale storage solution. The reverse reaction can be done using a fuel cell.

Moreover, with world records of renewable energy kilowatt-hour prices observed in recent years, the electrolysis sector via renewables is becoming more and more interesting compared with the conventional sector via the reforming of natural gas, especially when one puts into perspective their respective environmental costs. Hydrogen enters several industrial processes such as metallurgy, cement, and fertilizers.

The latest registered Power Purchase Agreement (PPA) is approximately \$0.0234/kWh for PV and \$0.03/kWh for wind. These figures are in perfect line with the IEA's estimates in its "Renewable Energy for Industry, IEA, 2017" report [2]. These estimations give a more competitive kilogram of green hydrogen than that obtained from conventional steam methane reforming (SMR), when electricity is around 0.03 degrees/kWh (30 degrees/MWh). Fig. 10.1 presents the price of a kilogram of hydrogen evolution according to the number of hours of charge of the electrolysis.

10.2 Hydrogen: Properties and fundamental

The hydrogen atom H consists of a single-proton P and a peripheral electron e^-. The atom H is the smallest, simplest, and lightest atom. It occupies the first place at the upper left entrance of Mendeleev's table. Equipped with a single peripheral electron, H is chemically the most reductive atom and can bind to practically all elements. It first binds to itself to form the dihydrogen molecule (H_2) whose two-electron peripheral layer possesses the stability of the first level of quantum equilibrium. However, the mass of the hydrogen is not sufficient for it to be retained by the earth's attraction. According to current knowledge, it is very rarely present as H_2 form in the crust (2/1000 of atoms) as in the

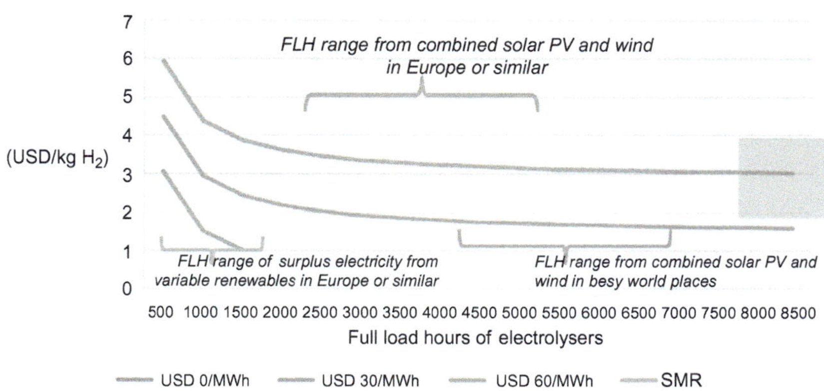

Fig. 10.1 Evolution of the price of a kilogram of hydrogen according to the number of hours of charge of the electrolysis (500–8500 h).

atmosphere (0.5 ppm), while it is the most common atom in the universe of which it constitutes more than 90% of atoms and 75% of mass.

On Earth, hydrogen as a reducing agent is present in several molecules, the most essential of which are very important to life such as with oxygen, it forms water; with carbon, it produces methane and the hydrocarbonate family, which make up organic chemistry; with nitrogen, it forms ammonia and its derivatives (fertilizer); and with halogens, it produces strong acids (such as hydrochloric acid).

Hydrogen is therefore not a directly available primary energy on earth, such as coal or oil. In H_2 form, it is only an energy vector that must be produced by extracting it from its compounds (water, methane, etc.) and that can then react powerfully with oxidants, especially oxygen, by releasing a lot of energy.

As an energy vector, hydrogen has several exceptional abilities:

- the highest mass density of energy, 1 kg of hydrogen that contains as much energy as about 3 kg of oil;
- high-energy combustion in oxygen with pure water production (about three times more than gasoline at constant weight);
- a deep duality with electricity, with a direct reciprocal transformation capacity by electrolysis or fuel cell;
- noncarbon combustion (no CO_2 emissions when it comes from renewable sources).

These major points make hydrogen, for some, the master asset of the decarbonization necessary, by substituting it for fossil fuels, to put an end to the current climate runaway. It is particularly seen as the intermittent remedy for wind and solar, of which H_2 would be the stocker/destocker, thus enabling a massive development of these sectors. However, the H_2 energy vector also has major drawbacks related to its low density of volume energy. To store, transport, and distribute hydrogen, it must either be liquefied at atmospheric pressure and at an extremely low temperature (−253°C) or compressed at very high pressure (700 bars) while controlling its risk of leakage, corrosion, and explosion.

10.3 Main chemical and oil uses of hydrogen

10.3.1 Hydrogen in the chemical industry

Hydrogen is one of the basic materials of the chemical and petrochemical industries. It can do the following:

- be manufactured specifically in units dedicated to its production (steam reforming methane, water electrolysis, etc.).

- be coproduced, as a byproduct, from other chemicals, such as ethylene or chlorine. Its use as a raw material is still far superior compared with its energy value.

In geographical areas where the chemical industry is strongly established (Benelux, Texas, etc.), these "hydrogen networks" of sites are interconnected by networks of "pipelines" connecting producer and consumer sites.

10.3.1.1 Ammonia

This sector is of strategic importance, especially in the context of food security and global warming. Fertilizers are one of the solutions to optimize agricultural production. At the base of this industry is ammonia, NH_3, as a product used in the manufacturing of fertilizers, which is the dominant chemical product in terms of hydrogen consumption. It is also responsible for about 1% of global greenhouse gas (GHG) emissions.

Thanks to the major developments in renewable energy projects, the "green ammonia" sector is experiencing a real enthusiasm on the part of international players, namely, Siemens and Yara. Research and development efforts are still investigating a number of ways to produce and consume ammonia in a sustainable way (use of concentrated thermal solar, electrochemical synthesis, ammonia as fuel, ammonia to drive turbines, etc.). On the other hand, there are preindustrial demonstration efforts that use mature technological building blocks, namely, water electrolysis for the production of hydrogen "connected" to the conventional "Haber-Bosch" process to generate ammonia.

The decline in the cost of generating renewable electricity, particularly solar and wind, is directly reflected in the price of a ton of green ammonia (Fig. 10.2), which is also becoming competitive with an electricity price of 30 USD/MWh.

If the amount of hydrogen used today in the manufacturing of chemicals such as amines, methanol, and oxygenated water is much lower than that required from the manufacturing of ammonia, its use has been developed in other sectors of the industry such as welding, the development of certain metals, the manufacture of glasses, semiconductors, or the food industry.

10.3.2 Hydrogen in the oil industry

Refineries strive to produce from a variety of crude oils the maximum amount of marketable products (petrol, diesel, kerosene, naphtha, etc.) that must meet specific standards. In the refining scheme a number of processing units subproduce hydrogen (thermal or catalytic cracking, catalytic reformer, etc.), while others consume it (hydrocracking, hydroprocessing,

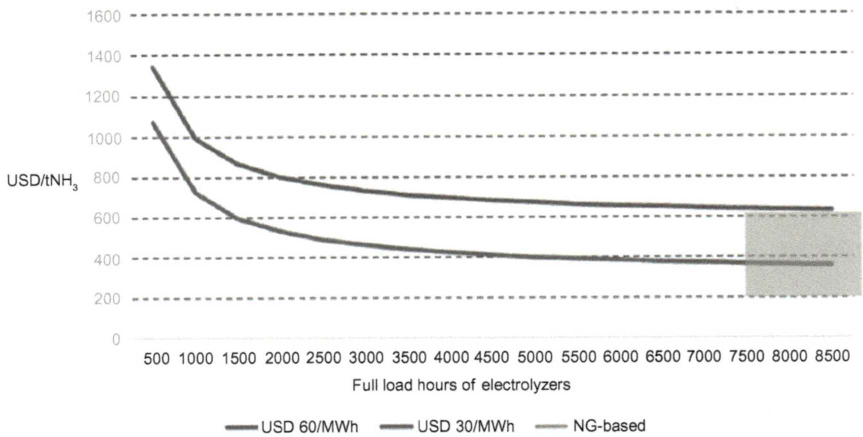

Fig. 10.2 Evolution of the price of a ton of ammonia based on the number of hours of electrolyzer's charging.

desulfurization, etc.). The trend toward increasingly stringent specifications for fuels and petroleum products is increasing the demand for hydrogen and leading to overall loss-making balance sheets. As a result, most refineries are required to produce the complementary hydrogen they need in "steam forming" units. Therefore the global need for the oil industry has become the largest source of hydrogen consumption in recent years. Table 10.1 presents the global breakdown of industrial consumption of hydrogen [3].

10.4 The different ways of producing hydrogen by electricity

The chemical process that transforms electricity into hydrogen is water electrolysis. It converts excess electricity into hydrogen. It is possible to pair an electrolyzer upstream with a renewable DC power supply (photovoltaic panels and wind turbines). Today, many small and medium capacity electrolytes (1–100 kW) have been developed. They can be used at the level of a neighborhood. The general principle of electrolysis is shown in Fig. 10.3.

The electrolyzer has two electrodes: an anode and a cathode. It is coupled upstream with a DC renewable energy source (wind turbines or photovoltaic panels). The supply of electrical energy depends mainly on the enthalpy (H −285 kJ/mol). Electrical (237.2 kJ/mol) and thermal (48.6 kJ/mol)

Table 10.1 Global breakdown of hydrogen consumption in the chemical industry and refining.

Use	Consumption (million tons per year)	%
Ammonia production	22.8	38
Other chemicals	4.8	8
Refining	26.4	44
Various (space, food industry, glass, etc.)	6	10
Total	60	100

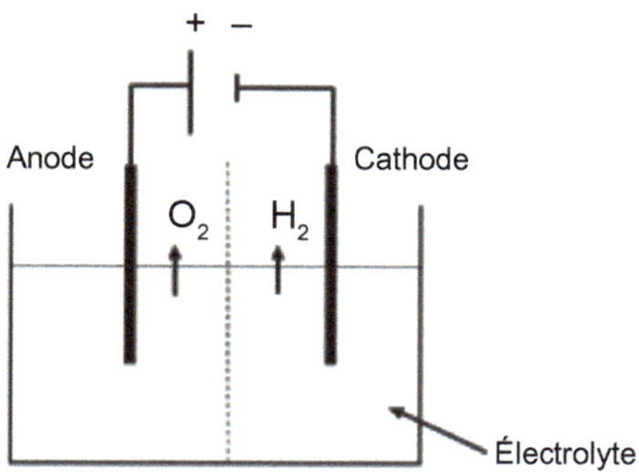

Fig. 10.3 General principle of electrolysis.

energy are converted into chemical energy. The variation in entropy for this reaction is 0.163 kJ/mol.

The minimum potential required is E −1.23 V (under normal temperature and pressure conditions). With cell potentials in the range of 1.7–2.1 V, electrolysis yields of 70%–85% can be achieved. The CEA has even developed an SOEC technology capable of achieving a return of more than 90%. The actual voltage of operation also depends on the intensity of the current, the internal resistance of the electrolyzers, and the polarization voltages. It is possible to perform a small to medium capacity electrolysis (1–100 kW). The power consumed is usually 4–5 kWh/Nm^3 of hydrogen produced. Electrolysis is weakly exothermic.

It may be interesting for a power-to-gas operator to combine different technologies to reap the benefits of each of these technologies. Thus, at the Jupiter 1000 pilot site, GRT Gaz combines alkaline electrolysis and PEM electrolysis. A blender ensures the stability of the mixture.

10.4.1 Alkaline electrolysis

Alkaline electrolysis is the most common technology for the production of electrolytic hydrogen. It thus enjoys a very high industrial maturity. In an alkaline electrolyzer the electrolyte is an aqueous solution of potassium hydroxide (KOH). The ion conduction is then provided by hydroxide ions (OH—) and potassium (K—). Fig. 10.4 presents a schematic of alkaline electrolysis. This type of electrolyzer works at relatively low current density (up; $0.4\,A\,cm^{-2}$). What favors the use of this type of electrolyzers is the fact that their operation does not use expensive catalysts based on platinum. This implies that their price is lower than that of membrane electrolysis.

- Electrodes: nickel/Fer
- Diaphragm: asbestos, sintered nickel or microperforated canvas
- Hydrogen pressure: up to 30 bars
- Life span: 50,000–90,000 h or 5.7–10.3 years continuously
- Power available: up to several megawatt
- Power consumed: 4.5–5 kWh/N m^3 of hydrogen produced

Alkaline electrolysis works at an average temperature (80–160°C) and a moderate pressure (3–30 bars) with potassium as a liquid electrolyte. Its good performance (60%–70%) is associated with high inertia that makes it unsuitable for rapid fluctuations in intermittent electrical sources. Improvements in the reactivity of these pressure-increasing electrolyzers have recently been achieved.

Currently, using mature bipolar technologies, the effort is focused on heavy modular architectures that could reach 100 MW (25 × 4 MW) capable of producing more than 40 tons per day of H_2 (and eight times more oxygen). These electrolyzers are intended for use in transportation (HGV stations and large fleets) and industry to compete with steam forming.

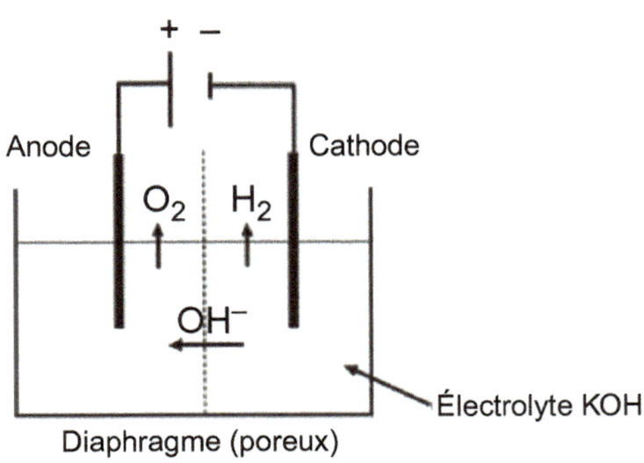

Fig. 10.4 Alkaline electrolysis.

However, to fill the cost gap, considerable efforts will be required. Alkaline electrolysis is constrained as the electrolyzer requires continuous heating.

10.4.2 PEM electrolysis (proton exchange membrane)

The main feature of this type of electrolyzers is its electrolyte (solid), compromised of a polymer membrane. This allows the conduction of hydronium ions (H_3O) generated in the anode and enables the separation of the produced oxygen and hydrogen. Fig. 10.5 shows a schematic of a proton exchange membrane (PEM) electrolysis. The benefits of this technology are the simplicity of the operating design, the limitation of corrosion problems, the compactness, and performance, which are higher than those of alkaline type. Since the electrolyte is not liquid, the system operates at a higher current density than the alkaline type (from 1 to 2 A cm^{-2}). They also have the ability to cope with current fluctuations, making PEM electrolysers the most suitable systems for electricity conversion from renewable sources such as wind and solar.

- Electrodes: platine, iridium, and ruthenium
- Temperature: up to 100°C
- Hydrogen pressure: up to 200 bar
- Life span: more than 40,000 h or 4.6 years continuously
- Power available: up to several megawatt
- Power consumed: 4.5–9 kWh/N m^3 hydrogen produced

PEM electrolysis is particularly suited to coupling a renewable energy source because it can withstand variations in available electrical power better than alkaline electrolysis. Indeed, electrolysis for power to gas must be able to absorb peak production, and, for this, they should have a significant peak power. However, PEM technology is much more expensive than alkaline technology.

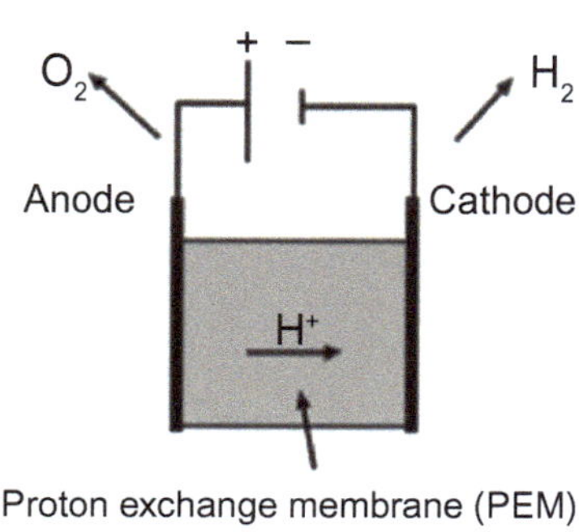

Fig. 10.5 PEM electrolysis.

10.4.3 SOEC electrolysis

Solid oxide electrolyzer cell (SOEC) technology performs electrolysis of water steam at a very high temperature (500–900°C). At this temperature the water breaks down into H_2. This reduces electrical requirements by about 16%. This technology is intended to be coupled with concentrated solar power system (CSP) and has the advantage of being able to operate in a reversible way: in an electrolyzer or in a fuel cell. With an efficiency of up to 90%, SOEC electrolysis is in the laboratory research phase. While the system efficiency is promising, accelerated cell degradation is the main obstacle facing the rapid development of this technology. Fig. 10.6 presents the general principle of a solid oxide electrolyzer cell.

10.4.4 The photoelectrolysis of water

This process of direct conversion of solar energy into chemical energy allows for decentralized "green" hydrogen production. It relies on technologies developed in the photovoltaic panel industry. The use of semiconductors as photocatalysts within a liquid electrolyte allows the capturing of photons to provide the energy needed to migrate electrons from the valence band to the conduction band within the semiconductor material. This illumination results in the reaction of electrons with water in the surface of photocatalysts, dissociating water into oxygen and hydrogen.

The main interest of this process lies in the direct conversion of solar energy, thus avoiding any cost of electrical installation or connection to the grid. In addition, unlike processes using thermal energy, photoelectrolysis is performed at low temperatures, reducing constraints on the installations. These positive aspects, as well as its strong association with solar

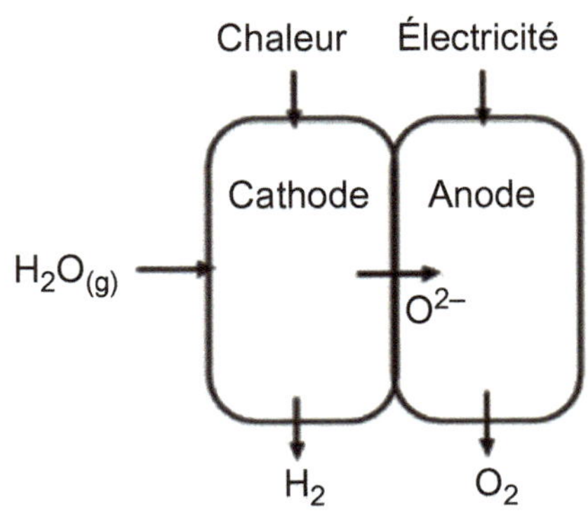

Fig. 10.6 SOEC electrolysis.

technologies, make it a solution with strong development potential. Fig. 10.7 presents an example of photoelectrolysis.

The activity of titanium dioxide, one of the most widely used photocatalysts, is limited to 4% of the solar spectrum. In addition, semiconductors must be protected, due to corrosion induced by the liquid electrolytes used. These constraints increase the already high share of capital in the total cost attributed to semiconductors. While the simplicity of the process reduces investment and maintenance costs, the materials currently used for electrodes do not results in a competitive hydrogen production price.

10.4.5 The thermochemical dissociation of water

This technology provides thermal decomposition energy using concentrated solar power or high-temperature nuclear power plants. However, the temperature required to achieve spontaneous decomposition is considerable. To reduce the energy expended to reach this temperature, increase the life span of the materials, reduce to high temperatures, and facilitate their manufacture, the process of thermochemical dissociation uses chemicals. The succession of reactions between them, as well as with water, ultimately allows the decomposition of water into oxygen and hydrogen at reduced temperatures of about 700°C. This process works according to chemical cycles, in which thermal energy enables the occurrence of the necessary reactions to obtain hydrogen but also those leading to the recycling of the substances introduced. This technology, therefore, proposes a clean hydrogen production solution consuming only water and heat from solar or nuclear sources. Fig. 10.8 presents a schematic of cerium oxide cycle coupled with concentrated solar power system.

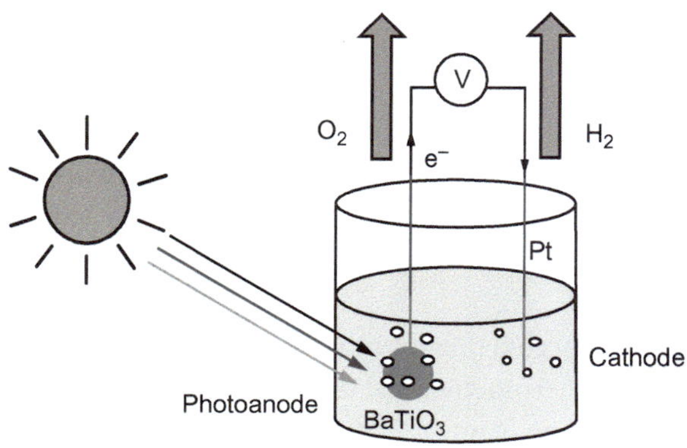

Fig. 10.7 Photoelectrolysis process.

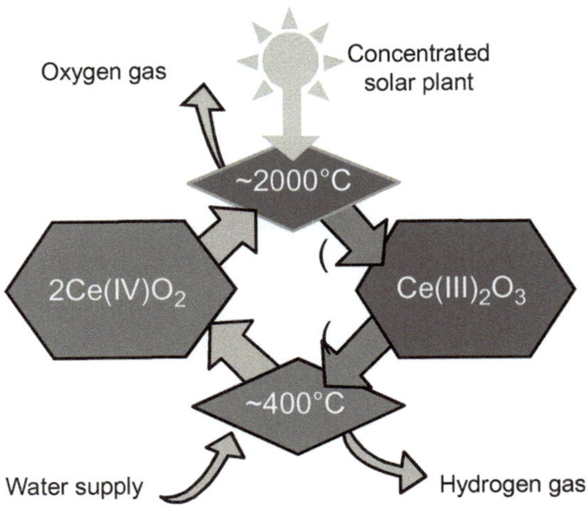

Fig. 10.8 Cerium oxide cycle coupled with a concentrated solar power system.

The thermochemical dissociation of water is in the early stages of development and is highly dependent on advances in concentrated solar technologies, whose most promising current cycles have an efficiency of around 20%–30%. While the potential of this technology is considerable, many barriers have been identified including the following:

- Poor knowledge of the thermochemical cycles used is an obstacle to the improvement of efficiency and to the reduction of costs associated with increased reactions.
- The level of technicality required is high at the temperatures reached, and the challenges concern both the life span of the materials and the corrosive substances, the manufacture of the reactor, its management in operation (heat and separation within certain chemical reactions), and concentrated solar technologies, which require progress.
- The ground surface required for the installation of such facilities and their degree of technicality entail a high investment cost.

The US Department of Energy conducted a study of hydrogen costs based on different factors. The simulation is based on a 13-km^2 concentrated solar power plant with a production capacity of 100 tons of hydrogen per day (see Table 10.2).

Each of these factors has the potential to reduce costs. This shows that the performance of the thermochemical dissociation of water is the most decisive element. Its increase from 6.2% to 34.3% reduces the cost of a kilogram of hydrogen by more than 87%.

Table 10.2 Hydrogen costs based on different factors.

Factors	Performance	Materials lifespan	Lower reactor cost	Cost of concentrated solar	Cost in $/kg	Cost in euros/kg
2015	6.2%	1 an	–	140$/m^2	35.22	27.1
2020	16.7%	5 years old	30% compared with 2015	75$/m^2	8.56	6.6
Beyond	34.3%	10 years old	30% compared with 2020	–	3.84	2.95

10.4.6 The use of photosynthetic microorganisms

The last "green" hydrogen production method mimics a process taking place in nature. Indeed, some microorganisms naturally produce hydrogen by photosynthesis. This technology harnesses the photosynthetic capabilities of green algae or cyanobacteria to produce oxygen and hydrogen. Solar photons are first captured by a pigment in organisms, chlorophyll, and then produce two enzymes. Chlorophyll then loses electrons, replaced by others during a water photolysis. The difference in the process of producing hydrogen occurs in the second phase, during which reduced carbon dioxide is replaced by protons to produce the carbohydrates necessary for the development of the organism. The induced reduction reaction produces molecular hydrogen.

Water and the sun are the only two sources of energy in this process. It, therefore, allows hydrogen production in complete dissociation with the actual infrastructure. In addition, the absence of the use of electrodes or semiconductors eliminates any constraints on water quality, which reduces the cost of the usually coupled devices and offers more opportunities for implantation.

This process requires a better understanding of chemical phenomena to "optimize" organisms. For example, in the second phase, the enzyme hydrogenase catalyst is the final reduction of protons, transforming the H-ions into hydrogen molecules. However, oxygen inhibits the functioning of this enzyme. Coproduced with hydrogen during this process, it limits the production time of hydrogen to a few minutes at most. One area of research is therefore to block the effects of oxygen by altering the structure of the enzyme. On the other hand the combined production of oxygen and hydrogen is a second obstacle, since an additional device is required. Finally the efficiency of this technology is naturally limited by the maximum solar energy absorbable by organisms. They use only a small part of the theoretically available energy and dissipate the rest in the form of heat.

Since the thermochemical dissociation of water is only at the very beginning of its development, there are few estimates of the cost of hydrogen production. The US Department of Energy estimates a cost per kilogram produced of about 6.9 euros in 2020, with a process efficiency of more than 5%.

10.5 Hydrogen in the energy transition

10.5.1 Hydrogen and the challenges of the energy transition

Hydrogen could address two key challenges in the energy transition. This includes the following:

- Decarbonization of the transportation sector. Electric vehicles equipped with a fuel cell (PAC) convert hydrogen into electricity and water vapor:
  - ✓ But this solution is only favorable in environmental terms if hydrogen is produced from decarbonized sources;
  - ✓ Hydrogen has a number of advantages over batteries, in terms of autonomy (500–700 km) and recharge time (up to 5 min).
- Compensation of the production variability of certain renewable energies with the possibility of storing hydrogen:
  - ✓ Hydrogen is produced by electrolysis of water using surplus electricity, wind, and photovoltaic production in particular.
  - ✓ Hydrogen can be stored and converted into electricity.
  - ✓ Storage is considered mainly in saline cavity.

This requires two conditions:

- the ability to produce decarbonized hydrogen or green hydrogen:
  - ✓ Hydrogen production does not emit CO_2 in the case of electrolysis, if electricity is itself decarbonized, that is, produced from renewable or nuclear energy.
  - ✓ In the case of natural gas reforming production, the emitted CO_2 must be captured and stored permanently underground to obtain decarbonized hydrogen.
- The ramp-up of decarbonized electricity generation, particularly from renewable energies, to meet the needs of hydrogen production by electrolysis.

Green hydrogen production has not become yet a reality. A transformation in the technicoeconomic context and the energy systems will be required to achieve this.

10.5.2 The valuation of the decarbonized hydrogen

There are generally four ways to enhance the decarbonized hydrogen. Fig. 10.9 presents the hydrogen value chain.

- **Power to industry**: direct sales to industries consuming green hydrogen (refining and chemistry) to decarbonize their industrial processes.
- **Power to gas**: valuation in the gas sector in two forms:
  - ✓ direct injection into gas networks for combustion,
  - ✓ by production of synthetic methane [according to the principle of methanation: conversion of monoxide (CO) or carbon dioxide (CO_2) in the presence of hydrogen], which can then be converted into heat, electricity, or fuel.
- **Power to power:** power generation from fuel cells.
- **Power to mobility—e-fuel**: transformation of hydrogen into another fuel via the process called Fischer-Tropsch. The aim is to produce another molecule from H_2 and CO_2 to obtain a fuel that can be used in today's engines.

10.5.3 Conditions for deploring decarbonized hydrogen

Large-scale deployment of hydrogen is not envisaged until 2030. The challenges are indeed multiple:

- The cost of producing hydrogen by electrolysis is still two to three times greater than that produced by natural gas reforming. Whether for fuel cell vehicles or electrolyzers, an important cost reduction across the value chain is necessary.
- At the same time a relatively high CO_2 price would reduce the cost gap with natural gas reforming. However, the increase in carbon taxation must be gradual and accompanied by public policies to support the poorest populations.

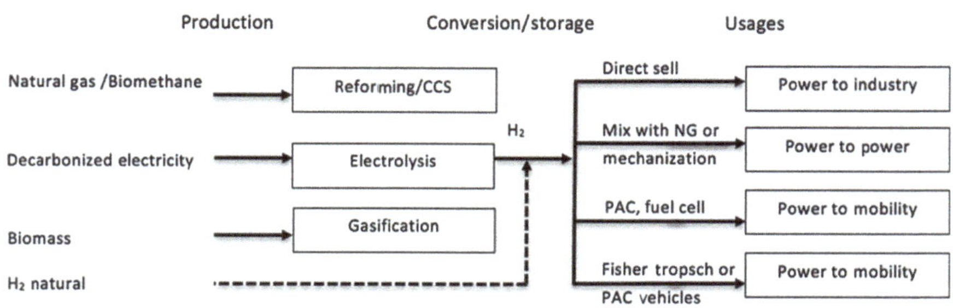

Fig. 10.9 Value chain of hydrogen.

- Access to cheap decarbonized electricity is needed to reduce the cost of electrolysis production. This condition requires further reduction in the cost of solar or wind energy.
- The various transformations involve, in some cases, cascades of energy-loss yields, which have the effect of increasing production costs.
- Some processing pathways (methanation and Fischer-Tropsch) involve access to CO_2, which leads to an additional cost and the need to develop CO_2 capture technologies.
- Hydrogen systems require the deployment of a new framework in the electricity sector incorporating renewable energy, associated grids, and adapted management modes (smart grid).
- The deployment of transport and distribution infrastructure requires significant investment and a relatively long implementation period.

10.6 Hydrogen for transportation and clean mobility

Hydrogen plays a major role in mobility. Apart from its production the three main challenges include the decarburization of energy, the development of economical vehicles, and the implementation of a charging infrastructure. Hydrogen mobility has a future only if the cost of fuel cells decreases drastically and the distribution chain is secure.

The use of an energy vector such as hydrogen coupled with renewable resources offers a variety of benefits on several scales. Hydrogen facilitates the exploitation of renewable resources in the transportation sector. It replaces fossil fuels, thus ensuring a reduction in polluting emissions. Hydrogen can then be a solution for today's energy challenges, but this can only be achieved by overcoming obstacles such as those related to the intermittent nature of renewable resources. Particular attention must be paid to the technical feasibility of the hydrogen supply chain, which is mainly surrounded by the intermittent nature of renewable resources.

In addition, the hydrogen infrastructure presents many challenges that must be addressed for a successful transition to a hydrogen-dependent economy. These difficulties are mainly due to purely economic problems, as well as the existence of numerous technological options for the production, storage, transport, and use of hydrogen. For this main reason, it is essential to understand and analyze the hydrogen supply chain in advance, to detect important factors that may play an imminent role in the development of an optimal configuration.

Hydrogen, stored and shipped in a compressed tank, is a reserve of energy for a traction or electric-powered vehicle. Hydrogen is converted into electricity and heat via a battery, which powers the electric motor, as well as the vehicle interior for heating. The traction chain of electric vehicles that implement a hydrogen fuel cell generally combines a battery, with a varied degree of hybridization according to the desired power and energy profiles. Battery and fuel cell technologies are now emerging as complementary in the field of electromobility. This complementarity is interesting for multiple usage profiles, such as passenger or goods transportation, for which one would like clean vehicles with a high level of service:

- Increased energy autonomy: Commercial vehicles for delivery, refrigerated vans, and home service company vehicles are vehicles with significant daily travel distances or are energy intensive.
- Load retention: In heavy vehicles, battery technology is not enough to provide energy and power under acceptable operating conditions. The combination of a battery and a hydrogen fuel cell allows the design of heavy goods vehicles, boats, locomotives, and electric buses, which can meet traditional use characteristics with liquid fuels.
- Vehicle availability: Filling the hydrogen tank in a matter of minutes frees itself from the constraints of stopping the vehicle due to the recharging of the batteries. Functional vehicles, road or nonroad, construction equipment, and logistics have intensive use profiles that require flexibility to recharge. This could be provided by hydrogen.
- Conditions of connection to the electricity grid: The multiplicity of electric vehicles for a fleet can lead to important contact power with the network or to problems of on-site congestion. The decoupling between hydrogen production and tank filling provides flexibility in this case. Moreover, in the field of rail, the use of hydrogen engines is an alternative to the electrification of high-cost catenary lines.

The deployment of battery electric vehicles will meet many mobility applications, including that of individuals for their daily travel. The use of hydrogen electric vehicles appears to be relevant for profiles that have not been achievable by these vehicles, especially in the field of professional, private, and public fleets.

Hydrogen-based solutions complement those based on "all-battery" electrification, especially for uses requiring fast charging times, large ranges of action, and heavy transportation. Technological advances make it possible to consider large-scale deployments.

10.7 Hydrogen for the storage of renewable energy

In the medium and long terms, renewable energy systems will necessarily rely on energy storage. In these cases the power-to-gas-to-power loop, via hydrogen, is an option to consider, despite its low energy efficiency (between 25% and 35% according to current technologies).

Technically the integration of large numbers of electrolyzers into the electrical sector translates into significant electricity consumption. From the point of view of power needs and security of supply, the integration of electrolyzers is not generally a cause for concern. This is linked on the one hand to the characteristics of the proposed electrical system, which must have significant margins at this time in link with the development of renewable energies and interconnections. In addition, electrolyzers are by nature flexible and can be disconnected during peak consumption.

The ability of electrolyzers to vary their level of electricity consumption in a matter of seconds gives them the technical ability to provide services to the electrical system, for supply-demand balance and for network operation. Except in very specific cases, however, the value associated with the provision of these services should be limited in relation to the costs of electrolyzers. In fact the provision of supply-demand-balanced services (frequency system services, fast and complementary reserves, etc.) can be remunerative but constitutes a small market where competition with other flexibilities is intense (active demand management and batteries). The participation of electrolyzers in these services is also associated with real constraints in availability that can affect the volume of hydrogen production.

In terms of services that can be provided to the grid, the value associated with the resolution of congestion remains low compared with other solutions (network development and localized cuttings), including areas with high renewable energy development.

The analysis of the models of actors currently envisaged for the production of decarbonized hydrogen around the world leads to consider several possible modes of operation for electrolyzers. Three modes of operation, deliberately very marked, are explored (see Fig. 10.10):

1. supply power to the market during periods of renewable surplus,
2. supply power to the electricity market on a base basis, excluding peak power situation,
3. coupling with renewable production (e.g., photovoltaic) under "local" models.

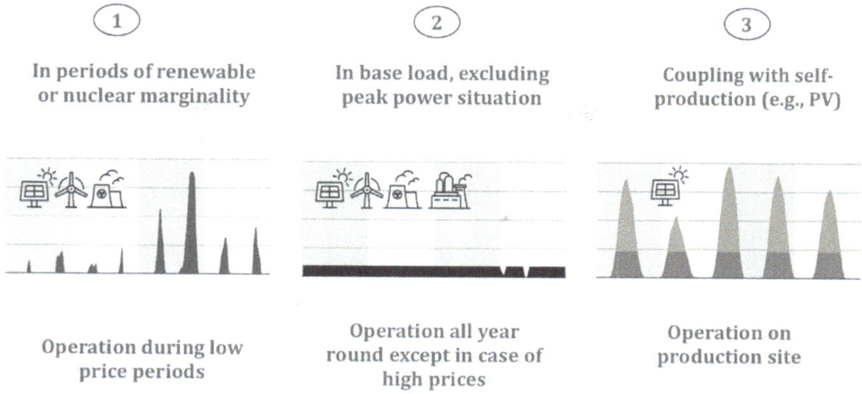

Fig. 10.10 Mode of operation.

These models lead to electrolyzer load factors and to very different technical and economic issues from one mode to another. For example, periods of decarbonized electricity generation surpluses, characterized by low or zero electricity prices, would be very unevenly distributed throughout the year and are likely highly variable. A model in which low-carbon hydrogen would be produced only during these periods would lead to irregular hydrogen production, which raises important issues for the organization of downstream chain (industrial integration and/or the need to develop dedicated hydrogen storage capabilities to ensure continuity in the supply of hydrogen).

These typical situations have an illustrative purpose, and it is likely in practice that different models will emerge, as evidenced by the great diversity of projects currently being set up, especially in the context of approaches supported by certain regions and metropolises, especially in Europe.

To achieve scale effects and large-scale energy targets, significant load factors appear to be required for at least some of the electrolysis installations (between 3000 and 6000 h per year).

From a community perspective, comparing the full cost of electrolysis with that of steam reforming is highly dependent on the recovery of CO_2 externality.

For a low CO_2 valuation hypothesis ($30/ton), the full cost of electrolysis appears to be much higher than that of steam reforming. This explains why the hydrogen used today is of fossil origin. On the other hand, by retaining a high value for environmental externality—for example, by considering the carbon's core value by 2035 ($375/ton)—electrolysis generally appears to be less expensive. This shows that it is socioeconomically relevant to replace electrolysis with steam reforming in the next 15 years.

In practice the actual development of the sector will be determined by the comparative competitiveness of the different modes of hydrogen production (carbon and decarbonized) from the point of view of economic actors, integrating all the economic signals they face.

The price of decarbonized hydrogen produced by electrolysis appears, in all three modes, higher than that of steam reforming, even taking into account significant cost reductions of electrolysis.

The development of the sector will therefore depend on the evolution of taxation and public support. However, the operating modes are characterized by different sensitivities:

- For mode 1 (market supply in renewable surplus situations), the price of electricity on the wholesale market has a small impact on the economic equation because the operation is centered on periods of low prices. Nevertheless, this mode of production implies reduced operating times, leading to increased sizing of electrolyzers for the same hydrogen production and eventually to the development of a downstream hydrogen distribution chain incorporating dedicated storage facilities necessary to compensate for the variability of the operation of the electrolyzers.
- For mode 2 (e.g., supply base on the European electricity market), the cost of electrolyzers is not determinative. The issue identified (e.g., in the European ETS market) is rather about access to low electricity prices, and the increase in the price of carbon on the European ETS market does not favor low-carbon hydrogen (compared with hydrogen from fossil fuels) in this mode. In fact the price of European electricity does not reflect a moderate cost and remains very dependent on the price of CO_2 on the market. Thus an increase in this price ultimately leads to penalizing the production of low-carbon hydrogen by electrolysis.
- Finally, for mode 3 (coupling with self-production), the main determinant of the business model is the full cost of renewable production facilities that are coupled with electrolyzers.

10.8 Conclusion

In the debate on energy transition, hydrogen is sometimes presented as a revolution for the transport and production of energy, as were, for example, electricity or natural gas in another era. Hydrogen is likely to eventually replace petroleum fuels used in transport (for trains, boats, cars, heavy goods vehicles, etc.) or to replace natural gas burned in boilers or power plants, without emitting greenhouse gases.

Four significant contributions to the energy transition are performed by hydrogen. This include the following:

- In the context of a future electricity mix combining renewable sources (solar, wind, and hydraulic), hydrogen provides solutions of flexibility and optimization to the energy networks. First, as a gas, hydrogen is storable and helps to offset the temporal variability of renewables and energy uses. In addition, through the "power-to-gas" process, it allows the gas networks and associated uses to be gradually decarbonized.
- Hydrogen as a storage option offers new opportunities for self-consumption on the scale of a building, island, or village. This storage is based on the complementarity between short-term storage (batteries) and a hydrogen chain that provides storage over several days, weeks, or months. The economic value of this solution is becoming real for some sites not interconnected to the continental electricity grid.
- Hydrogen provides new solutions for electromobility, especially for heavy vehicles, or for ensuring autonomy and availability for light commercial vehicles. This includes primarily vehicles for professional use, whether they are land, sea, river, and rail. Hydrogen mobility technologies are still expensive, and future experiments and predeployments require, like any emerging technology, support to initiate demand for vehicles and accelerate industrialization.

References

[1] Hydrogen Council. Hydrogen, scaling up, In: A sustainable pathway for the global energy transition. 2017, https://hydrogencouncil.com/.

[2] IEA. Renewable energy for industry, https://www.iea.org/reports/renewable-energy-for-industry; 2017.

[3] AFHYPAC. Production et Consomation d'hydrogene Aujourd'hui, http://www.afhypac.org/; 2016.

Index

Note: Page numbers followed by *f* indicate figures, and *t* indicate tables.

G

H

FSC
www.fsc.org
MIX
Papier | Fördert gute Waldnutzung
FSC® C083411

Kolibri 360 GmbH
Ferdinand-Jühlke-Straße 7
99095 Erfurt, Deutschland
produktsicherheit@kolibri360.de